STATISTICAL MECHANICS

STATISTICAL MECHANICS

Fundamentals and Modern Applications

RICHARD E. WILDE
SURJIT SINGH
Texas Tech University,
Lubbock, Texas

A Wiley-Interscience Publication

JOHN WILEY & SONS, INC.

New York • Chichester • Weinheim • Brisbane • Singapore • Toronto

Copyright © 1998 by John Wiley & Sons, Inc. All rights reserved.

Published simultaneously in Canada.

Library of Congress Cataloging in Publication Data:

Wilde, Richard E., 1931–
   Statistical mechanics : fundamentals and modern applications /
  Richard E. Wilde and Surjit Singh.
     p. cm.
  Includes bibliographical references and index.
  ISBN 0-471-16165-9 (cloth : alk. paper)
   1. Statistical mechanics. I. Surjit Singh, 1947–   II. Title.
QC174.8.W55   1997
530.13—dc21   97-3287
                                     CIP

Printed in the United States of America

10 9 8 7 6 5 4 3 2 1

CONTENTS

PREFACE

Determinism and the reductionistic approach to science are being replaced by the so-called emergent phenomena based on complexity and self-organization. This, in turn, is changing our traditional view of statistical mechanics. The new paradigms are affecting all fields of science and forcing their practitioners to deal with statistical principles. This book reflects these changing views at an intermediate level. Therefore, scientists not working in statistical mechanics can get an overview of the latest techniques and developments in this field from this book. Generally speaking, the book is intended for senior undergraduate and beginning graduate students of chemistry or physics, but the topics are basic and general enough to appeal to students in related subjects. It is assumed that the reader is familiar with the main ideas of classical and quantum mechanics and has had an introduction to thermodynamics and equilibrium statistical mechanics.

Because it is our belief that the serious reader needs to know how equations are derived in order to understand their limitations and the assumptions that must be made in their use, detailed steps are provided. If mathematical "tricks" are used in the derivation, these are pointed out. Occasionally, advanced mathematical concepts must be used. In all cases, these are fully explained and referenced. Although the mathematics is important, we do not want the reader to miss the physical significance of the theories and concepts. We have tried to explain the physical concepts behind the equations. As an aid to understanding, computer programs written in FORTRAN are used to illustrate the concepts in the text. These programs are listed in the Appendixes, and the readers are encouraged to explore these programs for themselves. In addition, more than 100 exercises and problems are provided.

The book is divided into three parts. Part I contains what we consider to be the essentials. The reader is urged to understand this part before proceeding. In any case, this part itself can serve as an introduction to or a refresher course in modern statistical mechanics. It can be used as a valuable "pocket guide" of definitions and formulas of statistical mechanics.

The next two parts contain modern applications of the basic ideas covered in Part I. Part II considers equilibrium statistical mechanics. It contains a discussion of phase transitions, critical phenomena, liquids, molecular dynamics and Monte Carlo techniques, polymers, protein folding, and spin glasses. Part III considers nonequilibrium statistical mechanics beginning with the Boltzmann equation. The book then proceeds to systems near enough to equilibrium that linear response can be used. Brownian motion, Zwanzig–Mori formalism, and the Kramers problem are then discussed. Nonlinear differential equations must be used to describe oscillating chemical reactions, which are systems far from equilibrium. These can be treated also by cellular automata, a relatively new research field, which holds the promise of future success in the understanding of emergent phenomena.

The following modern topics are brought together for the first time in this book, and many are rarely discussed in statistical mechanics textbooks of this level. Some of these ideas are so simple and versatile that it is surprising that they are not found in standard textbooks.

- Stochastic variables and processes.
- Dirac's bracket notation.
- Thermal wavelength and degeneracy.
- Poisson summation formula approach to ideal Bose gas.
- Broken symmetry, scaling, and universality.
- Renormalization group.
- Exact solutions of one-dimensional fluid models.
- A complete molecular dynamics simulation.
- Umbrella sampling.
- Monte Carlo renormalization group.
- Self-avoiding random walks.
- Spin glass models.
- Microscopic derivation of the Langevin equation.
- The optimized Rouse–Zimm model.
- Kramers and Grote–Hynes theories.
- The logistic equation.

- Cellular automata.
- Guide to the statistical mechanics resources on the Internet and the World Wide Web.

The book is derived from lecture notes developed by both of us in teaching a diverse group of students in the United States and India. We are indebted to many people, and it is a pleasure to thank them. In particular, one of us (S.S.) would like to thank Raj Pathria for introducing him to many wonderful techniques and ideas in statistical mechanics in his usual incisive manner and Wilse Robinson for many fruitful discussions on all kinds of topics and for his unyielding support.

We dedicate this book to our wives, Sophia and Harmesh, who patiently supported us throughout the writing of this book.

RICHARD E. WILDE
SURJIT SINGH

Lubbock, Texas, 1997

PART I

ESSENTIALS

1

CLASSICAL STATISTICAL MECHANICS

1.1 INTRODUCTION

As the name implies, statistical mechanics is the application of statistics and probability theory to the understanding of many-body problems in mechanics, both classical and quantum. This application works on two levels. First, statistical mechanics helps us to understand the macroscopic laws of thermodynamics. Second, it helps us to calculate macroscopic properties such as the specific heat, dielectric constant, and magnetic susceptibility. It does so by providing us with techniques that, starting from a microscopic description in terms of the Hamiltonian, lead to the calculation of experimentally observable macroscopic properties.

The laws of thermodynamics were obtained experimentally and are applicable to macroscopic, real systems, visible to the naked eye. To understand why these laws exist and to calculate macroscopic physical and chemical properties of a given system, one has to go beyond thermodynamics. Macroscopic or bulk systems are made of a large number (typically 10^{23}) of atoms or molecules, which are in constant motion. Because of these motions, the positions and velocities of these particles are rapidly varying. To extract the experimentally observed macroscopic quantities like the pressure or the specific heat, one has to perform statistical averages over these motions. The reader is no doubt familiar with the usual statistical argument for the derivation of the ideal gas law from kinetic theory, but this simple argument cannot be used when there are interactions among the particles. The most successful treatment applicable in all situations is statistical mechanics. In summary, statistical mechanics acts as a bridge between the microscopic world of atoms and the macroscopic world of observable physical and chemical properties.

The purpose of this chapter is to review the basic concepts of classical statistical mechanics. We begin with the random walk problem, especially as it relates to two of the most important distributions in physical systems, the Gaussian and Poisson distributions. Next, we consider the ensemble concept and the ergodic hypothesis, which lie at the heart of statistical mechanics. Three of the most important ensembles are then considered, and it is shown how thermodynamic averages and fluctuations around them are obtained from the given Hamiltonian of the system by evaluating the relevant partition functions. The Boltzmann definition of entropy is used to make the connection to thermodynamics. Due emphasis is laid on the importance of the thermodynamic limit.

Nonequilibrium statistical mechanics is then introduced by studying the time dependence of the statistical distribution function, which is given by Liouville's theorem. Boltzmann's H-theorem is then studied with the help of the concept of coarse graining. This study culminates in Boltzmann's definition of entropy, which is the most fundamental relation between the macroscopic, familiar world of thermodynamics and the microscopic, remote world of the Hamiltonian.

1.2   PROBABILITY THEORY

1.2.1   The Random Walk Problem

The random walk problem is also called the "drunkard's walk problem." In the one-dimensional version of this problem, a drunk starts out from an origin and takes a step of unit length either right or left. Then he takes a second step, again either right or left. His destination is at a distance d units away. He takes steps in the right or the left directions with the same fixed probabilities, r and l, respectively, with $r + l = 1$. Assume that N steps have been taken. The random walk problem is to calculate the probability that the drunk will have reached his destination d on his path in N steps. The problem can serve as a simple model of a Markovian process (Chapter 9) or a spin-$\frac{1}{2}$ paramagnet (Chapter 3).

The solution to the problem is an exercise in elementary probability theory. Since each step taken is an independent event, the probability of n_r steps to the right and $n_l \equiv N - n_r$ steps to the left in a single trial is the product of the individual probabilities, $r^{n_r} l^{N-n_r}$. In general, this probability can be achieved in several ways, each outcome increasing the probability of the event. We ask how N distinguishable steps can be distributed between right and left "boxes," the number of steps in each "box" being specified. This is $N!/[n_r!(N - n_r)!]$.

So, the probability of taking n_r steps away from the origin at the end of the total of N steps is

$$P(n_r) = \frac{N!}{n_r!(N - n_r)!} r^{n_r} l^{N-n_r} \tag{1.1}$$

To take a concrete example, consider a random walker who has taken a total of $N = 3$ steps, with $r = \frac{1}{4}, l - \frac{3}{4}$, to a destination at a distance $d = -1$. By definition

$$d = n_r - n_l = 2n_r - N \tag{1.2}$$

Therefore, $d = -1$ with $N = 3$ can be obtained only with $n_r = 1$ and $n_l = 2$, and the probability that the walker is at his destination is $P(1) = \frac{27}{64}$. Similarly, for other possible destinations ($d = 3, 1$, and -3) the probabilities are $P(3) = \frac{1}{64}, P(2) = \frac{9}{64}$, and $P(0) = \frac{27}{64}$, respectively. The fact that all these probabilities add up to unity is an expression of the assumption that the walker ends up somewhere and is not "absorbed" in the process. More general random walks with creation and absorption can, of course, be considered. This "conservation" of the random walker can be shown generally by using Newton's binomial theorem:

$$(x + y)^N = \sum_{n=0}^{N} \frac{N!}{n!(N - n)!} x^n y^{N-n} \tag{1.3}$$

Summing Eq. (1.1) over n_r and using this theorem shows that

$$\sum_{n_r=0}^{N} P(n_r) = (r + l)^N = 1 \tag{1.4}$$

Since $P(n_r)$ tells us how the total probability of unity is distributed among different values of n_r, it is called a *probability distribution*. Because of its relation to the binomial theorem, $P(n_r)$ given by Eq. (1.1) is called the *binomial distribution*. So, the probability that the drunk has taken n_r steps away from the origin is given by the binomial distribution. After a total of N steps, how far away is the drunk from the origin on the average? The answer to this question depends on the concept of the expectation, also called the expected value, mean value, or average value of the distribution. It is denoted by angular brackets and is defined by

$$\langle n_r \rangle = \sum_{n_r=0}^{N} n_r P(n_r) \tag{1.5}$$

Once $\langle n_r \rangle$ is obtained

$$\langle n_l \rangle = N - \langle n_r \rangle \tag{1.6}$$

and from Eq. (1.2)

$$\langle d \rangle = \langle n_r - n_l \rangle = 2\langle n_r \rangle - N \tag{1.7}$$

The mean value in Eq. (1.5) is also called the first moment of n_r. In general, one can calculate the kth moment, defined by

$$\langle (n_r)^k \rangle = \sum_{n_r=0}^{N} (n_r)^k P(n_r) \tag{1.8}$$

The various moments can be obtained conveniently from the *moment generating function*

$$G(t) = \langle \exp(n_r t) \rangle \tag{1.9}$$

as discussed in Appendix B. For the binomial distribution,

$$G(t) = \sum_{n_r=0}^{N} \exp(n_r t) P(n_r) = [r \exp(t) + l]^N \tag{1.10}$$

where the binomial theorem [Eq. (1.3)] has been used. Note that

$$G(0) = 1 \tag{1.11}$$

as expected because of the normalization of the distribution. The first moment of the binomial distribution is

$$\langle n_r \rangle = Nr \tag{1.12}$$

and thus

$$\langle d \rangle = N(2r - 1) \tag{1.13}$$

from Eq. (1.7).

For a symmetric distribution, where $r = \frac{1}{2}$, $\langle d \rangle$ is zero, so that the drunk did not get anywhere on the average! However, he certainly made an effort to reach his destination. This effort is reflected in the *second central moment* of the distance d, defined as $\langle (d - \langle d \rangle)^2 \rangle$. The second central moment of d is 4 times the second central moment $\langle (n_r - \langle n_r \rangle)^2 \rangle$ of n_r, where

$$\langle (n_r - \langle n_r \rangle)^2 \rangle = \sum_{n_r=0}^{N} (n_r - \langle n_r \rangle)^2 P(n_r) \tag{1.14}$$

The kth *central moment* is defined as

$$\langle (n_r - \langle n_r \rangle)^k \rangle = \sum_{n_r=0}^{N} (n_r - \langle n_r \rangle)^k P(n_r) \tag{1.15}$$

In general, kth central moments can be defined for any discrete distribution and can be obtained from the moment generating function. The second central moment of n_r is called the *variance*, $\text{Var}(n_r)$, and its square root is the root-mean-square deviation, called the *standard deviation*, $\sigma(n_r)$. In the case of the random walker, it is found that

$$\langle n_r^2 \rangle = Nr + N(N - 1)r^2 \tag{1.16}$$

whence

$$\text{Var}(n_r) = \langle n_r^2 \rangle - \langle n_r \rangle^2 = Nr(1 - r) \tag{1.17}$$

$$\sigma(n_r) = N^{1/2}[r(1 - r)]^{1/2} \tag{1.18}$$

and

$$\text{Var}(d) = 4Nr(1 - r) \tag{1.19}$$

$$\sigma(d) = 2N^{1/2}[r(1 - r)]^{1/2} \tag{1.20}$$

The random walk problem is often used to model processes such as diffusion (see Chapter 9). The drunk is the diffusing particle and the random walk of the particle is due to collisions with other particles. The variable d becomes the position along the x axis, which is continuous, so that the total number of steps N essentially goes to infinity. The probability distribution $P(n_r)$ is then replaced by the *probability density* $p(x)$, defined in such a way that $p(x)\,dx$ gives the probability of finding the particle in a small range between x and $x + dx$. The kth moment of x is defined as

$$\langle x^k \rangle = \int_{-\infty}^{\infty} x^k p(x)\,dx \tag{1.21}$$

where the particle is assumed to move along the whole x axis. The zeroth moment is unity because of normalization. The kth order central moment is defined to be

$$\langle (x - \langle x \rangle)^k \rangle = \int_{-\infty}^{\infty} (x - \langle x \rangle)^k p(x)\, dx \tag{1.22}$$

Again, the second-order central moment is called the variance, Var(x), and its square root is called the standard deviation, $\sigma(x)$.

1.2.2 The Gaussian Distribution

The binomial distribution depends on two parameters, N and $r \equiv 1 - l$. In physical applications, N is usually very large. Therefore, it is important to study the binomial distribution in the limit $N \to \infty$. Two different distributions are obtained in this limit depending on the value of r. These are the *Gaussian* and *Poisson distributions*. We discuss the approach of the binomial distribution to the Gaussian distribution in this section.

For simplicity, let us assume a symmetric binomial distribution, where $r = l = \frac{1}{2}$, and take N to be an even number. More general cases are discussed in textbooks on probability theory (e.g., Feller, 1968, Vol. I, Chapter VII). The symmetric distribution is given by

$$P(n) = \frac{N!}{n!(N-n)!} \left(\frac{1}{2} \right)^N \tag{1.23}$$

where we have replaced n_r by n for convenience. The maximum value of $P(n)$ occurs at $N/2$, which is the mean value of n according to Eq. (1.12). The ratio of the standard deviation of n [Eq. (1.18)] to the mean of n is called the relative standard deviation, which in the limit of large N goes to zero as $1/N^{1/2}$. It is useful to expand $P(n)$ around its mean value in terms of the distance d.

$$d = 2n - N = 2 \left(n - \frac{N}{2} \right) \tag{1.24}$$

For the symmetric distribution, Eqs. (1.13) and (1.20) give

$$\langle d \rangle = 0 \qquad \sigma(d) = N^{1/2} \tag{1.25}$$

Both the variable d and its standard deviation have dimensions of distance. It is convenient to introduce a dimensionless variable

$$y = \frac{n - N/2}{N^{1/2}} \tag{1.26}$$

The variable y has an advantage over the original variable n in that it is dimensionless, its average is zero, its standard deviation is fixed at 0.5, and it becomes continuous as N tends to ∞. We now substitute Eq. (1.26) in Eq. (1.23),

$$P(n) = \frac{N!}{(N/2 + y\sqrt{N})!(N/2 - y\sqrt{N})!} \left(\frac{1}{2}\right)^N \tag{1.27}$$

To improve the convergence to the $N \to \infty$ limit, it is found useful to take the log of this equation and use Stirling's formula (Pathria, 1996, Appendix B), which is

$$\ln N! \approx N \ln N - N + \ln \sqrt{2\pi N} \tag{1.28}$$

This approximation can be used for all values of y except for $y \simeq \pm\sqrt{N}/2$, which corresponds to $n \simeq N$ or $n \simeq 0$, respectively. In most applicable cases the variable n lies near its average $N/2$, so in those cases the approximation is quite good. In the equation resulting from the application of Stirling's formula, we use the Taylor's series expansion

$$\ln(1 + z) = z - \frac{z^2}{2} + \frac{z^3}{6} + \cdots \tag{1.29}$$

Retaining terms to order $1/N$, we obtain

$$\ln P(n) \approx \frac{1}{2} \ln 2 - \frac{1}{2} \ln(\pi N) - 2y^2 + \frac{2y^2}{N} \tag{1.30}$$

Assuming $|y| \ll \sqrt{N}/2$ and taking the antilog of Eq. (1.30), we get

$$P(n) = \left(\frac{2}{N\pi}\right)^{1/2} \exp(-2y^2) \tag{1.31}$$

Since y becomes a continuous variable in the large N limit, we define a probability density $p(y)$ such that

$$p(y)\, dy \approx P(n) \left(\frac{\Delta n}{\Delta y}\right) dy \tag{1.32}$$

where from Eq. (1.26) it is seen that if n changes to $n + 1$, the change in y is $\Delta y = 1/N^{1/2}$. Comparing Eqs. (1.31) and (1.32) in the large N limit,

$$p(y) = \left(\frac{2}{\pi}\right)^{1/2} \exp(-2y^2) \tag{1.33}$$

In terms of n, Eq. (1.31) is

$$P(n) = \left(\frac{2}{N\pi}\right)^{1/2} \exp\left[-\frac{2(n - \langle n \rangle)^2}{N}\right] \qquad (1.34)$$

and, using Eq. (1.18), one obtains

$$P(n) = \frac{1}{\sigma\sqrt{2\pi}} \exp\left[-\frac{(n - \langle n \rangle)^2}{2\sigma^2}\right] \qquad (1.35)$$

where the simplified notation σ has been adopted for the standard deviation of n.

This completes the limiting process, and the resulting distribution [Eq. (1.35)] is called a *Gaussian* (or *normal*) *distribution*. In the case of a continuous variable x, the Gaussian probability density is

$$p(x) = \frac{1}{\sigma\sqrt{2\pi}} \exp\left[-\frac{(x - \mu)^2}{2\sigma^2}\right] \qquad (1.36)$$

where $\mu = \langle x \rangle$. Equation (1.33) represents a Gaussian probability density with $\mu = 0$ and $\sigma = \frac{1}{2}$. The Gaussian distribution is completely specified by the mean μ and by the standard deviation σ. Figure 1.1 shows a Gaussian distribution for the random walk problem and $N = 100$. From the ordinate scale, it is seen that the probabilities are very small.

The maximum value of the Gaussian distribution occurs at the mean. Both the maximum value and the width depend on σ. When σ approaches zero, the maximum value increases and the width decreases, so that the distribution is sharply peaked at the mean, in effect becoming a *Dirac delta function* (Pathria, 1996, Appendix B) centered at the mean. This function, denoted by $\delta(x - \alpha)$, has the property

$$\int_a^b f(x)\delta(x - \alpha)\,dx = f(\alpha) \qquad (1.37)$$

if the variable α lies in the range of integration; otherwise the integral is 0. If α lies exactly at one of the end points, the integral is $f(\alpha)/2$.

This behavior of the Gaussian distribution has important consequences for statistical mechanics. In particular, it explains the observation that macroscopic measurements are sharply peaked at their average values in the thermodynamic limit, as we shall see in Section 1.6.2.

Gaussian distributions are of great importance in the application of statistics to physical situations. There are two main reasons for this:

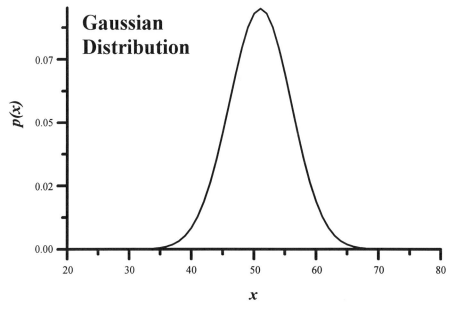

Figure 1.1   The random walk Gaussian distribution.

- When a physical event is composed of a series of independent events randomly distributed according to any distribution with a finite mean and a finite variance, the probability density is Gaussian. This result is a consequence of the *central limit theorem* (Feller, 1968, Vol. II, Chapter VIII).

- Linear operations on Gaussian distributions, such as multiplying the variables by a constant or adding two variables, or combinations of these two operations, yield Gaussian distributions (Feller, 1968, Vol. II, Chapter III).

Because the Gaussian distribution is symmetric, it is a good approximation to the binomial distribution only if $r \approx l \simeq \frac{1}{2}$. If either r or l is much smaller than $\frac{1}{2}$, the Poisson distribution becomes a better approximation to the binomial distribution.

1.2.3   Poisson Distribution

We now study the limit of the binomial distribution as N becomes very large and r goes to zero in such a way that $\lambda = \langle n \rangle = Nr$ is fixed. Clearly,

$l = 1 - \frac{\lambda}{N}$. With these substitutions, Eq. (1.1) becomes

$$P(n) = \frac{(1 - \lambda/N)^N \lambda^n}{n!} \left[\frac{N!}{N^n(N - n)!(1 - \lambda/N)^n} \right] \qquad (1.38)$$

It can be shown that the expression in square brackets in Eq. (1.38) approaches 1 as $N \to \infty$. Also, noting that $\lim_{N \to \infty}(1 - \lambda/N)^N = \exp(-\lambda)$, Eq. (1.38) becomes

$$P(n) = \frac{\lambda^n}{n!} \exp(-\lambda) \qquad (1.39)$$

This approximation to the binomial distribution is called the *Poisson distribution*. It can be shown by the moment generating function that the standard deviation of the Poisson distribution is $\sqrt{\lambda}$. For large values of λ, it is useful to define a new variable $z = (n - \lambda)/\sqrt{\lambda}$, and to consider the limit $N \to \infty$. For $|z| \ll \sqrt{\lambda}$, we get the Gaussian distribution

$$P(z) = \frac{1}{\sqrt{2\pi}} \exp\left(-\frac{z^2}{2}\right) \qquad (1.40)$$

with zero mean and unit variance.

The Poisson distribution is used to describe many processes such as noise in vacuum tubes, radioactive decay, genetic mutations, and epidemics. All these processes are characterized by the fact that the events are very rare, that is, $r \ll 1$. The Poisson distribution is completely determined by the single parameter λ. In Figure 1.2, we show it for $N = 100$ and $P(r) = \lambda/N = \frac{1}{20}$. Once again, note the very small probabilities on the ordinate.

1.3   RANDOM VARIABLES AND STOCHASTIC PROCESSES

1.3.1   Introduction

A *random variable* or a *stochastic variable* is like a function in calculus in that the variable takes on a discrete set of values. The difference is that a function takes its values with perfect certainty, whereas the random variable takes its values with a given set of probabilities. Examples are the number of steps that a random walker takes to the right, the number of heads in a coin-tossing game, and the energy eigenvalue of a one-dimensional harmonic oscillator in quantum mechanics. In each of these examples, the variable takes on its possible values with probabilities that can be calculated. The totality

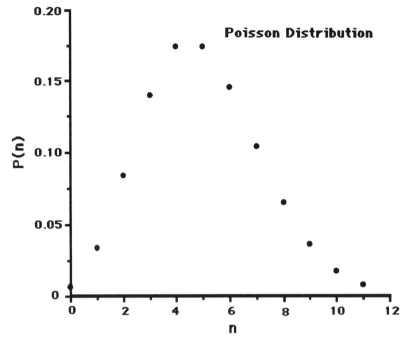

Figure 1.2   The Poisson distribution.

of the values of the random variable and its associated probabilities is called the probability distribution of the variable. We have already studied three well-known probability distributions.

The possible values of the random variable may be continuous rather than discrete. Let $X(t)$ be a stochastic or random variable, which is a function of time. The change of X with time is called a *stochastic* or a *random process*. Thus, a stochastic process involves a stochastic variable that takes random values whose development in time is not governed by a deterministic equation. Only the probabilities that the random variable will take certain values at given times are known. When a random variable attains a certain value, we say that the underlying physical system is in a certain "state" and its time development is thought of as a progression from one state to another. We now describe two broad classifications of stochastic processes.

First, one may consider how the probability of going from one state to another depends on time. If it depends on the actual times at which the system occupies the two states, we have a *nonstationary* process. On the other hand, if the probability depends only on the difference in the times corresponding to the two states, it is called a *stationary* process.

Second, one may consider how many different past states the future evolution depends on. If the probability corresponding to a present state is the only one that determines the probability corresponding to the immediate next state, then the process is *Markovian*. On the other hand, if the future depends on how the present state was arrived at or on the history or memory of states other than the present, we have a *non-Markovian* process.

Brownian motion, which is an example of a Markovian process, is studied in detail in Chapter 9. Here we study general random variables.

1.3.2   Moments of Random Variables

We consider a random variable X that takes values x_i with probability distribution $P(x_i)$. The kth moment of the variable X is exactly the same as the kth moment of the probability distribution. The same is true of central moments such as the expectation value, variance, standard deviation, and so on. If we have two random variables, the *joint probability distribution* is a function that depends on values of both variables. In general, a joint probability distribution can be a function of the values of several random variables. For example, one may consider a two-dimensional random walk on a square lattice in which the random walker may take steps in any of the four directions with given probabilities.

Consider two random variables, X and Y, with probability distributions, $P(x_i)$ and $Q(y_j)$, respectively. Let their joint probability be denoted by $J(x_i, y_j)$. If the joint probability is summed over all values of y_j for a given x_i, we must obtain $P(x_i)$. In other words,

$$\sum_{y_j} J(x_i, y_j) = P(x_i) \tag{1.41}$$

The joint probability is defined when the two variables take their values irrespective of each other. What if one of the variables, say Y, has taken a given value y_j? What is the probability then that the variable X will take the value x_i? This probability, which is equal to $J(x_i, y_j)/Q(y_j)$, is called the *conditional probability* and is denoted by $C(x_i \mid y_j)$:

$$C(x_i \mid y_j) = \frac{J(x_i, y_j)}{Q(y_j)} \tag{1.42}$$

Sometimes it may happen that

$$C(x_i \mid y_j) = P(x_i) \tag{1.43}$$

which, on using Eq. (1.42), may be rewritten as

$$J(x_i, y_j) = P(x_i)Q(y_j) \tag{1.44}$$

In this case, when the joint distribution of two variables is simply the product of their individual distributions, the two random variables are said to be *stochastically independent*. As an example, one may think of the random walk on a square lattice, where the probabilities of taking steps in different directions are mutually independent.

The random variables can be added and multiplied. The expectation of the sum of random variables is equal to the sum of the expectations of the individual random variables. However, the same is not true of higher moments. Similarly, the expectation of the product of two random variables is not equal to the product of their expectations, unless they are independent. Covariance of two variables is defined by

$$\text{Cov}(X, Y) = \langle (X - \langle X \rangle)(Y - \langle Y \rangle) \rangle = \langle XY \rangle - \langle X \rangle \langle Y \rangle \tag{1.45}$$

If two variables are independent, the covariance is zero. The converse is not true. For a discussion of this point, see Feller (Feller, 1968, Vol. I, Chapter IX). In physics and chemistry, it is more common to consider the dimensionless *correlation function* of two variables defined as

$$\text{Corr}(X, Y) = \frac{\langle (X - \langle X \rangle)(Y - \langle Y \rangle) \rangle}{\langle X \rangle \langle Y \rangle} \tag{1.46}$$

We shall see many examples of these throughout the book.

To illustrate these ideas, consider the random walk discussed in Section 1.2.1. Let X_i be the number of steps to the right at the ith step in the random walk. It is a random variable that has only two values 0 and 1 with probabilities $1 - r$ and r, respectively. Therefore, its expectation is

$$\langle X_i \rangle = 0 \times (1 - r) + 1 \times r = r \tag{1.47}$$

Now consider a random variable $\sum_{i=1}^{N} X_i$, which is the total number of steps to the right after a total number of N steps. This random variable was denoted by n_r in Section 1.2.1. Its expectation is given by

$$\langle n_r \rangle = \sum_{i=1}^{N} \langle X_i \rangle = Nr \tag{1.48}$$

where use has been made of Eq. (1.47). In the same way,

$$\langle X_i^2 \rangle = 0^2 \times (1 - r) + 1^2 \times r = r \tag{1.49}$$

whence

$$\langle n_r^2 \rangle = \sum_{i=1}^{N} \sum_{j=1}^{N} \langle X_i X_j \rangle = \sum_{i=1}^{N} \langle X_i^2 \rangle + \sum_{i \neq j} \langle X_i \rangle \langle X_j \rangle = Nr + N(N-1)r^2 \quad (1.50)$$

where mutual independence of individual steps has been used.

Equations (1.48) and (1.50) are in agreement with with Eqs. (1.12) and (1.16) derived previously. The simplicity of the present derivation is obvious, as generating functions or calculations of binomial coefficients were not needed. Here we focused on the random variables, whereas in Section 1.2.1 importance was given to their probability distributions. This duality between variables and their distributions occurs commonly in physics and chemistry. For example, we mention the Heisenberg and the Schrödinger pictures in quantum mechanics (Chapter 2), the Langevin and the Fokker–Planck approaches to Brownian motion (Chapter 9), and the operator and distribution approaches in the Zwanzig–Mori formalism (Chapter 10).

1.4   EQUILIBRIUM STATISTICAL MECHANICS

1.4.1   Ergodic Hypothesis

In order to study classical statistical mechanics, it is convenient to use the Hamiltonian approach to mechanics as opposed to Newton's equations of motion. This approach is more general, puts the coordinates and momenta on an equal footing, has the conservation laws built in, and affords a relatively smooth transition to quantum statistical mechanics. We begin with a summary of the Hamiltonian approach.

Newton's equations, including constraints, if any, are first replaced by the equations of motion for the n generalized coordinates of the Lagrangian formulation. These coordinates constitute the *configuration space*. A system is characterized by a point in this hyperspace, and in time this point traces out a curve in configuration space. Generalized momenta are now defined for each coordinate, and the Hamiltonian is obtained in the standard way by eliminating the velocities. Thus, in the Hamiltonian formulation, we have a $2n$-dimensional space formed from the n generalized position coordinates, collectively denoted by q, and their n conjugate generalized momentum coordinates, collectively denoted by p. This space is called *phase space*, and in time the system traces out a curve in phase space. According to Tolman (Tolman, 1979, Section 19), it was Boltzmann who pointed out the importance of using position and momentum coordinates in classical statistical mechanics. As we develop the ensemble approach to statistical mechanics, we shall see why this is so.

We begin our considerations with an isolated system in equilibrium. As an example, consider a sealed cylinder full of nitrogen gas at room temperature and pressure. Because the gas is in equilibrium, its temperature, pressure, volume, and other thermodynamic properties are fixed. So the system is in a unique thermodynamic state, also called a *macrostate*. Despite the outward calm implied by the macrostate, the nitrogen molecules in the cylinder are forever moving in different directions with different velocities and colliding with one another and with the cylinder walls. At a given time, the system is represented by a point in its phase space and is said to be in a *microstate*, which is determined by the values of its generalized position and momentum coordinates.

Because a typical physical or chemical system contains on the order of 10^{23} molecules, it is impossible to determine the generalized coordinates of every particle in order to apply Hamilton's equations of motion. Since complete knowledge of the system is denied us, we are forced to use statistics and to measure average values of the system's physical observables. Typically, various quantities are obtained experimentally by taking time averages. The *time average* of the physical observable G is

$$\langle G \rangle_t = \frac{1}{\tau} \int_0^\tau G(p, q) \, dt. \tag{1.51}$$

As $\tau \to \infty$, the system settles down into an equilibrium state, even if it was in a nonequilibrium state in the beginning. In practice, G is measured at time intervals $\Delta t \ll \tau$, the measured G values are added, and then divided by the number of measurements

$$\langle G \rangle_t \approx \frac{\Delta t}{\tau} \sum_{t=0}^\tau G(t) \tag{1.52}$$

where G is an implicit function of time owing to its dependence on the generalized coordinates. It is sufficient to choose τ to be much larger than the time needed to reach equilibrium. The average obtained in this way will agree with the theoretical average if, during the course of time, the system passes through all possible microstates consistent with the given macrostate. The assumption that this is indeed so is called the strong form of the *ergodic hypothesis*. If it is valid, the two averages [Eqs.(1.51) and (1.52)] are equal. It turns out that the strict form of the hypothesis is not valid except for the simplest systems. To remedy the situation, a *quasiergodic* hypothesis, also called the weak form of the ergodic hypothesis, has been proposed. This weak form assumes that the system will pass arbitrarily close to a given microstate as time passes. We shall not study the complex subject of the validity of either hypothesis (see Tolman, 1979; Ma, 1985 for details). As physicists and chemists are wont to do, we shall simply assume the strong hypothesis and proceed to calculate various quantities. The justification for this is that, most of the time, there is agreement between theory and experiment. The disagreements, if any, are usually traced to an oversimplified theory rather than to the failure of the ergodic hypothesis.

In the case of a system undergoing phase transitions, there is a possibility of the breakdown of the ergodic hypothesis. This phenomenon is closely related to *broken symmetry* and is discussed in detail in Section 3.2.4. We shall just mention here that one has to take ensemble averages very carefully to obtain results consistent with the ergodic hypothesis.

For classical systems it is possible, in principle at least, to determine the position and momentum of every particle in the system at a given instant of time. This provides one point in phase space. This process is repeated at the next instant of time to provide a second point. After a sufficiently long time has elapsed, an essentially continuous distribution of phase points is available from which an average value of a physical observable can be obtained. If the particles in the system are few and of macroscopic size, a camera can be used to carry out this process. For example, let the system be a pendulum executing simple harmonic motion. The pendulum is photographed with a movie camera. Each frame of movie film records a state of the system. The position of the pendulum is measured directly on each frame. The change of position between frames gives the velocity and, hence, the momentum, since the mass of the pendulum is known. When the data from the movie film are plotted in phase space, an ellipse is obtained as shown in Figure 1.3.

In the simple one-dimensional pendulum example, there is no doubt that the ergodic hypothesis is valid, and in time all points on the ellipse will be visited. However, for a chemical system composed of Avogadro's number of molecules, it is impossible to be certain that all accessible regions of phase space have been visited. The problem was solved by J. Willard Gibbs, who proposed the ensemble concept, which is independent of the ergodic hypothesis.

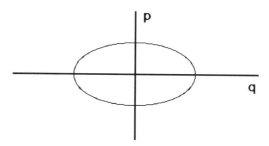

Figure 1.3   Trajectory of a pendulum in phase space.

1.4.2   Ensemble Theory

The Gibbsian-ensemble idea is the following theoretical proposal. Construct macroscopically identical copies of the system with one provision. Each copy should correspond to the same macrostate but a different microstate! The number of copies needed is equal to the number of microstates to which the given thermodynamic state corresponds. The collection of all such copies needed for a given macrostate is called an *ensemble*.

As an explicit example, consider the system of N one-dimensional anharmonic oscillators described by the Hamiltonian,

$$\mathcal{H} = \sum_i \left(\frac{p_i^2}{2m_i} + \frac{1}{2} m_i \omega_i^2 x_i^2 \right) + \sum_{i>j} \lambda x_i^7 x_j^{11} \tag{1.53}$$

where x_i and p_i are the position and linear momentum coordinates of the particles with masses m_i and angular frequencies ω_i, and λ is a coupling constant between different oscillating particles. Let us fix the number of particles at N and the volume of the system at V. If the system is isolated, its energy is a constant, say E_c, which restricts the allowed values of x_i and p_i. For a given set of $\{x_i\}$ and $\{p_i\}$ from this constant-energy-restricted set, we get one microstate and one member of our ensemble. Changing the values of x_i and p_i so that they still lie in the constant-energy-restricted set gives another member of the ensemble, and so on. These members of the ensemble have different values of the x_i and p_i, which puts them in different microstates. However, the values of the x_i and p_i for each member are such that the energy is equal to E_c. This statement implies that all members are in the same macrostate, because thermodynamically a macroscopic system of given total energy, number of particles and volume is in a unique macrostate, irrespective of its microscopic properties. Here, we have given a prescription to construct a *microcanonical ensemble*, which will be discussed in detail in Section 1.4.3.

Remember that these copies of the system, that is, the members of the ensemble, are mental copies. Therefore, they are all independent of one another. To obtain the average of any physical quantity, we simply take the average over the members of the ensemble. Because the members are all identical under the given constraints (i.e., same number of particles, temperature, pressure, etc.), only one phase space need be considered, and the ensemble appears as a "cloud" of points in phase space. As a practical matter, we must work with a finite number of phase-space points N_e. The number of points N_e must be such that there is an essentially continuous distribution of points in phase space. There is then associated with the ensemble a probability density $\rho(p, q)$, which represents the fraction of phase points per unit "volume" at the point (p, q).

A natural question arises as to the weights given to different microstates when the averages are taken over the ensemble. For an isolated system, there is no reason to prefer one member of the ensemble over another. Hence, we invoke the *postulate of equal* a priori *probabilities*, which states that, for isolated systems at equilibrium, all accessible regions of phase space have equal *a priori* probabilities for equal volumes. As with the ergodic hypothesis, there is no derivation of this postulate. Its justification lies in its phenomenal success as the cornerstone of statistical mechanics. In terms of our pendulum example, this postulate can be implemented by assigning one and exactly one point to each movie frame, each point given equal weight without prejudice. We shall soon see how this hypothesis is used to derive various ensembles.

1.4.3 Statistical Distribution Function

The probability dw of finding the system in the infinitesimal volume element $dp\,dq = \prod_{i=1}^{n} dp_i\,dq_i$ is proportional to $dp\,dq$.

$$dw \propto dp_1\,dp_2 \cdots dp_n\,dq_1\,dq_2 \cdots dq_n \tag{1.54}$$

The constant of proportionality is $\rho(p, q)$.

$$dw = \rho(p, q)\,dp\,dq \tag{1.55}$$

In statistics, ρ is called a *probability density*; in classical statistical mechanics, it is called the *statistical distribution function*.

The volume of the phase space element $dp\,dq$ cannot be infinitesimal, since then we should have an infinite number of microstates to contend with. Just as we are forced to follow the motion of a system at finite time intervals Δt to get a time average, we must work with finite volumes $\Delta p \Delta q$ to get an ensemble average. The volume $\Delta p \Delta q$ is arbitrary. Obviously, if we know the value of ρ for each microstate, we have complete information about the statistical distribution function (often referred to as simply the distribution function). As we take a larger volume $\Delta p \Delta q$, we have a ρ averaged over the microstates in $\Delta p \Delta q$. If $\Delta p \Delta q$ is taken still larger, we must consider an average over even more microstates. If the basic volume $\Delta p \Delta q$ is too large, we lose information about the distribution function. To circumvent this, we assume that the region $\Delta p \Delta q$ contains many microstates, the number of which is small compared to the total number of states. In this case, we require that the volume $\Delta p \Delta q$ be such that the distribution function can be treated as changing continuously in phase space, and the *ensemble average* of the physical observable G can be written

$$\langle G \rangle = \int G(p,q)\rho(p,q)\,dp\,dq \qquad (1.56)$$

By the ergodic hypothesis, this average is equal to the experimental time average given by Eq. (1.52).

To summarize, for a given macroscopic system it is impossible to know the positions and momenta of all particles. This means that we do not have complete information about the state of the system. The best we can do is to know the probability of finding the system at a phase point in a region $\Delta p \Delta q$ of phase space that is microscopic but still contains a large number of points. In this case, the basic principle is that of equal a priori probabilities. The validity of this principle must rest with experiment.

1.5 TYPES OF ENSEMBLES

If we are to use an ensemble average, it is crucial that the distribution function be known. It is possible to choose a distribution function subject to certain constraints on the ensemble. The simplest ensemble is one whose distribution function is a constant, $\rho(p,q) = C$. This equation says that each member of the ensemble is equally probable. Physically, it corresponds to a system whose energy (E), number of particles (N), and volume (V) can vary unconditionally. Thus, for the purpose of constructing averages, we have no need of the ensemble but can consider just one member. Because this member can be in any microstate, we know nothing about the system. Hence, the *uniform ensemble* is an ensemble of a system about which we have no information. It is the simplest ensemble but is not very useful.

1.5.1 Microcanonical Ensemble

Next, we consider a system under the constraints of almost fixed energy, volume, and number of particles, which defines an almost isolated system. A completely isolated system is an idealization, which is impractical. The point is that, to study any system, we have to interact with it. For example, in order to measure its temperature we have to let a little bit of energy ΔE flow to the thermometer. The concept of an almost isolated system is useful if none of the thermodynamic properties of the system depends on the magnitude of ΔE. The ensemble corresponding to this system is called a *microcanonical ensemble* and its distribution function is given by

$$\rho(p,q) = 1/\Omega, \quad E < \mathcal{H}(p,q) < E + \Delta E \qquad (1.57)$$

and zero otherwise. Here $\mathcal{H}(p,q)$ is the Hamiltonian of the system, and the quantity $\Omega(E,V,N)$ is the total number of microstates satisfying the given constraints and is given by

$$\int_{\mathcal{R}} dp\,dq = \Omega \tag{1.58}$$

where the integral is a $2n$-dimensional integral over a region $\mathcal{R}$ of the phase space satisfying the constraints of constant volume, constant number of particles, and $E < \mathcal{H}(p,q) < E + \Delta E$. For those systems of almost fixed energy E, this ensemble is like the uniform ensemble. The difference is that ρ is mostly zero here. The microcanonical ensemble is the basic ensemble for almost isolated systems.

Two questions remain now. How is the quantity Ω related to physical properties, and how is Ω evaluated? It was shown by Boltzmann that Ω is related to entropy according to the famous formula, first written by Planck,

$$S = k_{\rm B} \ln \Omega(E,V,N) \tag{1.59}$$

where $k_{\rm B}$ is *Boltzmann's constant*. A nice derivation of this result has been given by Pathria (Pathria, 1996, Chapter 1). Since a relation between a statistical and a thermodynamic quantity has been found, one can calculate all other properties. Let us write the differential of S from thermodynamics,

$$dS = \frac{dE}{T} + \frac{PdV}{T} - \frac{\mu dN}{T} \tag{1.60}$$

where P and μ are the pressure and the chemical potential of the system, respectively. Relations like this one are commonplace in thermodynamics. A well-known text written from a modern point of view is by Callen (Callen, 1985), to which the reader is referred for thermodynamics. From Eq. (1.59),

$$dS = k_{\rm B}\frac{\partial \ln \Omega}{\partial E}dE + k_{\rm B}\frac{\partial \ln \Omega}{\partial V}dV + k_{\rm B}\frac{\partial \ln \Omega}{\partial N}dN \tag{1.61}$$

By comparing this equation with Eq. (1.60), we get

$$\frac{\partial \ln \Omega}{\partial E} = \frac{1}{k_{\rm B}T} \qquad \frac{\partial \ln \Omega}{\partial V} = \frac{P}{k_{\rm B}T} \qquad \frac{\partial \ln \Omega}{\partial N} = -\frac{\mu}{k_{\rm B}T} \tag{1.62}$$

When we differentiate partially with respect to one variable in these equations, it should be remembered that the other variables are to be held constant. The variables E and V are usually thought of as continuous variables, so differentiating with respect to them is acceptable. Since the variable N is assumed to be discrete, how is the derivative with respect to it possible? The answer is that, in a typical situation, N is so large that a fractional change obtained by changing one particle in a macroscopic quantity $S(N)$,

$$S(N + 1) - S(N) = \Delta S/\Delta N \tag{1.63}$$

may be thought of as the derivative $\partial S/\partial N$.

1.5.2 Canonical Ensemble

The microcanonical ensemble is the simplest ensemble with a physical interpretation. Unfortunately, because of the microcanonical constraints, the integral (1.58) is very difficult to evaluate mathematically, except for very simple systems. [Two examples of physical systems in which it is evaluated and shown to be independent of ΔE can be found in Pathria (Pathria, 1996, Chapter 2).] In view of the independence of the results from ΔE, from now on we assume $\Delta E = 0$. Instead of the microcanonical ensemble, we now consider a "small" system in contact with a large thermal reservoir, with which energy can be exchanged, as being one large isolated system. The ensemble corresponding to the "small" system is a *canonical ensemble*. It should be kept in mind that the small system studied here is small only relative to the reservoir, but it is still a very large macroscopic system. The total, isolated system can be described by a microcanonical ensemble. As seen above, the microcanonical distribution function is $\rho = 1/\Omega(E)$, where we do not display the number of particles and volume because they are fixed. Let the instantaneous energies of the system and reservoir be E_s and E_r, respectively. The total energy is fixed at $E = E_r + E_s$. Treating the system and the reservoir as weakly interacting, they can be considered to be almost independent systems so that $\rho = \rho_r \rho_s = 1/(\Omega_r \Omega_s)$ and

$$\rho_s(E_s) = \Omega_r(E_r)/\Omega(E) \tag{1.64}$$

since $\Omega = \Omega_r \Omega_s$.

Because $E_r \gg E_s$, Ω_r is in general a very large number, which varies as energy is exchanged between system and reservoir. One way of handling this situation is to expand $\Omega_r(E_r)$ in a Taylor's series about E. However, Ω_r is large and rapid convergence of the series cannot be assured. The way around this problem is to expand the slowly varying function $\ln \Omega_r(E_r)$ about $E_r = E$:

$$\ln \Omega_r(E_r) = \ln \Omega_r(E) + (E_r - E) \left(\frac{\partial \ln \Omega_r}{\partial E_r} \right)_{E_r = E} + \cdots$$

$$= \ln \Omega_r(E) - E_s \left(\frac{\partial \ln \Omega_r}{\partial E_r} \right)_{E_r = E} + \cdots \tag{1.65}$$

Ignoring terms beyond the linear, we have

$$\ln \Omega_r(E_r) = \ln \Omega_r(E) - \beta E_s \tag{1.66}$$

where

$$\beta = \left(\frac{\partial \ln \Omega_r}{\partial E_r} \right)_{E_r = E} \tag{1.67}$$

The antilog of Eq. (1.66) gives

$$\Omega_r(E_r) = \Omega_r(E) \exp(-\beta E_s) \tag{1.68}$$

By using Eq. (1.64), we get

$$\rho_s(E_s) = C \exp(-\beta E_s) \tag{1.69}$$

where C is a normalization constant, independent of E_s. Normalization requires

$$\sum_{s'} \rho_{s'} = C \sum_{s'} \exp(-\beta E_{s'}) = 1 \tag{1.70}$$

where the sum is over all states of the system, so that

$$C = \left[\sum_{s'} \exp(-\beta E_{s'}) \right]^{-1} \tag{1.71}$$

Here the sum is symbolic only. In reality, in classical mechanics the sum over microstates means an integral over the coarse-grained cells of the phase space. Finally,

$$\rho_s = \frac{\exp(-\beta E_s)}{\sum_{s'} \exp(-\beta E_{s'})} \tag{1.72}$$

It can be shown for a fixed E_s that

$$\left(\frac{\partial \ln \Omega_r}{\partial E_r}\right)_{E_r = E} = \left(\frac{\partial \ln \Omega}{\partial E}\right)_{N,V} \tag{1.73}$$

Therefore, from Eqs. (1.62) and (1.67), it follows that

$$\beta = 1/k_B T \tag{1.74}$$

The proof that the terms beyond the linear can be neglected in Eq. (1.65) is an exercise at the end of this chapter. These terms become very small when the reservoir becomes large and can be neglected entirely in the case of an infinite reservoir. It is important to mention here that Eq. (1.72) does not depend on the detailed nature of the reservoir. It depends only on a single parameter, the parameter β, which is inversely proportional to the temperature. Equation (1.72) is the distribution function for the canonical ensemble. The sum over microstates in the denominator of Eq. (1.72) is called the *canonical partition function* and is denoted by $Z(N, V, T)$. Clearly,

$$Z(N, V, T) = \sum_{s'} \exp(-\beta E_{s'}) \tag{1.75}$$

The sum here is over all the microstates. Denoting the number of microstates corresponding to the energy $E_{s'}$ as $\Omega_{s'}$, we can write it as a sum over possible energy levels in quantum mechanics.

$$Z(N, V, T) = \sum_{\{\text{levels}\}} \Omega_{s'} \exp(-\beta E_{s'}) \tag{1.76}$$

The normalization requirement for the microcanonical distribution function requires that $\sum(1/\Omega) = 1$, where the sum is over all microstates Ω. In this sense, Ω is the partition function of the microcanonical ensemble. Thus, Eq. (1.76) can be looked upon as a weighted sum of microcanonical partition functions with a weight $\exp(-\beta E_{s'})$. The relationship of $Z(N, V, T)$ with thermodynamics will be established in Section 1.6.

1.5.3   Grand Canonical Ensemble

We now relax the requirement that the system be closed, so that the system and reservoir can exchange particles. We let the instantaneous number of particles in the system and reservoir be N_s and N_r, respectively, and we let the total number of particles be N. The derivation of the *grand canonical distribution function* follows the same reasoning that was used for the distribution function of the canonical ensemble. Now, however, $\ln \Omega_r(E_r, N_r)$ is expanded about E and N, where $N = N_s + N_r$. The distribution function is

$$\rho_s = \frac{\exp(-\alpha N_s - \beta E_s)}{\sum_{N_{s'}} \sum_{s'} \exp[-\alpha N_{s'} - \beta E_{s'}(N_{s'})]} \tag{1.77}$$

The double sum is required in the denominator of Eq. (1.77), since both the energy $E_{s'}$ and number of particles $N_{s'}$ are variable now. Also, $E_{s'}$ depends on $N_{s'}$, a fact explicitly indicated in Eq. (1.77). Here, the quantity α is defined as

$$\alpha = \left(\frac{\partial \ln \Omega_r}{\partial N_r}\right)_{E_r = E, N_r = N} \tag{1.78}$$

and it can be shown that

$$\alpha = -\mu/k_B T \tag{1.79}$$

The quantity in the denominator of Eq. (1.77) is called the *grand canonical partition function* and is denoted by $\Xi(\mu, V, T)$:

$$\Xi(\mu, V, T) = \sum_{N_{s'}} \sum_{s'} \exp[+\beta \mu N_{s'} - \beta E_{s'}(N_{s'})] \tag{1.80}$$

The quantity $\exp(\beta \mu)$ is called the *absolute activity* (also called *fugacity* by some authors) and is denoted by z. Equation (1.80) can also be written as

$$\Xi(\mu, V, T) = \sum_{N_{s'}} z^{N_{s'}} Z(N_{s'}, V, T) \tag{1.81}$$

which makes it clear that the grand canonical partition function is a weighted sum of canonical partition functions with weight $z^{N_{s'}}$.

Other ensembles can be defined similarly (see McQuarrie, 1976, Chapter 3), but these will not concern us here. The *isothermal-isobaric* (or the constant T–constant P) *ensemble* will be needed in Chapter 4 and is left as an exercise.

Once the ensembles are defined, averages of physical quantities may be calculated analytically, at least in principle. For many systems, the averages must be obtained numerically. Two broad methods are available for numerical calculations. In the first method, called the *molecular dynamics method*, one solves Hamilton's or Newton's equations of motion for the given system to take the system through a series of microstates, and one then takes time averages. This deterministic scheme is the subject of Chapter 5. In the second method, called the *Monte Carlo method*, the system is taken from one microstate to another in a probabilistic manner, generating enough microstates consistent with the given macrostate so that ensemble averages can be determined. This stochastic method is discussed in detail in Chapter 6. Do the two methods give the same results for the averages? In a wide variety of cases, they do, and, to that extent the ergodic hypothesis may be considered to have been verified numerically.

1.6 THERMODYNAMICS

1.6.1 Canonical and Grand Canonical Partition Functions

The Boltzmann relation between entropy and the microcanonical distribution function, Eq. (1.59), can be generalized to other ensembles. The generalization is

$$S = -k_B \langle \ln \rho \rangle \tag{1.82}$$

where the average is taken over the appropriate ensemble corresponding to ρ. For the microcanonical ensemble, ρ is a constant equal to $1/\Omega$, so that Eq. (1.59) is obtained, as expected.

For the canonical ensemble, in which N, V, and T are fixed, but the energy of the system is allowed to vary, ρ is given by Eq. (1.72). Substituting this in Eq. (1.82) gives

$$S = -k_B \langle \ln[(1/Z)\exp(-\beta E)] \rangle = k_B \ln Z + \langle E \rangle / T \tag{1.83}$$

where the relations (1.74) and (1.75) have been used, and $Z \equiv Z(N, V, T)$. Equation (1.83) can be rewritten to give

$$-k_B \ln Z = (\langle E \rangle - TS)/T = A/T \tag{1.84}$$

where A is the Helmholtz free energy. Therefore,

$$A = -k_B T \ln Z \tag{1.85}$$

Once the Helmholtz free energy has been found, other quantities can be obtained using thermodynamics. For example, using the relation

$$dA = -S\,dT - P\,dV + \mu\,dN \tag{1.86}$$

one may obtain S, P, and μ. Then the average energy may be obtained by using

$$\langle E \rangle = A + TS \qquad (1.87)$$

The average energy can also be obtained directly from its definition

$$\langle E \rangle = \frac{1}{Z} \sum_s E_s \exp(-\beta E_s) = -\frac{\partial}{\partial \beta} \ln Z \qquad (1.88)$$

Clearly, the partition function, which plays a major role in the evaluation of thermodynamic functions, is a direct result of the normalization of the statistical distribution function [Eq. (1.72)]. The partition function sum is over all the accessible microstates. The classical partition function can be written as an integral over phase space:

$$Z(N, V, T) = \frac{1}{N! h^{3N}} \int \exp[-E(p, q)/k_B T] \, dp \, dq \qquad (1.89)$$

where the factors $N!$ and h^{3N}, h being Planck's constant, cannot be obtained within classical statistical mechanics. They result when the classical limit of the quantum mechanical partition function is taken (see Chapter 2). Briefly, the factor $N!$ arises from the fact that, in reality, identical particles of a given system are indistinguishable from one another, whereas classical mechanics treats them as distinguishable. The $1/h^{3N}$ correction may be thought of as a factor needed to make the partition function dimensionless, as it should be. If these factors are not included, paradoxes arise, a discussion of which may be found in Pathria (Pathria, 1996, Chapter 1).

In the *grand canonical ensemble*, both the energy and the number of particles of the system are allowed to change. The partition function is defined by

$$\Xi(\mu, V, T) = \sum_{N_s} \sum_s \exp[\beta\{\mu N_s - E_s(N_s)\}] \qquad (1.90)$$

Corresponding to Eq. (1.85), we get,

$$G - A = k_B T \ln \Xi \qquad (1.91)$$

where $G = \mu \langle N \rangle$ is the Gibbs free energy. Noting that,

$$G = A + PV \qquad (1.92)$$

it is seen that

$$PV = k_B T \ln \Xi \qquad (1.93)$$

By using the thermodynamic relationship

$$d(PV) - P\,dV + S\,dT + \langle N \rangle\,d\mu \qquad (1.94)$$

the quantities S and $\langle N \rangle$ can be obtained. For future reference, we quote the result

$$\langle N \rangle = k_B T \left(\frac{\partial \ln \Xi}{\partial \mu} \right)_{V,T} \qquad (1.95)$$

1.6.2   Fluctuations and the Thermodynamic Limit

By using the results in Section 1.5, it can be shown (see Exercises at the end of this chapter) that, in the vicinity of the average energy, $\Omega(E, V, N)$ becomes a Gaussian distribution as a function of energy. The mean of the distribution is equal to the average energy $\langle E \rangle$. The standard deviation is a measure of the *fluctuations* of the energy around the mean. To see how sharply peaked the thermodynamic functions are, we now calculate the relative fluctuations in the internal energy and particle number. For the energy in the canonical ensemble, the variance can be obtained by differentiating Eq. (1.88) with respect to β. We have

$$\frac{\partial \langle E \rangle}{\partial \beta} = \frac{1}{Z} \left[\sum_s -\beta E_s^2 \exp(-\beta E_s) \right]$$
$$- \frac{1}{Z^2} \left[\sum_s -\beta E_s \exp(-\beta E_s) \right] \left[\sum_s E_s \exp(-\beta E_s) \right] \qquad (1.96)$$

The right-hand side of this equation is clearly equal to

$$- \beta \langle E^2 \rangle + \beta \langle E \rangle^2 \qquad (1.97)$$

Therefore,

$$\sigma(E)^2 = \langle E^2 \rangle - \langle E \rangle^2 = k_B T^2 \left(\frac{\partial \langle E \rangle}{\partial T} \right)_{N,V} = k_B T^2 C_V \qquad (1.98)$$

where C_V is the heat capacity at constant volume. For example, for 1 mol of an ideal monatomic gas, $C_V = (\frac{3}{2})N_0 k_B$, where N_0 is Avogadro's number, and at 300 K, $\sigma(E) = [(\frac{3}{2})N_0 k_B^2 T^2]^{1/2} \approx 3.9 \times 10^{-9} J$. Since $\langle E \rangle = (\frac{3}{2})N_0 k_B T \approx 3.7 \times 10^3 J$, the percent relative fluctuation, $(\sigma(E)/\langle E \rangle) \times 100$, is about $1 \times 10^{-10}\%$, an extremely small number. This means that, in spite of the fact that the sum in Eq. (1.72) is over all possible energy values, by far the largest contribution to the sum comes from the energy range near the average energy. Hence, the results for the canonical ensemble approach those of the microcanonical ensemble. Similar results follow for systems other than the ideal gas.

If the limit $N \to \infty$ and $V \to \infty$ is taken in a given system with reasonably well-behaved interactions, the following results are found:

- All the extensive quantities of the system go to infinity proportionally to the number of particles.
- All the intensive quantities approach a limit that is independent of the number of particles, but, of course, may depend on the number density, equal to $\lim_{N \to \infty} N/V$.
- The values of thermodynamic quantities found from the microcanonical and canonical distribution are exactly identical.
- The identity of results from the two ensembles remains valid even in the region of phase transitions, where the various thermodynamic quantities develop mathematical singularities (see Chapter 3).

The limit involved here is called the *thermodynamic limit* and was introduced into statistical mechanics by Kramers. An excellent discussion of this limit and many other rigorous results is one by Griffiths (Griffiths, 1972, pp. 7–109).

The percent relative fluctuation in the particle number can be obtained exactly in the same way as the fluctuation in energy above. It can be shown that (see McQuarrie, 1976, p. 61)

$$\sigma(N)^2 = \langle N^2 \rangle - \langle N \rangle^2 = \langle N \rangle^2 k_B T \kappa_T / V \tag{1.99}$$

where $\kappa_T = -(1/V)(\partial V/\partial P)_{N,T}$ is the isothermal compressibility. For an ideal gas, $\kappa_T = \langle N \rangle k_B T/(P^2 V) = 1/P$, and $\sigma(N)^2 = \langle N \rangle$. Thus, $\sigma(N) = \langle N \rangle^{1/2}$ and $\sigma(N)/\langle N \rangle = \langle N \rangle^{-1/2}$. For 1 mol of an ideal gas, $\sigma(N)/\langle N \rangle \approx 1.3 \times 10^{-12}$ or $1.3 \times 10^{-10}\%$. Similar results can be shown to hold for other systems.

These results show that the relative fluctuation in the number of particles in the grand canonical ensemble is very small for macroscopic systems. Therefore, in Eq. (1.77) the sum over the variable number of particles is dominated very strongly by the term involving the average number of particles, thereby making the grand canonical ensemble equivalent to the canonical ensemble. In the thermodynamic limit, all the statements made above remain valid, particularly the last two items, which now include the grand canonical ensemble.

1.6.3   Boltzmann Distribution

Consider a system in a canonical ensemble whose N constituent molecules are independent. An example is a classical ideal gas. The phase space of this system is sometimes called Γ space (gas space). Now focus attention on just one molecule. The phase space of one molecule is called μ space (molecule space). If the instantaneous values of the positions and momenta of all N molecules are determined and plotted in one μ space, a cloud of points is obtained in μ space. This cloud of points in μ space provides just one point in Γ space. In what follows we work in μ space.

In analogy to Eq. (1.72) and the discussion of the canonical ensemble, a one-molecule distribution function in μ space can be written as

$$f_j = \frac{\exp(-\epsilon_j/k_B T)}{\sum_j \exp(-\epsilon_j/k_B T)} = \frac{\exp(-\epsilon_j/k_B T)}{Z_1} \tag{1.100}$$

where ϵ_j is the energy of the jth molecular energy state, $Z_1 \equiv Z(1, V, T)$ is called the *molecular partition function*, and the subscript 1 is a reminder that it is for one molecule. In Eq. (1.100), f_j is the fraction of molecules that are in the jth energy state. This fraction is not the fraction of molecules having an energy ϵ_j. To obtain that fraction we have to know the degeneracy factor g_j, which indicates the number of states per energy level. Once g_j is known, the fraction of particles having an energy ϵ_j can be written as

$$\frac{n_j}{N} = g_j \frac{\exp(-\epsilon_j/k_B T)}{Z_1} \tag{1.101}$$

where Z_1 can be written as a sum over energy levels:

$$Z_1 = \sum_i g_i \exp(-\epsilon_i/k_B T) \tag{1.102}$$

Equations (1.100) and (1.101) are alternate forms of the *Boltzmann distribution*; $\exp(-\epsilon_i/k_B T)$ is called the *Boltzmann factor*. The Boltzmann distribution is also known as the Maxwell–Boltzmann distribution in honor of

J. C. Maxwell, who first derived an expression for the distribution of molecular velocities. Later, Boltzmann obtained the same distribution as part of a general treatment of the nonequilibrium behavior of gases (see Chapter 8). As derived from kinetic molecular theory, the Maxwell–Boltzmann distribution of molecular velocities $\mathbf{v}$ for molecules of mass m is

$$f(\mathbf{v}) = \frac{N}{V}\left(\frac{m}{2\pi k_B T}\right)^{3/2}\exp\left(\frac{-m\mathbf{v}^2}{2k_B T}\right) \tag{1.103}$$

which is normalized such that

$$\int f d^3 v\, d^3 r = N \tag{1.104}$$

The Boltzmann distribution can be obtained also by use of the microcanonical ensemble. This utilizes the method of the most probable distribution. A discussion of the method can be found in Pathria (Pathria, 1996, Section 6.1). The method involves distributing N particles into various energy levels with n_j particles in the energy level ϵ_j and maximizing the entropy. Stirling's formula is used to simplify the factorials such as $n_j!$. This produces an inconsistency, because Stirling's formula is valid for $n_j \gg 1$, a condition that is violated for high energies or low temperatures (see Pathria, 1996, Figure 6.2).

The derivation of the Boltzmann distribution from the grand canonical ensemble, as a limiting case of the quantum mechanical distributions, is discussed in Chapter 2.

1.6.4   Classical Ideal Gas

For a classical ideal polyatomic gas, the canonical partition function Z is the product of the individual molecular partition functions. Division by $N!$ corrects for the indistinguishability of the molecules, which affects only the translational motion and has no effect on the internal motions. Thus, $Z = Z_1^N/N!$. If the translational, rotational, vibrational, and electronic molecular degrees of freedom are independent,

$$
\begin{aligned}
Z_1 &= \sum_{t,r,v,e}\exp[-\beta(\epsilon_t + \epsilon_r + \epsilon_v + \epsilon_e)] \\
&= \sum_t \exp(-\beta\epsilon_t)\sum_r \exp(-\beta\epsilon_r)\sum_v \exp(-\beta\epsilon_v)\sum_e \exp(-\beta\epsilon_e) \\
&= Z_t Z_r Z_v Z_e
\end{aligned}
\tag{1.105}
$$

In this case,

$$Z = \frac{1}{N!}Z_t^N Z_r^N Z_v^N Z_e^N \tag{1.106}$$

The evaluation of the molecular partition functions for the individual degrees of freedom is discussed in undergraduate physical chemistry texts (e.g., Levine, 1995, Section 22.6) and so will not be discussed here. We shall sketch the derivation for the monatomic gas called the *Boltzmann gas* and quote the results for future reference. The Hamiltonian for the translational motion of the *i*th particle is given by

$$\mathcal{H}_i = \frac{p_i^2}{2m} \tag{1.107}$$

where m is the mass of the particle. In applying Eq. (1.89) for $N = 1$, the position integrals give V and the momentum integrals for the three components are Gaussian integrals. We find that

$$Z_t = V\left(\frac{2\pi m k_B T}{h^2}\right)^{3/2} \tag{1.108}$$

so that

$$Z = \frac{1}{N!}V^N\left(\frac{2\pi m k_B T}{h^2}\right)^{3N/2} \tag{1.109}$$

With this, the sum over N in the grand canonical partition function [Eq. (1.81)] is the series expansion of the exponential function. We obtain

$$\Xi(\mu, V, T) = \exp\left[zV(2\pi m k_B T/h^2)^{3/2}\right] \tag{1.110}$$

The pressure and the average number of particles are obtained by using Eqs. (1.93) and (1.95), respectively. We get

$$PV = k_B TzV(2\pi m k_B T/h^2)^{3/2} \tag{1.111}$$

$$\langle N \rangle = zV(2\pi m k_B T/h^2)^{3/2} \tag{1.112}$$

so that

$$PV = \langle N \rangle k_B T \tag{1.113}$$

the familiar ideal gas equation. Other thermodynamic quantities can be obtained by using standard thermodynamics.

Following the procedure in Section 1.6.2, one may calculate the fluctuations in the energy of a single molecule. We get

$$\sigma(\epsilon)^2 = \langle \epsilon^2 \rangle - \langle \epsilon \rangle^2 = k_{\mathrm{B}} T^2 \left(\frac{\partial \langle \epsilon \rangle}{\partial T} \right)_V \tag{1.114}$$

For monatomic molecules, $\langle \epsilon \rangle = (\frac{3}{2}) k_{\mathrm{B}} T$ and $\sigma(\epsilon) = \sqrt{3/2} k_{\mathrm{B}} T$. The percent deviation is $\sigma(\epsilon)/\langle \epsilon \rangle \times 100 \approx 82\%$. This value says that if we choose a molecule at random, the energy distribution in a single molecule is not peaked at the average value and fluctuates wildly. Previously we saw that, in the space of the whole gas, the Γ space, the energy of the system is extremely close to the average. The differences between μ and Γ space make clear the fact that the ideas of thermodynamics are not applicable to individual molecules but to a large collection of molecules. As mentioned before, thermodynamics is rigorously obtained from statistical mechanics only if the thermodynamic limit is taken, in which case the number of molecules becomes infinite.

1.7 NONEQUILIBRIUM STATISTICAL MECHANICS

1.7.1 Liouville's Theorem

For a system in equilibrium, the statistical distribution function ρ does not change in time. The reason is that, in equilibrium, the values of thermodynamic quantities, which are averages over the distribution function, are independent of time. Of course, we learned previously that things are "quiet" only on the macroscopic level. On a microscopic level, the system is in a constant state of motion, and the generalized coordinates keep changing according to Hamilton's equations of motion. Why is this motion not reflected in the equilibrium distribution function? It is because of our drastic but essential assumption that ρ depends on the generalized coordinates through strictly conserved quantities, such as the total energy or the total number of particles.

Just as for equilibrium, a system will trace out a trajectory in $2n$-dimensional phase space under nonequilibrium conditions. In order to study how a system, not in equilibrium, reaches a state of equilibrium, we need to study the time dependence of the distribution function $\rho(p, q, t)$. It is a point function whose total differential is

$$d\rho = \frac{\partial \rho}{\partial t} dt + \sum_i \frac{\partial \rho}{\partial p_i} dp_i + \sum_i \frac{\partial \rho}{\partial q_i} dq_i \tag{1.115}$$

and hence

$$\frac{d\rho}{dt} = \frac{\partial \rho}{\partial t} + \sum_i \left(\frac{\partial \rho}{\partial p_i} \dot{p}_i + \frac{\partial \rho}{\partial q_i} \dot{q}_i \right) \tag{1.116}$$

where $d\rho/dt$ is the change in the distribution function with time for a given volume of phase space and $\partial \rho/\partial t$ is the time variation of the distribution function at a fixed point, say (p_0, q_0), in phase space. As usual, the dots over the positions and momenta indicate time derivatives. The motions of the generalized coordinates are governed by Hamilton's equations of motion.

$$\dot{p}_i = -\frac{\partial \mathcal{H}}{\partial q_i} \qquad \dot{q}_i = +\frac{\partial \mathcal{H}}{\partial p_i} \tag{1.117}$$

For a fixed and closed volume V_Γ in phase space having a surface area A_Γ, the number of points N_V in volume V_Γ at time t is

$$N_V = N_e \int_{V_\Gamma} \rho \, dp \, dq \tag{1.118}$$

and hence

$$\frac{dN_V}{dt} = N_e \int_{V_\Gamma} \frac{\partial \rho}{\partial t} \, dp \, dq \tag{1.119}$$

where N_e is the total number of points in the ensemble. Equation (1.118) follows from the definition of ρ, and Eq. (1.119) represents the change in the number of points per unit time in the volume V_Γ. The number of points passing into the volume V_Γ per unit time can be written

$$\frac{dN_V}{dt} = -N_e \int_{A_\Gamma} \rho \mathbf{v} \cdot d\mathbf{A}_\Gamma \tag{1.120}$$

where $\mathbf{v} = (\dot{p}_1, \dot{p}_2, \ldots, \dot{p}_n, \dot{q}_1, \dot{q}_2, \ldots, \dot{q}_n)$, and the negative sign arises because $d\mathbf{A}_\Gamma$ is a vector directed outward from the surface along the unit normal vector. The integral in Eq. (1.120) is over those phase points on the surface of the closed volume V_Γ. This equation can be found in texts on vector analysis as can a discussion of Gauss' divergence theorem (e.g., Margenau and Murphy, 1956, Section 4.18), which we now use to rewrite Eq. (1.120) in terms of the *del operator*

$$\nabla = \left(\frac{\partial}{\partial p_1}, \frac{\partial}{\partial p_2}, \ldots, \frac{\partial}{\partial p_n}, \frac{\partial}{\partial q_1}, \frac{\partial}{\partial q_2}, \ldots, \frac{\partial}{\partial q_n} \right) \tag{1.121}$$

as

$$\frac{dN_V}{dt} = -N_e \int_{V_\Gamma} \nabla \cdot (\rho \mathbf{v}) \, dV_\Gamma \tag{1.122}$$

If there are no sources or sinks in the volume V_Γ, so that the number of phase points is conserved, Eqs. (1.119) and (1.122) can be equated to give

$$N_e \int_{V_\Gamma} \frac{\partial \rho}{\partial t} \, dp \, dq = -N_e \int_{V_\Gamma} \nabla \cdot (\rho \mathbf{v}) \, dV_\Gamma \tag{1.123}$$

Equation (1.123) can be valid for an arbitrary volume V_Γ only if

$$\frac{\partial \rho}{\partial t} + \nabla \cdot (\rho \mathbf{v}) = 0 \tag{1.124}$$

or

$$\frac{\partial \rho}{\partial t} + \text{div}(\rho \mathbf{v}) = 0 \tag{1.125}$$

where div is the familiar divergence of vector analysis. This is the well-known *equation of continuity*, which states that if there is no sink or source (i.e., an almost isolated system), the number of phase points entering the volume V_Γ must also leave in order to conserve their number. Notice that this requires that an ensemble (equilibrium or nonequilibrium) must include all accessible points in phase space, even if some points may not be accessible at equilibrium, otherwise there will appear to be sources and/or sinks.

It should be noted that, unless q is an orthogonal Cartesian coordinate, operators of the form $\partial/\partial q$ are symbolic only. If q is not an orthogonal Cartesian coordinate, the differential operator $\partial/\partial q$ must be determined by the chain rule according to the equations relating the Cartesian and new coordinate systems. Such transformations are adequately discussed in most vector analysis and quantum mechanics books (e.g., Levine, 1983, Chapter 5).

By using the standard relation (Margenau and Murphy, 1956, p. 153)

$$\nabla \cdot (\rho \mathbf{v}) = \rho(\nabla \cdot \mathbf{v}) + (\nabla \rho) \cdot \mathbf{v} \tag{1.126}$$

Eq. (1.124) can be written as

$$\frac{\partial \rho}{\partial t} + \rho \sum_i \left(\frac{\partial \dot{p}_i}{\partial p_i} + \frac{\partial \dot{q}_i}{\partial q_i} \right) + \sum_i \left(\frac{\partial \rho}{\partial p_i} \dot{p}_i + \frac{\partial \rho}{\partial q_i} \dot{q}_i \right) = 0 \tag{1.127}$$

Now we use Hamilton's equations of motion [Eq. (1.117)] on both sums. The first sum vanishes because

$$\frac{\partial \dot{p}_i}{\partial p_i} = -\frac{\partial^2 \mathcal{H}}{\partial q_i \partial p_i} = -\frac{\partial \dot{q}_i}{\partial q_i} \tag{1.128}$$

The second sum is recognized as a *Poisson bracket*, which is defined as

$$[\rho, \mathcal{H}] = \sum_i \left(\frac{\partial \rho}{\partial q_i} \frac{\partial \mathcal{H}}{\partial p_i} - \frac{\partial \rho}{\partial p_i} \frac{\partial \mathcal{H}}{\partial q_i} \right) \tag{1.129}$$

Therefore, Eq. (1.127) reduces to

$$\frac{\partial \rho}{\partial t} = -[\rho, \mathcal{H}] \tag{1.130}$$

On looking at the sum in Eq. (1.116), it is seen that this sum is also the Poisson bracket. Consequently, Eq. (1.116) can be written as

$$\frac{d\rho}{dt} = \frac{\partial \rho}{\partial t} + [\rho, \mathcal{H}] \tag{1.131}$$

which, on using Eq. (1.130), becomes

$$d\rho/dt = 0 \tag{1.132}$$

which is *Liouville's theorem*. It states that the distribution function is conserved in time. Another way of stating this is that phase space is incompressible. In this chapter, we do not consider dissipative systems to which Liouville's theorem does not apply. Such systems are considered in Chapters 12 and 13.

The statistical distribution function is associated with a volume or "cell" $\Delta p \Delta q$ in phase space. The phase points within the cell each trace out a trajectory in time. Because the solutions of Hamilton's equations of motion are unique, trajectories cannot cross, and points within and on the boundary of a phase cell can never escape from the cell. The total differential $d\rho/dt$ represents the change in the distribution function of the cell as it moves through phase space. From Liouville's theorem, $d\rho/dt = 0$, and the volume of the cell remains constant, although the cell is distorted as it streams through phase space. Because the shape of the cell is distorted in time, the density of phase points is not uniform in the given cell but is continually changing such that the average density is constant. The simplicity of Liouville's theorem and other results discussed here are a direct consequence of choosing a phase space composed of conjugate coordinates. Otherwise Eq. (1.128) would not be true, and the final results would not be so simple.

Liouville's equation (1.132) is the basic equation that must be solved in order to understand the behavior of the distribution function for irreversible processes. On the other hand, for reversible processes, which are obtained by letting systems go from one thermodynamic state to another through almost-equilibrium states, ρ must be explicitly independent of time, giving rise to a *stationary* ensemble. Otherwise, the thermodynamic averages obtained using it would depend on time, which would be contrary to the definition of equilibrium averages. The change in time of the statistical distribution function at a fixed point in phase space is given by Eq. (1.130). Therefore, a simple way to assure that we have a stationary ensemble is to choose ρ to satisfy

$$\left[\rho, \mathcal{H}\right] = 0 \tag{1.133}$$

This condition is satisfied by the equilibrium ensembles introduced in this chapter. In the microcanonical ensemble, ρ is constant or zero; in the canonical ensemble, ρ is a function of $\mathcal{H}$ only, and in the grand canonical ensemble it is a function of $\mathcal{H}$ and another conserved quantity N, the total number of particles. In all these cases, $[\rho, \mathcal{H}] = 0$ as required.

1.7.2 The Joule Expansion

To help clarify the somewhat formal and mathematical treatment given above, we consider the Joule expansion; that is, the expansion of an ideal gas into a vacuum. We must first set up an ensemble. We choose an ensemble whose independent members consist of an ideal gas at fixed internal energy, number of molecules, and volume. The physical situation consists of two identical bulbs connected by a stopcock. Initially, one bulb contains the gas and the other is empty. The volume consists of the total volume of both bulbs. The ensemble must consist of all accessible configurations; this includes, upon opening the stopcock, the configurations consisting of (1) all molecules in the one bulb, (2) all molecules in the other bulb, and (3) all configurations in between.

There are two ways of viewing the Joule expansion. One is to consider an equilibrium ensemble and let the system move to equilibrium. The other is to consider a nonequilibrium ensemble and observe its behavior as the system traces out a trajectory in phase space. We use the first method and choose the microcanonical ensemble. At equilibrium, we expect $\rho(p, q)$ to be small for all but equal numbers of molecules in each bulb. When the stopcock is opened, the system is in a region of phase space where $\rho(p, q)$ is vanishingly small. Thermodynamics tells us that the system will spontaneously move to a region of high probability (see Section 1.7.3). Therefore, the system will seek the equilibrium distribution of equal numbers of molecules in each bulb.

The second way to view the Joule expansion is to let the statistical distribution function $\rho(p, q, t)$ be a function of time and to observe how it changes in time as the system traces out a trajectory in phase space. In this case, Liouville's theorem tells us that $\rho(p, q, t)$ does not change. However, the distribution function at each point of phase space must change in such a way that the final ensemble is an equilibrium ensemble.

1.7.3   Boltzmann's H-Theorem

In thermodynamics, entropy can be calculated only for a reversible path. If the entropy of an isolated system increases for a given process, the process is said to be irreversible in accordance with the second law of thermodynamics. The concept of entropy can be broadened by a generalization of Eq. (1.82) to include nonequilibrium situations. To this end, we seek a function that provides a measure of the extent to which a system deviates from macroscopic equilibrium. The function we seek was enunciated in Boltzmann's famous H-theorem. This theorem states that the approach of a system toward equilibrium may be regarded as corresponding to a decrease of a function H toward a minimum value. The proof of Boltzmann's theorem traditionally depends on the Boltzmann equation (see Chapter 8) and the associated distribution functions in μ space. Instead, we follow the argument of Tolman (Tolman, 1979, Section 51), which is entirely in Γ space. The discussion by Tolman is a mathematical version of qualitative arguments presented by Gibbs earlier. The main idea is that, according to Liouville's theorem, ρ does not change in time, and thus it cannot be used to define entropy for nonequilibrium processes. If, however, ρ is averaged, or *coarse-grained* over a suitable volume, a definition of entropy consistent with thermodynamics becomes possible.

The coarse-grained H function is *defined* as

$$H = \ln P \tag{1.134}$$

where P is a coarse-grained probability density given by

$$P = \frac{1}{\delta V} \int_{\delta V} \rho \, dp \, dq \tag{1.135}$$

Clearly, if the H function were defined with ρ, which may be called the *fine-grained distribution*, instead of P, it would never change in time for either equilibrium or nonequilibrium processes because of Liouville's theorem. On the other hand, the time dependence of P is not so clear cut. In the definition of P, $\delta V = dp\,dq$ is smaller than macroscopic dimensions but much larger than molecular dimensions. We are interested in the average of the H function,

$$\langle H \rangle = \int \rho \ln P \, dp \, dq \tag{1.136}$$

By dividing the whole phase space into the set of cells $\{\delta V\}$, this average becomes a sum over this set:

$$\langle H \rangle = \sum_{\{\delta V\}} \left(\int_{\delta V} \rho \, dp \, dq \right) \ln P \tag{1.137}$$

By using the definition of P [Eq. (1.135)], this can be rewritten as

$$\langle H \rangle = \sum_{\{\delta V\}} P \ln P \delta V \tag{1.138}$$

On letting the cell size become small, the sum becomes an integral so that we get

$$\langle H \rangle = \int P \ln P \, dp \, dq \tag{1.139}$$

By comparing Eqs. (1.136) and (1.139), we see that

$$\langle H \rangle = \int \rho \ln P \, dp \, dq = \int P \ln P \, dp \, dq \tag{1.140}$$

so that the average of H is the same whether P or ρ is used to calculate it.

We now consider a system that has been disturbed from its initial equilibrium state by changing the external conditions at time $t = t_0$. We wait till it reaches a new equilibrium under these altered conditions and hold it there. Clearly, with respect to the initial state, the system is not in equilibrium in the new altered state. We release the constraints at time $t = t_1$ and let the system relax till it reaches the initial equilibrium state at a later time $t = t_3$, say. We observe the system at time t_2 soon after t_1, before it has reached equilibrium. Hence, the system is also in a nonequilibrium state at $t = t_2$. Let the average H value associated with the nonequilibrium state at $t = t_1$ be $\langle H_1 \rangle$ and the one at $t = t_2$ be $\langle H_2 \rangle$. We aim to show that $\langle H_1 \rangle > \langle H_2 \rangle$.

As explained above, we must work with coarse-grained probabilities in order to avoid the consequences of Liouville's theorem. Let the systems at times t_1 and t_2 be characterized by the coarse-grained distributions P_1 and P_2, respectively. Let the corresponding fine-grained distributions be ρ_1 and ρ_2, respectively. Since the system at time t_1 is in equilibrium with respect to the altered conditions, we can use an equilibrium ensemble to characterize it. Let us choose the microcanonical ensemble in which ρ_1 is a constant. Then, from the definition given by Eq. (1.135), $P_1 = \rho_1$. Therefore at time t_1,

$$\langle H_1 \rangle = \int \rho_1 \ln \rho_1 \, dp \, dq \qquad (1.141)$$

At time t_2,

$$\langle H_2 \rangle = \int P_2 \ln P_2 \, dp \, dq \qquad (1.142)$$

The difference $\langle H_1 - H_2 \rangle$ is

$$\langle H_1 - H_2 \rangle = \int (\rho_1 \ln \rho_1 - P_2 \ln P_2) \, dp \, dq = \int (\rho_2 \ln \rho_2 - P_2 \ln P_2) \, dp \, dq \qquad (1.143)$$

since according to Liouville's theorem $d\rho/dt = 0$, and from this it follows that $d(\int \rho \ln \rho \, dp \, dq)/dt = 0$. At this point, we add and subtract 1 to get

$$\langle H_1 - H_2 \rangle = \int (\rho_2 \ln \rho_2 - P_2 \ln P_2 - \rho_2 + P_2) \, dp \, dq \qquad (1.144)$$

since both ρ and P are normalized over the entire phase space, implying

$$\int \rho_2 \, dp \, dq = \int P_2 \, dp \, dq = 1 \qquad (1.145)$$

By using Eq. (1.140), Eq. (1.144) can be written as

$$\langle H_1 - H_2 \rangle = \int (\rho_2 \ln \rho_2 - \rho_2 \ln P_2 - \rho_2 + P_2) \, dp \, dq \qquad (1.146)$$

As shown by Tolman (Tolman, 1979, Section 51), the integrand is greater than or equal to 0, with equality occurring only if $\rho_2 = P_2$. Therefore, $\langle H_1 \rangle$ is greater than or equal to $\langle H_2 \rangle$, and H is a function whose average decreases or remains unchanged as equilibrium is approached.

Liouville's equation (1.132), Hamilton's equations, and indeed all the equations of mechanics are reversible equations. By this is meant that, for any process described by these equations, the reverse process is as likely to occur as the forward process. In other words, in microscopic systems there is no direction to time. How then to account for an irreversible process such as the expansion of a gas into a vacuum? As will be seen in Chapter 8, Boltzmann's treatment of the μ space distribution function for gas-phase collisions introduces irreversibility. A proof of the H-theorem can be given based on the Boltzmann equation. However, there are several objections to basing the proof of the H-theorem on the Boltzmann equation (see Tolman, 1979, Chapter 6; Prigogine, 1980, Chapter 7; McQuarrie, 1976, Section 18-5).

The approach we use is the concept of coarse graining, first suggested by Gibbs. It states that irreversibility is observed only in macroscopic systems; this is because the coarse-grained probability density differs from the fine-grained probability density as time progresses. This is not to say that microscopic equations are invalid. Liouville's equation is valid for any conservative process. Thus, if you were able to observe an irreversible process at a given point in phase space, the probability density would evolve according to Liouville's equation, but you would not be aware that the process had occurred irreversibly. From your perspective, a movie of the process would make just as much sense when run backward as forward in time. The same movie of the macroscopic process, say an egg being smashed when dropped, would appear absurd when run backward.

There are also objections to the concept of coarse graining (see, e.g, Prigogine, 1980; Prigogine and Stengers, 1984, p. 285; Hill, 1987, Section 17). The problem of getting irreversibility from reversible equations is a continuing one that no doubt will occupy future theorists. However, we know that time does have a direction in everyday problems and that direction can be based on the second law of thermodynamics. The second law tells us that entropy may be used as time's arrow. Many other processes, such as the expansion of the universe, have been used occasionally to define time's arrow.

We have shown that the average of Boltzmann's H function seeks a minimum value at equilibrium. Therefore, we postulate that entropy is proportional to the negative of $\langle H \rangle$. Because $\langle H \rangle$ is dimensionless and S has the dimensions of the Boltzmann constant k_B, the constant of proportionality between $-\langle H \rangle$ and S can be chosen as k_B. Thus, under equilibrium conditions, we may write

$$S = -k_B \langle H \rangle \tag{1.147}$$

and use of Eq. (1.136) gives

$$S = -k_B \int \rho \ln P \, dp \, dq = -k_B \langle \ln P \rangle \tag{1.148}$$

This definition is precisely what was used in previous sections to relate statistical mechanics to thermodynamics.

Our aim in this section was to introduce the reader to the ideas of nonequilibrium statistical mechanics and, in particular, derive Boltzmann's relation between entropy and his H function. A variety of ways to treat systems in nonequilibrium states are studied in detail in Part III.

1.8 EXERCISES AND PROBLEMS

1. Derive Eq. (1.12) using a generating function.

2. Derive Eq. (1.18) using a generating function.

3. Consider an even more drunk random walker such that each time he takes a step to the right, to the left, or takes no step, all with equal probability $\frac{1}{3}$. Calculate the mean of the distance traveled and its standard deviation for a total of N steps taken.

4. Show that the average value of a continuous random variable is also the most probable value for a Gaussian distribution.

5. For the Gaussian distribution in Figure 1.1, determine μ, Var(x), and $\sigma(x)$. By comparing Var(x) and $\sigma(x)$ with the figure, suggest meanings for these statistical quantities.

6. Show that

$$\lim_{N \to \infty} \frac{N!}{N^{n_r}(N - n_r)!(1 - \lambda/N)^{n_r}} = 1$$

7. Use the Maclaurin's series expansion of e^x to show that the Poisson distribution is normalized to unity. For the Poisson distribution, use the Maclaurin's series expansion of xe^x to find the first moment μ_1 from the moment generating function.

8. Show that the next term in Eq. (1.65)

$$\frac{1}{2}E_s^2 \left(\frac{\partial^2 \ln \Omega_r}{\partial E_r^2} \right)_{E_r = E}$$

can be neglected in the limit when the number of particles in the reservoir becomes very large.

9. Derive Eq. (1.73), which implies that the temperature of the reservoir is equal to that of the whole system, as expected for equilibrium.

10. Derive an expression for the distribution function of the isothermal-isobaric ensemble (constant N, P, T) by using the approach of Section 1.5. (*Hint:* The thermodynamic function of interest for a constant pressure process is the enthalpy.)

11. Show that the total number of states accessible to an isolated large system (composed of a reservoir and a relatively smaller system) near the average energy is a Gaussian function of energy. (*Hint:* Expand $\ln(\Omega_s \Omega_r)$ in a Taylor's series and use the result of Exercise 9.)

12. Boron is 20 mol% ^{10}B and 80% ^{11}B, while chlorine is 76% ^{35}Cl and 24% ^{37}Cl. Calculate the partial pressure of all the isotopic species of BCl_3 at 1.0-atm total pressure assuming ideal gas behavior.

13. Calculate and plot the molar heat capacity of a system of harmonic oscillators as a function of temperature. Assume that the vibrational wavenumber is 300 cm^{-1}.

14. Look up the rotational constant of HBr and then calculate the fraction of molecules in the $J = 0$, $J = 5$, and $J = 10$ rotational levels at 300 K.

15. Use the discussion in Section 1.6.4 to obtain an expression for the entropy of particles in a cubic box. This expression is known as the Sackur–Tetrode equation. Use this equation to find the entropy of mercury gas at 298 K and 1 atm.

2

QUANTUM STATISTICAL MECHANICS

2.1 INTRODUCTION

Many systems can be treated quite adequately by classical statistical mechanics. However, factors such as $N!$ and h^{3N}, which are needed in the expression for the classical partition function [Eq. (1.89)] for consistency with experiments, cannot be explained within the classical theory. Even more serious are the discrepancies between the classical theory and experiments at low temperatures. As examples, we recall the Rayleigh–Jeans law for spectral distribution of energy for black-body radiation and the Dulong–Petit law of the specific heat of solids. Both laws were derived using classical statistical mechanics and disagreed with experiments. These discrepancies ultimately led to the birth of quantum mechanics. The focus of this chapter is quantum statistical mechanics, based on quantum mechanics.

First, the essentials of quantum mechanics, including dynamics, are reviewed using Dirac's bra and ket notation. The quantum mechanical distribution function, called the *density matrix*, which allows the ensemble concept to be extended to quantum mechanical systems, is introduced and studied. It will be seen that introduction of the quantum concept changes the nature of ensembles. It is then explained how the fundamental principle of indistinguishability of identical particles gives rise to two types of quantum statistics.

The partition functions of ideal quantum gases are evaluated in terms of measurable quantities and, from these partition functions, thermodynamic functions of quantum gases are calculated. Corrections are obtained to classical behavior at high temperatures and/or low densities. For low temperatures and/or high densities, new quantum effects arise. One example of a new effect is that of *Bose–Einstein condensation*, which is an active area of research at present. The chapter closes with two more applications of quantum statistics.

2.2   ESSENTIALS OF QUANTUM MECHANICS

2.2.1   Dirac Bra and Ket Vectors

This section is a review of quantum theory using Dirac's bra and ket vector notation (Dirac, 1958; Sakurai, 1985). There are many advantages to working in the Dirac notation. These will become apparent as the notation is used in the sequel. For now, three reasons will be mentioned. The notation allows a very concise formulation of quantum mechanics, applicable to both discrete states as well as to the continuum. Second, it allows one to view quantum mechanics from a very general standpoint of a *Hilbert space* (Jauch, 1968). Finally, most of the work in modern physical chemistry and physics uses this notation, so that the student will do well to become familiar with it.

In wave mechanics, a quantum mechanical system is described by complex-valued functions of position coordinates and/or spin variables. Observables are represented by combinations of functions of position, derivatives with respect to position, and spin matrices. The time-independent Schrödinger equation appears as a partial differential equation or a matrix equation. Calculating the expectation values of observables involves integrals over position coordinates or sums over spin indexes. By making suitable transformations, one may go to different representations as convenient, which may leave the student with a bewildering array of formulas, derivations, and interpretations.

This bewildering array can be unified into a general point of view and given a mathematical basis using Dirac's formalism. According to Paul Dirac, a state in quantum mechanics is a member of a set of vectors in an abstract Hilbert space (Jauch, 1968), which is a generalization of the usual three-dimensional Euclidean space. Position wave functions, momentum wave functions, and spin column vectors are concrete representations of these abstract quantum mechanical states. In Hilbert space, a state is specified by a *ket vector* $|\psi\rangle$. If the ket vector is multiplied by an arbitrary complex number c, the resulting ket $c|\psi\rangle$ still represents the same state. Another way of saying it is that the state is determined by the direction of the ket and not by its magnitude. The vector space is linear, so that general linear combinations such as $a|\psi_1\rangle + b|\psi_2\rangle$, where a and b are complex numbers, are also members of the vector space.

To each ket vector there corresponds a unique *bra vector*, which is the *Hermitian conjugate* of the ket, and for the ket $|\psi\rangle$ is denoted by $\langle\psi|$, or, symbolically,

$$|\psi\rangle^{\dagger} = \langle\psi| \tag{2.1}$$

where the superscript † denotes the Hermitian conjugate. The inner product, or scalar product, of a bra and a ket is in general a complex number and is denoted by the notation $\langle\psi_1|\psi_2\rangle$. Notice that the order in which the Dirac vectors occur in the inner product is bra–ket, which forms a "bracket." The bracket notation $\langle\cdots|\cdots\rangle$ replaces the $\int\cdots d\tau$ notation of wave mechanics and the matrix product notation of matrix mechanics. The normalization condition is $\langle\psi|\psi\rangle = 1$. The normalization condition determines the ket up to a *phase factor* $\exp(i\phi)$ or a *phase angle* ϕ. The state of the system does not depend on either of these. In other words, the absolute phase of a ket has no physical significance. However, if two different kets are superimposed, the resulting *phase difference* can give rise to interesting quantum interference effects, not seen in classical mechanics.

In Hilbert space, there are defined linear operators, which operate on ket vectors and which are denoted by putting a *wide hat* over them. To each such operator $\widehat{Q}$ there corresponds a Hermitian conjugate denoted by $\widehat{Q}^{\dagger}$, which is defined by the equation

$$\langle\psi_1|\widehat{Q}^{\dagger}|\psi_2\rangle = \langle\psi_2|\widehat{Q}|\psi_1\rangle^* \tag{2.2}$$

where the superscript * denotes the complex conjugate, as usual.

A physical observable is represented by a *Hermitian* operator, which is defined to be one whose Hermitian conjugate is equal to itself. The most important Hermitian operator is the Hamiltonian operator $\widehat{\mathcal{H}}$, in terms of which the time-independent Schrödinger equation is

$$\widehat{\mathcal{H}}|\mathcal{E}_i\rangle = E_i|\mathcal{E}_i\rangle \tag{2.3}$$

where the unit or the identity operator is understood on the right-hand side. Equation (2.3) is an *eigenvalue* equation, where $|\mathcal{E}_i\rangle$ is now taken to be a special ket, the eigenket of the Hamiltonian operator. Research in atomic, molecular, solid state, and to a large extent, nuclear physics may be thought of as exercises in proposing Hamiltonian operators and solving their eigenvalue equations! The energy eigenvalues $E_i(i = 1, 2, \ldots)$, which generally form a discrete set, are the only possible energies the system is allowed to have. The eigenkets of the Hamiltonian are called *stationary states*.

Let us take a general Hermitian operator corresponding to some physical observable G that is explicitly independent of time. In general, its eigenvalue equation can be written

$$\hat{G}|g_j\rangle = g_j|g_j\rangle \tag{2.4}$$

where g_j are the eigenvalues and $|g_j\rangle$ are the eigenkets. The eigenvalue equation (2.4) implies that any measurement of the physical observable G can result only in the real numbers g_j. Any other value, for example, $(g_2 + 11g_3)/7$, where g_2 and g_3 are different eigenvalues, can never occur as a result of a measurement of G. The set of eigenkets $\{|g_j\rangle\}$ is *orthonormal*, which means that, for any two eigenkets $|g_j\rangle$ and $|g_k\rangle$,

$$\langle g_j|g_k\rangle = \delta_{jk} \tag{2.5}$$

where the notation δ_{jk} is a generalized Kronecker delta, so that it is equal to the Dirac delta function $\delta(j - k)$ in case the labels j and k are continuous. The orthonormal set of eigenkets may be chosen as a *basis* for the vector space, implying that any given normalized ket $|\psi\rangle$ can be expressed as a linear combination of basis kets.

$$|\psi\rangle = \sum_j c_j|g_j\rangle \tag{2.6}$$

The complex numbers c_j may be found by multiplying Eq. (2.6) on the left by a particular eigenbra $\langle g_k|$ of the basis set

$$\langle g_k|\psi\rangle = \sum_j c_j\langle g_k|g_j\rangle = c_k\langle g_k|g_k\rangle = c_k \tag{2.7}$$

where the orthonormality relation [Eq. (2.5)] has been used. By using Eq. (2.7), Eq. (2.6) can be written in the compact form

$$|\psi\rangle = \sum_j |g_j\rangle\langle g_j|\psi\rangle \tag{2.8}$$

The state ket $|\psi\rangle$ is said to be in the representation formed by the $|g_j\rangle$ basis kets. Equation (2.6) [or its equivalent Eq. (2.8)] is the foundation of the *superposition principle* of quantum mechanics. This principle can be traced ultimately to the fact that the Schrödinger equation [Eq. (2.3)] is a linear equation.

What happens if the system is in the superposition state $|\psi\rangle$ and a measurement of G is performed? The measured values in any such experiment will be one of the eigenvalues g_j. No other values can result. However, every g_j value has an a priori probability of occurring given by $|c_j|^2 \equiv c_j^* c_j$. This feature of the physical observable G makes it a random or a stochastic variable in the sense of Section 1.3. The probabilities must add to unity.

$$\sum_k c_k^* c_k = 1 \tag{2.9}$$

By substituting for c_k from Eq. (2.7), Eq. (2.9) becomes

$$\sum_k \langle \psi | g_k \rangle \langle g_k | \psi \rangle = 1 \tag{2.10}$$

The normalization of the ket $|\psi\rangle$, $\langle \psi | \psi \rangle = 1$, is consistent with Eq. (2.10) only if

$$\sum_k | g_k \rangle \langle g_k | = 1 \tag{2.11}$$

where both sides of this equation are obviously operators, the right-hand side being the unit operator. For simplicity, the wide hat will not be shown explicitly on the unit operator. This equation is called the *completeness* or the *closure* condition. The completeness relation holds for any complete set of kets that span the vector space. Incidentally, using the closure relation, Eq. (2.11), the expansion of a superposition ket in terms of the eigenkets gives Eq. (2.8) immediately.

Consider a given term in Eq. (2.11), say

$$\widehat{P}_k = | g_k \rangle \langle g_k | \tag{2.12}$$

This operator has the property

$$\widehat{P}_k^2 = \widehat{P}_k \tag{2.13}$$

which is the defining property of *projection* operators. It can be shown from Eq. (2.13) that projection operators are Hermitian and have eigenvalues 0 and 1. Projection operators are at the heart of the *Zwanzig–Mori* formalism used in nonequilibrium statistical mechanics, which is discussed in Chapter 10.

Once a particular measurement of G has been made, the system is forced to choose one of the eigenstates, because only one of the values g_j can result. If many measurements are made, the weighted average, called the expectation value and given by

$$\langle G \rangle = \langle \psi | \widehat{G} | \psi \rangle \tag{2.14}$$

is obtained. This equation can be written more explicitly by using the closure condition and inserting a complete set of orthonormal states, not necessarily eigenstates.

$$\langle G \rangle = \sum_{m,n} \langle \psi|m\rangle\langle m|\widehat{G}|n\rangle\langle n|\psi\rangle \tag{2.15}$$

The complex numbers in the middle, $G_{mn} \equiv \langle m|\widehat{G}|n\rangle$, are the *matrix elements* of the operator $\widehat{G}$ in the representation of the $|m\rangle$ *general basis kets*. The off-diagonal elements ($m \neq n$) have the natural interpretation that the squares of their moduli give the probabilities of transition from one state to another (state $|n\rangle$ to state $|m\rangle$). What happens if the system is in one of the eigenstates, say $|g_j\rangle$, of the operator $\widehat{G}$? In this case, the double sum in Eq. (2.15) collapses and we get $\langle g_j|\widehat{G}|g_j\rangle = g_j$. This means that the operator $\widehat{G}$ is diagonal in the representation of its own eigenkets with the eigenvalues g_j ($j = 1, 2, \ldots$) as the diagonal elements. Also, if the system is in an eigenstate $|g_j\rangle$, the expectation value is equal to the eigenvalue g_j itself.

Since the Hamiltonian operator is diagonal in the *energy representation*, no transitions are possible between energy eigenkets, which explains the name stationary states for its eigenkets. In Section 2.2.2, we consider nonstationary states.

2.2.2 Schrödinger and Heisenberg Pictures

To study the dynamics of a quantum mechanical system, two points of view called *pictures* are available. In the *Schrödinger picture*, the states are explicit functions of time satisfying the time-dependent Schrödinger equation

$$\widehat{\mathcal{H}}|\psi, t\rangle = i\hbar\frac{\partial|\psi, t\rangle}{\partial t} \tag{2.16}$$

which has the formal solution

$$|\psi, t\rangle = \exp(-i\widehat{\mathcal{H}}t/\hbar)|\psi\rangle \tag{2.17}$$

Here $\hbar \equiv h/2\pi$ is Dirac's constant and the exponential operator is defined in terms of its Taylor's series about $t = 0$. The exponential operator is sometimes called a *time propagation operator* (see Chapter 10). For energy eigenkets $|\mathcal{E}\rangle$, Eq. (2.16) has the solution

$$|\mathcal{E}, t\rangle = \exp(-iEt/\hbar)|\mathcal{E}\rangle \tag{2.18}$$

which shows the time dependence of the ket. Since the time dependence is in the form of a complex number of modulus unity, the state is unchanged in time. Again, we see why an energy eigenstate is called a stationary state.

In the *Heisenberg picture* the situation is reversed, and states are time independent, whereas the operators evolve in time. Consider an operator $\widehat{G}$ that does not explicitly depend on time. The matrix elements of $\widehat{G}$ are time dependent in the Schrödinger picture because the kets evolve in time. A typical matrix element is given by

$$\langle \psi, t | \widehat{G} | \psi, t \rangle = \langle \psi | \exp(i\widehat{\mathcal{H}}t/\hbar) | \widehat{G} | \exp(-i\widehat{\mathcal{H}}t/\hbar) | \psi \rangle \qquad (2.19)$$

where Eq. (2.17) has been used. Therefore, the time dependence of the operator $\widehat{G}$ is given by

$$\widehat{G}(t) = \exp(i\widehat{\mathcal{H}}t/\hbar) | \widehat{G} | \exp(-i\widehat{\mathcal{H}}t/\hbar) \qquad (2.20)$$

By differentiating this equation with respect to time and using the definition of the exponential operator, one obtains

$$\frac{d\widehat{G}(t)}{dt} = \frac{i}{\hbar}[\widehat{\mathcal{H}}, \widehat{G}(t)] \qquad (2.21)$$

which is the equation of motion of the operator $\widehat{G}$ in the Heisenberg picture. Here $[\widehat{\mathcal{H}}, \widehat{G}(t)] \equiv \widehat{\mathcal{H}}\widehat{G}(t) - \widehat{G}(t)\widehat{\mathcal{H}}$ is a quantum mechanical commutator bracket, which is the quantum mechanical analogue of the classical Poisson bracket, Eq. (1.129).

2.2.3 Pure and Mixed States

Quantum statistical mechanics differs fundamentally from classical statistical mechanics in several respects. First, in quantum theory the Heisenberg uncertainty principle tells us that both position and momentum cannot be known exactly simultaneously. Therefore, the concept of phase space is not as important in quantum mechanics as it is classically. Instead of representing a state by the phase volume $\Delta p \Delta q$, quantum mechanics specifies the microstate by a wave function or by a state vector in Hilbert space.

Second, the ability to measure position and momentum exactly in classical mechanics means that in each state of a classical ensemble the physical observable G is known exactly. This statement is not true in quantum mechanics, where only the expectation value in any microstate $|\psi\rangle$ is known. So, in quantum mechanics double averaging is required to calculate the statistical mechanical average values of the observables. One averaging is due to the probabilistic nature of quantum mechanics and the other is due to the lack of knowledge of the microscopic properties of the system.

Third, identical particles are considered *indistinguishable* in quantum mechanics. In classical mechanics, each particle carries a label, its position coordinate, which makes it distinct from other particles, even though it may be identical to them in all other respects. In quantum mechanics this is not so. In quantum mechanics, indistinguishability gives rise to new effects not seen in classical mechanics. For example, in a two-particle state $|\psi(1)\psi(2)\rangle$, if the two identical particles are exchanged, the Hamiltonian cannot change and must be a symmetric function of the two particles. Another way of stating this is that it is impossible, by any experiment, to determine exactly the energy of either of the two particles. This finding gives rise to two possibilities. Either the two-particle state is unchanged or it changes sign when the particles are exchanged. The former case leads to *Bose–Einstein statistics* and the latter to *Fermi–Dirac statistics*. These cases will be discussed in detail in Section 2.3.

Fourth, the nature of an ensemble is changed in quantum mechanics because of possible *interference* between two different states of a system. This makes it necessary to consider the phase angles of wave functions. To understand this, one needs to realize that, in quantum mechanics, it is possible to have two kinds of ensembles called *pure* and *mixed states*. In a pure state, all the members of the ensemble are in the same microstate. In a mixed state, there are many microstates accessible to the system.

We illustrate the difference by a simple example. Consider a beam of silver atoms emerging from a Stern–Gerlach experiment in which they have been aligned along the $+Z$ direction. If they are then put through an inhomogeneous magnetic field aligned along the $+X$ direction, the probability that any given silver atom will emerge in the spin $+\frac{1}{2}$ or spin $-\frac{1}{2}$ beam is $\frac{1}{2}$ for each beam. Before passage through the second magnet, a silver atom is in a superposition state made up of spin $\pm\frac{1}{2}$ eigenstates of the operator for spin in the X direction. After passage through the second magnet the atom is in one of the two X-direction eigenstates and its eigenvalue is then known exactly. In either case, before and after passage through the second magnet, the silver atom is said to be in a pure state. A pure state possesses the maximum information available in a quantum system.

In contrast to a pure state, a system can be in a mixed state. A mixed state occurs when two or more pure states are accessible to the system with different weights assigned to them. An example of a mixed state is a beam of silver atoms, two-thirds of which have been prepared with spins in the $+Z$ direction and one-third of which have their spins in the $+X$ direction. There is no way to write a state vector for this system, which is an incoherent mixture of pure states. By incoherent is meant that there exists no phase relation (see below) between the two pure-state vectors.

Let us consider a mixed state consisting of $\mathcal{N}$ pure states labeled by the index α, the pure state $|\psi^\alpha\rangle$ carrying a weight w_α such that $\sum_{\alpha=1}^{\mathcal{N}} w_\alpha = 1$. The average of the observable G is

$$\langle G \rangle = \sum_{\alpha=1}^{\mathcal{N}} w_\alpha \langle \psi^\alpha | \widehat{G} | \psi^\alpha \rangle \tag{2.22}$$

By definition, mixed microstates cannot be represented by kets. They have to be represented by a *density operator* or *density matrix*, a concept to which we now turn.

2.2.4   The Density Matrix

We first introduce the density matrix for pure states. In the Schrödinger picture, the expectation value of G is

$$\langle G \rangle = \langle \psi, t | \widehat{G} | \psi, t \rangle \tag{2.23}$$

By using a complete set of eigenkets $|E_i\rangle$ of the Hamiltonian and introducing the closure relation, Eq. (2.11), twice in Eq. (2.23), the expectation value may be written as

$$\langle G \rangle = \sum_{i,j} \langle \psi, t | E_i \rangle \langle E_i | \widehat{G} | E_j \rangle \langle E_j | \psi, t \rangle$$

$$= \sum_{i,j} \langle E_j | \psi, t \rangle \langle \psi, t | E_i \rangle \langle E_i | \widehat{G} | E_j \rangle \tag{2.24}$$

where a slight rearrangement has been made. In Eq. (2.24), the last factor in the sum is recognized as G_{ij}, which is the (i, j) matrix element of $\widehat{G}$ in the energy representation. The first two factors also look like a matrix element. We introduce a new operator called the density operator, which is defined by its matrix elements

$$\rho_{ij} = \langle E_i | \widehat{\rho} | E_j \rangle = \langle E_i | \psi, t \rangle \langle \psi, t | E_j \rangle \tag{2.25}$$

Use the density operator to rewrite Eq. (2.24) as

$$\langle G \rangle = \sum_{i,j} \rho_{ji}(t) G_{ij} = \sum_{j} [\rho(t)G]_{jj} = \mathrm{tr}[\widehat{\rho}(t)\widehat{G}] \tag{2.26}$$

where tr indicates the trace, which is defined as the sum of the diagonal matrix elements of a given operator.

From Eq. (2.25), which gives the matrix elements of the density operator in the energy representation, the density operator is given by

$$\widehat{\rho} = |\psi, t\rangle\langle\psi, t| \tag{2.27}$$

This density operator is for the pure state $|\psi, t\rangle$ and is seen to be a projection operator [Eq. (2.12)]. It satisfies the normalization condition

$$\operatorname{tr}\widehat{\rho} = 1 \tag{2.28}$$

which can be proved by substituting the unity operator for $\widehat{G}$ in Eq. (2.26).

With this introduction to the density operator for pure states, the generalization to mixed states is straightforward. For a system of mixed states, we define

$$\widehat{\rho} = \sum_{\alpha=1}^{\mathcal{N}} w_\alpha |\psi^\alpha, t\rangle\langle\psi^\alpha, t| \tag{2.29}$$

or, in the energy representation,

$$\rho_{ij} = \sum_{\alpha=1}^{\mathcal{N}} w_\alpha \langle \mathcal{E}_i | \psi^\alpha, t\rangle\langle\psi^\alpha, t | \mathcal{E}_j\rangle \tag{2.30}$$

The expectation value of the observable G is written as

$$\langle G\rangle = \operatorname{tr}(\widehat{\rho}\widehat{G}) \tag{2.31}$$

or more explicitly,

$$\langle G\rangle = \sum_{i,j} \sum_{\alpha=1}^{\mathcal{N}} w_\alpha \langle \mathcal{E}_j | \psi^\alpha, t\rangle\langle\psi^\alpha, t | \mathcal{E}_i\rangle G_{ij} \tag{2.32}$$

which is seen to be the same as Eq. (2.22). Notice the appearance of the double average in Eq. (2.32), as mentioned in Section 2.2.3. First, there is the ensemble average over α, as in classical systems. But then, there is also the quantum mechanical average over i and j, which gives the expectation value for each microstate. The latter average is not necessary in classical mechanics, where any observable can be measured, in principle, to any degree of accuracy with probability one.

In contrast with the density matrix for pure states, the density matrix for mixed states, Eq. (2.29), does not satisfy Eq. (2.13) in general, and hence is not a projection operator. In the special case, $w_\alpha = 1$ for some particular α and zero otherwise, the general density matrix reduces to the one for the pure state α and does satisfy Eq. (2.13).

In the density matrix, we have found the analogue of the classical statistical distribution function studied in Chapter 1. Now we prove the quantum mechanical version of Liouville's theorem.

2.2.5   Quantum Mechanical Liouville's Theorem

To derive Liouville's theorem we need the time dependence of the bra and ket vectors. The time dependence of the ket vectors is given by Eq. (2.16) and that of the bra vectors is given by

$$- i\hbar \frac{\partial \langle \psi, t |}{\partial t} = \langle \psi, t | \widehat{\mathcal{H}} \tag{2.33}$$

which is obtained by taking the Hermitian conjugate of Eq. (2.16). If we differentiate Eq. (2.29) and use the time derivatives of the bra and ket vectors, we obtain

$$\frac{\partial \widehat{\rho}}{\partial t} = \frac{1}{i\hbar} \sum_{\alpha=1}^{\mathcal{N}} w_\alpha \widehat{\mathcal{H}} | \psi^\alpha, t \rangle \langle \psi^\alpha, t | - \frac{1}{i\hbar} \sum_{\alpha=1}^{\mathcal{N}} w_\alpha | \psi^\alpha, t \rangle \langle \psi^\alpha, t | \widehat{\mathcal{H}} \tag{2.34}$$

from which it readily follows that

$$\frac{\partial \widehat{\rho}}{\partial t} = \frac{i}{\hbar} \left[\widehat{\rho}, \widehat{\mathcal{H}} \right] \tag{2.35}$$

Equation (2.35) is the quantum mechanical analogue of Eq. (1.130) and is the basic equation that must be solved to follow the time evolution of the density matrix. This equation is the quantum mechanical expression of Liouville's theorem.

2.2.6   The Fundamental Postulates

Ensembles in quantum mechanics are constructed in the same way as in classical mechanics. Once again, it is necessary to invoke certain postulates regarding averages. To describe the equilibrium properties of an isolated system, the density matrix is chosen to be independent of time, so that the ensemble corresponding to it is stationary. Accordingly, the quantum Liouville theorem, Eq. (2.35), requires that the density matrix commute with the

Hamiltonian. The commutation can be assured by making the density matrix either a constant or an explicit function of the Hamiltonian or a function of the operators that commute with the Hamiltonian.

We now show how the microcanonical ensemble is obtained. Other ensembles can be obtained similarly (Pathria, 1996, Chapter 5). Just as in classical statistical mechanics, the postulate of equal a priori probabilities is invoked for lack of information. In the quantum mechanical case, this postulate says that at equilibrium all accessible quantum microstates are equally probable. This means that the weights w_α are equal to one another. To be consistent with the notation used in classical statistical mechanics, we denote the total number, $\mathcal{N}$, of microstates $|\psi^\alpha\rangle$ in the microcanonical ensemble by Ω. Then normalization dictates that $w_\alpha = 1/\Omega$. Using this information, Eq. (2.29) reduces to

$$\hat{\rho} = \frac{1}{\Omega} \sum_{\alpha=1}^{\Omega} |\psi^\alpha, t\rangle\langle\psi^\alpha, t| \tag{2.36}$$

The matrix elements of the density matrix in a representation of general orthonormalized basis kets are

$$\rho_{mn} = \frac{1}{\Omega} \sum_{\alpha=1}^{\Omega} \langle m|\psi^\alpha, t\rangle\langle\psi^\alpha, t|n\rangle \tag{2.37}$$

Since the kets $|\psi^\alpha, t\rangle$ form a complete orthonormal set, the completeness relation can be used to give

$$\rho_{mn} = \frac{1}{\Omega}\delta_{mn} \tag{2.38}$$

The density matrix is a multiple of the unit operator and commutes with the Hamiltonian, as required for a stationary ensemble. Equation (2.38) also says that the density matrix is diagonal in any representation, since the basis kets were taken to be completely general in Eq. (2.37).

Equation (2.38) can also be obtained by writing the representative $\langle\psi^\alpha, t|n\rangle$ in the form

$$\langle\psi^\alpha, t|n\rangle = r \exp[i\theta_n^\alpha(t)] \tag{2.39}$$

where r and $\theta_n^\alpha(t)$ are the modulus and the phase angle of the complex number $\langle\psi^\alpha, t|n\rangle$. Use this equation to write Eq. (2.37) as

$$\rho_{mn} = \frac{r^2}{\Omega} \sum_{\alpha=1}^{\Omega} \exp[i(\theta_n^\alpha - \theta_m^\alpha)] \tag{2.40}$$

where the time dependence has been suppressed for simplicity. If the density matrix is to be diagonal,

$$\langle \exp[i(\theta_n^\alpha - \theta_m^\alpha)]\rangle = \frac{1}{\Omega}\sum_{\alpha=1}^{\Omega}\exp[i(\theta_n^\alpha - \theta_m^\alpha)] = \delta_{mn} \tag{2.41}$$

The only way this can be assured is to postulate that at equilibrium there exist random a priori phases for each member of the ensemble. Then, expressing the exponential function as $\cos\{\theta_n^\alpha - \theta_m^\alpha\} + i\sin\{\theta_n^\alpha - \theta_m^\alpha\}$ and allowing random values for each phase angle, the sum over all α will cause the cosine and sine terms to vanish unless $m = n$. Thus,

$$\rho_{mn} = r^2\delta_{mn} \tag{2.42}$$

The normalization of ρ ensures that $r^2 = 1/\Omega$, so that this equation is identical to Eq. (2.38).

2.3   QUANTUM STATISTICS

2.3.1   Ensembles for Independent Particle Systems

A real system poses a complicated many-body problem. However, in certain situations, it may be possible to model it by a system of independent particles or *quasiparticles* moving in an external potential. In this model, the mutual interactions between particles have been replaced by some kind of average single-particle potential or mean field. Technically, one says that correlations between the particles have been neglected. Examples are electrons in metals or semiconductors, phonons, the quanta of sound waves in solids, and orbitals in atomic or molecular Hartree–Fock type theories. For simplicity, we shall call all these entities particles.

Consider an isolated system having a fixed energy E, a fixed volume V, and composed of a fixed number N of identical, independent particles. The appropriate equilibrium ensemble is the microcanonical ensemble. Let each particle have accessible to it a set of energy eigenstates $|\epsilon_i\rangle$ and eigenvalues ϵ_i, where $i = 1, 2, \ldots, r$ labels the single-particle states. Following the convention used in chemistry, we call these single-particle states *orbitals*. (It is to be noted that this term is taken in a general sense and is not restricted to electron orbitals.) The total number of states, r, can be finite or infinite and the eigenvalues may be discrete or continuous or both. For example, for a spin-1 system, $r = 3$ corresponding to spin up, down, or zero. For a particle moving in a Coulomb field, r represents an infinite number of discrete and continuum orbitals. In the simplest case, even the mean field can be neglected and one has an ideal quantum gas. In that case the energy eigenvalues are simply the eigenvalues of the kinetic energy. We shall consider this in Section 2.3.2. For now, we retain the general mean-field case.

Since the particles are independent, the total Hamiltonian operator separates into the sum of N one-particle operators

$$\widehat{\mathcal{H}} = \sum_{k=1}^{N} \widehat{\mathcal{H}}_k \tag{2.43}$$

such that, for the kth particle,

$$\widehat{\mathcal{H}}_k |\epsilon_i\rangle = \epsilon_i |\epsilon_i\rangle \tag{2.44}$$

where ϵ_i is a single-particle energy. Because the particles are identical, the N Hamiltonians $\widehat{\mathcal{H}}_k$ are formally the same with formally the same eigenvalues and eigenstates. The total Hamiltonian $\mathcal{H}$ is a symmetric function of the N particles under *permutations*, which are processes in which two or more particles are interchanged. It can be shown that because of the permutation symmetry of the total Hamiltonian, the many-body states of the system are either symmetric or antisymmetric under permutations.

At any given instant, there will be n_i particles in the ith orbital, so that

$$N = \sum_i n_i \tag{2.45}$$

and the total energy of the system is

$$E = \sum_i \epsilon_i n_i \tag{2.46}$$

The energy E is an eigenvalue of the Hamiltonian operator with the energy eigenvector $|\mathcal{E}\rangle \equiv |\psi^\alpha\rangle$ so that

$$\widehat{\mathcal{H}} |\psi^\alpha\rangle = E |\psi^\alpha\rangle \tag{2.47}$$

where

$$|\psi^\alpha\rangle = |\epsilon_1, \epsilon_2, \epsilon_3, \ldots, \epsilon_r\rangle \tag{2.48}$$

is the many-body eigenstate. Since the many-body eigenstate has to be symmetric or antisymmetric under permutations, it cannot be just a product of orbitals. It has to be a sum of products of orbitals with appropriate signs.

Those particles whose wave function is symmetric under the interchange of two or more particles are called *bosons*. Examples are ^{4}He, H_2, and ^{12}CH$_4$. In this case, the many-body eigenstate $|\psi^\alpha\rangle$ is symmetric under interchange of any two particles and is the sum of the products of orbitals. It is given by

$$|\psi^\alpha\rangle = \frac{1}{\sqrt{N!}} \sum_{\{p\}} |\epsilon_1^a\rangle|\epsilon_2^b\rangle \cdots |\epsilon_r^N\rangle \qquad (2.49)$$

where the particles are labeled with the superscripts $a, b, \ldots, N$, the sum is over all $N!$ permutations $\{p\}$ of the particles, and the factor $1/\sqrt{N!}$ results from the normalization of the many-body eigenket. The ket given by Eq. (2.49) refers to N particles and total energy E, and the right-hand side of the equation tells us how many bosons are in each orbital, thus satisfying the constraints (2.45) and (2.46). For example, for a system of two bosons in the orbitals $|\epsilon_1\rangle$ and $|\epsilon_2\rangle$, the two-boson eigenstate is $(1/\sqrt{2})(|\epsilon_1^a\rangle|\epsilon_2^b\rangle + |\epsilon_1^b\rangle|\epsilon_2^a\rangle)$. Evidently, a given orbital labeled by i may have n_i bosons with $n_i = 0, 1, 2, \ldots, N$.

On the other hand, if the wave function of the system is antisymmetric under the interchange of two or more particles, then the particles are called *fermions*. Some examples of fermions are ^{3}He, HD, and ^{14}NH$_3$. In this case, the many-body eigenstate $|\psi^\alpha\rangle$ is antisymmetric under the interchange of two particles and is given by

$$|\psi^\alpha\rangle = \frac{1}{\sqrt{N!}} \sum_{\{p\}} (\pm 1)|\epsilon_1^a\rangle|\epsilon_2^b\rangle \cdots |\epsilon_r^N\rangle \qquad (2.50)$$

where the sum is once again over all $N!$ permutations $\{p\}$ of the particles. In contrast with the boson case, a given term in the sum has a positive or a negative sign depending on whether it is an even or an odd permutation. An even (odd) permutation is one reached by an even (odd) number of pair exchanges from a given starting permutation. The right-hand side of Eq. (2.50) is called the *Slater determinant*. This form of the many-body ket automatically satisfies *Pauli's exclusion principle*, according to which at most one fermion may occupy a given orbital. Therefore, for any i, $n_i = 0, 1$ for fermions. A system of two fermions in the orbitals $|\epsilon_1\rangle$ and $|\epsilon_2\rangle$ has a two-fermion eigenstate given by $(1/\sqrt{2})(|\epsilon_1^a\rangle|\epsilon_2^b\rangle - |\epsilon_1^b\rangle|\epsilon_2^a\rangle)$.

Because there are Ω ways of arranging the N particles among the r orbitals $|\epsilon_i\rangle$, there exists an Ω-fold degeneracy of the energy E (i.e., $\alpha = 1, 2, \ldots, \Omega$), and Ω is the number of states accessible to the system consistent with its maintaining a fixed energy E and number of particles N. The density matrix for the microcanonical ensemble is diagonal with each diagonal element given by $1/\Omega(E, V, N)$.

If the system is closed but no longer isolated, the canonical ensemble must be used. In this case, the energy eigenstates $|\psi^\alpha\rangle$ are no longer all degenerate, each eigenstate now having an eigenvalue E_α.

$$\widehat{\mathcal{H}}|\psi^\alpha\rangle = E_\alpha|\psi^\alpha\rangle \tag{2.51}$$

Each eigenvalue E_α has associated with it Ω_α states, each state having equal a priori probability. The canonical partition function can be derived in the same way as in classical mechanics (see Section 1.5) and is given by

$$Z = \sum_\alpha \exp(-\beta E_\alpha) \tag{2.52}$$

where the sum is over all accessible energy eigenstates, and $\beta = 1/k_B T$. For a sum over energy levels,

$$Z = \sum_{\{levels\}} \Omega_\alpha \exp(-\beta E_\alpha) \tag{2.53}$$

As in the case of the classical distribution function [Eq. (1.72)], the density matrix operator in the canonical ensemble is

$$\widehat{\rho} = \frac{\exp(-\beta\widehat{\mathcal{H}})}{Z} \tag{2.54}$$

The partition function in Eq. (2.52) can be written also as

$$Z = \text{tr}[\exp(-\beta\widehat{\mathcal{H}})] \tag{2.55}$$

The trace has to be taken with respect to the many-body ket, Eq. (2.49), in the case of bosons and to the ket (2.50) in the case of fermions. Since both ket and bra vectors are involved in the trace, a factor of $N!$ naturally comes in the denominator. This factor comes from the indistinguishability of the particles and remains even in the classical limit, which explains the occurrence of this factor in the expression for the classical canonical partition function given by Eq. (1.89).

If the system is open, the number of particles can change. The treatment of the grand canonical ensemble is the same as the canonical ensemble except for a variable number of particles. The grand canonical partition function is

$$\Xi = \sum_{N=0}^{\infty} \sum_\alpha \exp[\beta\{\mu N - E_\alpha(N)\}] \tag{2.56}$$

The double summation is to be understood to mean that we first sum over all $E_\alpha(N)$ for a fixed N and repeat the process for another N, until all possible values of N have been summed over.

2.3.2 Quantum Gases of Independent Particles

In order to study the thermodynamic properties of quantum gases of independent particles introduced in Section 2.3.1, the partition functions have to be evaluated. This evaluation can be done in any convenient ensemble. However, it turns out that the grand canonical ensemble is the most suitable; for other ensembles, see Pathria (Pathria, 1996, Chapter 6).

The partition function is given by Eq. (2.56), where N and E are given by Eqs. (2.45) and (2.46), respectively. By using these equations, the partition function can be written

$$\Xi = \sum_{N=0}^{\infty} \sum_{\alpha} \exp\left[\beta\mu \sum_i n_i - \beta \sum_i n_i \epsilon_i\right] \tag{2.57}$$

where the sum over α depends on the distribution of particles among the energy eigenstates. The main complications in evaluating the partition function of Eq. (2.57) come from the two constraints, and these can be bypassed by focusing our attention on the orbitals (Kittel, 1969, Chapter 9).

Consider the boson case first. Even though the particles are indistinguishable, the orbitals are not. Take a given orbital labeled by quantum numbers collectively denoted by i. Each such orbital can be occupied by an arbitrary number of particles. Therefore, the contribution of the orbital labeled by i to the partition function [Eq. (2.57)] is

$$\sum_{n_i=0}^{N} \exp[\beta n_i(\mu - \epsilon_i)] \tag{2.58}$$

which is a geometric series and can be evaluated readily to give

$$\frac{1 - x_i^{N+1}}{1 - x_i} \tag{2.59}$$

where

$$x_i = \exp[\beta(\mu - \epsilon_i)] \tag{2.60}$$

In the thermodynamic limit, when $N \rightarrow \infty$, the quantity x_i^{N+1} will vanish if $x_i < 1$. For now, we assume this to be true, in which case

$$\sum_{n_i=0}^{N} \exp[\beta n_i(\mu - \epsilon_i)] = \frac{1}{1 - x_i} \tag{2.61}$$

Since the orbitals are distinguishable and independent, the partition function can be obtained by multiplying the contributions of all the orbitals to give

$$\Xi = \prod_{i=1}^{r} \frac{1}{1 - x_i} \tag{2.62}$$

The fermion case is much easier because the sum analogous to Eq. (2.58) is for $n_i = 0, 1$. Therefore, the partition function is given by

$$\Xi = \prod_{i=1}^{r} (1 + x_i) \tag{2.63}$$

The two Eqs. (2.62) and (2.63) can be combined to give

$$\Xi = \prod_{i=1}^{r} (1 \pm x_i)^{\pm 1} \tag{2.64}$$

where the upper and lower signs correspond to fermions and bosons, respectively.

Once the partition function has been calculated, one may obtain all the thermodynamic quantities by using the results of Section 1.6.1, which involve the log of the partition function. Of particular interest is the average $\langle n_i \rangle$, which denotes the mean number of particles in the ith single-particle state. This average may be obtained by taking the log of both sides of Eq. (2.57) and differentiating both sides with respect to a given ϵ_k. We get

$$\frac{\partial \ln \Xi}{\partial \epsilon_k} = \frac{1}{\Xi} \sum_{N=0}^{\infty} \sum_{\alpha} (-\beta n_k) \exp \left(\beta \mu \sum_i n_i - \beta \sum_i n_i \epsilon_i \right) = -\beta \langle n_k \rangle \tag{2.65}$$

On the other hand, using Eq. (2.64), we get

$$\ln \Xi = \pm \sum_{i=1}^{r} \ln(1 \pm x_i) \tag{2.66}$$

By differentiating this equation with respect to ϵ_k and using Eq. (2.60), we get

$$\frac{\partial \ln \Xi}{\partial \epsilon_k} = \sum_{i=1}^{r} \frac{x_i}{1 \pm x_i}(-\beta\delta_{ik}) \tag{2.67}$$

where we have used the fact that $(\pm \times \pm) = 1$ and the Kronecker delta enters because

$$\frac{\partial \epsilon_i}{\partial \epsilon_k} = \delta_{ik} \tag{2.68}$$

The sum in Eq. (2.67) collapses into a single term, and, by comparing the result with Eq. (2.65), we obtain

$$\langle n_k \rangle = \frac{1}{\exp[\beta(\epsilon_k - \mu)] \pm 1} \tag{2.69}$$

where the upper (lower) sign corresponds to fermions (bosons), as usual. Equation (2.69) for the average number of particles in any given energy state is known as the *Fermi–Dirac distribution* for fermions and the *Bose–Einstein distribution* for bosons. The total number of particles is given by

$$\langle N \rangle = \sum_{k=1}^{r}\langle n_k \rangle = \sum_{k=1}^{r} \frac{1}{\exp[\beta(\epsilon_k - \mu)] \pm 1} \tag{2.70}$$

It is possible to obtain limits on the values of the chemical potential μ using simple physical arguments. The largest value of $\langle n_k \rangle$ is that for the lowest possible energy state, the ground state, for which we assume $\epsilon_0 = 0$. Therefore, $\langle n_k \rangle \leq \langle n_0 \rangle$, where

$$\langle n_0 \rangle = \frac{1}{\exp(-\beta\mu) \pm 1} \tag{2.71}$$

For the case of bosons, the physical requirement that $\langle n_0 \rangle \geq 0$ implies that $\exp(-\beta\mu) \geq 1$, which assures that $\mu \leq 0$. By using Eq. (2.69), this reduces to the requirement that $\exp[\beta(\epsilon_k - \mu)] \geq 1$, which on using Eq. (2.60) implies that $x_i \leq 1$. We assumed in the derivation of Eq. (2.62) that $x_i < 1$. The above physical argument justifies this assumption post facto. The case $x_i = 1$ can arise only for the ground orbital, and in that case the sum for the ground orbital in Eq. (2.58) is equal to N, which goes to ∞ in the thermodynamic limit. This predominant occupancy of the ground orbital gives rise to the phenomenon called *Bose–Einstein condensation*, which will be discussed in Section 2.4.2.

2.3.3 Thermal Wavelength and Degeneracy

We have mentioned before that new quantum effects arise at low temperatures and/or high densities. Let us quantify this statement. At a temperature T, the thermal energy of a particle is of the order of $k_B T$. Considering this as the maximum uncertainty in energy gives the maximum uncertainty in the momentum as $(2mk_B T)^{1/2}$. By using the Heisenberg uncertainty principle, this translates into an uncertainty in position $h/(2mk_B T)^{1/2}$. This may be considered as the de Broglie wavelength of the particle or the extent to which it can be localized. Customarily, the thermal wavelength is defined as

$$\lambda = \frac{h}{(2\pi m k_B T)^{1/2}} \tag{2.72}$$

The thermal wavelength gives one length scale in the problem. Another length scale is the mean distance between the particles estimated by $v^{1/3}$, where $v = V/\langle N \rangle$ is the volume per particle. A natural dimensionless ratio is

$$R_d = \lambda^3 / v \tag{2.73}$$

Depending on the value of this ratio, different results will ensue.

Consider the case $R_d \ll 1$ first, which can happen if either the temperature is high or the interparticle spacing is large or both. The particles are localized such that their wave packets do not overlap. We can distinguish the particles from one another. The case $R_d \ll 1$ is the regime of *weak degeneracy*. In the limit $R_d \to 0$, classical statistical mechanics will be obtained and for small R_d there will be small corrections to the classical limit. It may be recalled that this limit formally corresponds to taking Planck's constant to be zero, a well-known procedure to obtain Newtonian mechanics. In the opposite case, when $R_d \gg 1$, the wave packets of the particles overlap so much that we cannot distinguish between them. This case is the regime of *strong degeneracy* in which the classical approximation fails completely and new effects that are quantum mechanical in nature arise.

If we keep the density of the gas fixed and let the temperature vary, we can go from the weakly degenerate to the strongly degenerate case. The quantal effects start to become important at the temperature at which $R_d \simeq 1$. From Eqs. (2.72) and (2.73), we see that this temperature is of the order of

$$h^2 / (2\pi m k_B v^{2/3}) \tag{2.74}$$

It will be seen later that both the *critical temperature* of a boson gas and the *Fermi temperature* of a fermion gas are of the same order of magnitude.

These are the temperatures near and below which these two quantum gases cannot be treated correctly by classical statistics.

All the thermodynamic quantities for quantum gases can be obtained using the Bose–Einstein and the Fermi–Dirac distributions (Kittel, 1969, Chapters 14–17). Of course, they may be obtained also from the partition function, Eq. (2.64), or its log, Eq. (2.66). This approach is taken in Section 2.3.4.

2.3.4 Partition Functions

We now specialize to ideal quantum gases, so that the energy levels ϵ are obtained solely from the kinetic energy. We take the particle-in-a-box model. The energy levels are denoted by ϵ rather than ϵ_k for simplicity and are given by

$$\epsilon = \frac{\hbar^2 k^2}{2m} \tag{2.75}$$

where m is the mass of a particle, and the allowed values of the wave vector $\mathbf{k}$ depend on the boundary conditions. We choose *periodic boundary conditions* in which

$$|\psi(x + L, y, z)\rangle = |\psi(x, y + L, z)\rangle = |\psi(x, y, z + L)\rangle = |\psi(x, y, z)\rangle \tag{2.76}$$

where L is the length of the cubic box in each of the three directions. The allowed values of $\mathbf{k}$ are given by

$$\mathbf{k} = \frac{2\pi\mathbf{s}}{L} \tag{2.77}$$

where $\mathbf{s}$ is a vector each of whose components is a quantum number given by $0, \pm 1, \pm 2, \ldots$. If the particles have spin S, and the system is in a magnetic field, the allowed energy levels also depend on S. In the absence of the magnetic field, the case we adhere to, inclusion of spin simply makes the energy levels $(2S + 1)$-fold degenerate so that all the sums and integrals over energy are multiplied by $2S + 1$. By using all this information in Eq. (2.66), we can rewrite $\ln \Xi$ in the form

$$\ln \Xi = \pm(2S + 1) \sum_{\{s_i\}=-\infty}^{\infty} \ln\{1 \pm \exp[\beta(\mu - \epsilon)]\} \tag{2.78}$$

with

$$\epsilon = \frac{h^2}{2mL^2}(s_1^2 + s_2^2 + s_3^2) = \frac{h^2}{2mL^2}s^2 \tag{2.79}$$

The standard method of performing the triple sums in Eq. (2.78) is as follows (Pathria, 1996, Sections 7.1, 8.1, and Appendix A). In the thermodynamic limit, the number of particles and the volume of the box go to infinity, and one replaces the triple sum by an energy integral by calculating the *density of states*, $g(\epsilon)$, which is defined as the number of quantum states per unit energy in a thin spherical shell between ϵ and $\epsilon + d\epsilon$. In general, this quantity is quite complicated. In the thermodynamic limit, when the volume of the system goes to infinity, the energy levels given by Eq. (2.79) crowd together and an almost continuous distribution of allowed points in the ϵ space is obtained. For a given ϵ, the points lie in a sphere of radius

$$s = (2mL^2\epsilon/h^2)^{1/2} \tag{2.80}$$

The total number of points in the thin shell between ϵ and $\epsilon + d\epsilon$ is given by $g(\epsilon)\,d\epsilon$, by definition, and is equal to the volume of the thin shell, which is given by the product of the area of the shell, $4\pi s^2$, and its thickness, ds. Therefore,

$$g(\epsilon)\,d\epsilon = 4\pi s^2\,ds = (8\pi mL^2\epsilon/h^2)\,ds = 2\pi(2m/h^2)^{3/2}V\epsilon^{1/2}\,d\epsilon \tag{2.81}$$

where Eq. (2.80) has been used to change variables from s to ϵ on the right-hand side, and $V = L^3$ is the volume of the system. Using this, one obtains (Pathria, 1996, Sections 7.1, 8.1, and Appendix A)

$$\ln \Xi = \pm(2S + 1)\frac{2\pi V}{h^3}(2m)^{3/2}\int_0^\infty \epsilon^{1/2}\ln\{1 \pm \exp[\beta(\mu - \epsilon)]\}\,d\epsilon \tag{2.82}$$

An expression for the average number of particles can be obtained by using Eq. (1.95). One obtains

$$\langle N \rangle = (2S + 1)\frac{2\pi V}{h^3}(2m)^{3/2}\int_0^\infty \frac{\epsilon^{1/2}\,d\epsilon}{\exp[\beta(\epsilon - \mu)] \pm 1} \tag{2.83}$$

We remind the reader that the upper sign in the denominator of Eq. (2.83) applies to fermions and the lower sign to bosons.

It is well known that the solution of a differential equation, such as the Schrödinger equation, depends on the boundary conditions. Under boundary conditions other than periodic, the allowed values of the **k** vector will change, and this change will in turn affect the density of states. However, in the infinite volume limit, the final results are found to be independent of boundary conditions (Pathria, 1996, Appendix A).

Once the partition function has been calculated, thermodynamic quantities can be obtained using the results in Section 1.6.1. We do this in Section 2.4, where we shall see how classical statistics emerges naturally in the high-temperature/low-density limit and how new quantum effects arise in this limit.

2.4 IDEAL BOSE–EINSTEIN AND FERMI–DIRAC GASES

2.4.1 Weak Degeneracy

In the classical limit, the absolute activity, by definition, is

$$z = \exp(\mu/k_B T) = \langle N \rangle (h^2/2\pi m k_B T)^{3/2}/V$$
$$= \langle N \rangle \lambda^3/V = R_d \tag{2.84}$$

where Eqs. (1.112) and (2.73) have been used. For weak degeneracy, $R_d \ll 1$ so that $z \ll 1$, which implies that μ is large and negative. Since $\epsilon \geq 0$, it means that $\exp[\beta(\mu - \epsilon)] \ll 1$. It is legitimate in this case to expand the log on the right-hand side of Eq. (2.82) using

$$\ln(1 \pm x) = \pm \left(x \mp \frac{x^2}{2} + \cdots \right) \tag{2.85}$$

which gives

$$\ln \Xi = (2S + 1)\frac{2\pi V}{h^3}(2m)^{3/2}$$
$$\times \int_0^\infty \epsilon^{1/2} \left[\exp\{\beta(\mu - \epsilon)\} \mp \frac{1}{2} \exp\{2\beta(\mu - \epsilon)\} + \cdots \right] d\epsilon \tag{2.86}$$

By changing variables to $y = \epsilon^{1/2}$ and evaluating the resulting Gaussian integrals, we get

$$\ln \Xi = (2S + 1)\frac{Vz}{\lambda^3} \left(1 \mp \frac{z}{2^{5/2}} + \cdots \right) \tag{2.87}$$

where Eq. (2.72) and the definition of absolute activity have been used. The average number of particles follows by using Eq. (1.95) or by a similar treatment of Eq. (2.83). In either case, we obtain

$$\langle N \rangle = (2S + 1)\frac{Vz}{\lambda^3} \left(1 \mp \frac{z}{2^{3/2}} + \cdots \right) \tag{2.88}$$

By using the relation $PV = k_B T \ln \Xi$ and eliminating z between Eqs. (2.87) and (2.88), we get

$$PV = Nk_B T \left(1 \pm \frac{R_d}{2^{5/2}} + \cdots \right) \tag{2.89}$$

where Eq. (2.73) has been used.

The series expansion in powers of the absolute activity or in powers of R_d is a technique that can be used to obtain expressions for other thermodynamic quantities such as the entropy, free energy, specific heat, and so on. The first term in these series expansions gives the same results that were obtained for the Boltzmann gas, discussed in Section 1.6.4. However, unlike the discussion of Section 1.6.4, it was not necessary in the present section to introduce the factors $N!$ and h^{3N} on an ad hoc basis. These factors automatically come out of the quantum mechanical treatment. In this sense, the quantum statistical mechanics is more fundamental than the classical statistical mechanics.

We obtained the leading quantum correction to the pressure of a Boltzmann gas. Higher order corrections are not shown here; the interested reader is referred to other texts (e.g., Pathria, 1996, Sections 7.1 and 8.1). For a weakly degenerate quantum gas, the ideal gas equation is no longer obeyed. Hence, Maxwell's law of distribution of velocities is no longer valid. The correct law shows deviation from Gaussian behavior. The expansion in terms of the small variable R_d is valid for low densities at a fixed temperature. This low-density expansion is called the *virial expansion* in the theory of interacting gases. Here, the expansion is for noninteracting quantal systems. Therefore, one can think of quantum mechanics as introducing effective statistical interactions among the particles. From the sign of the correction to leading order in Eq. (2.89), one can see that fermions have an effective repulsive interaction, whereas bosons have an effective attractive potential.

As the temperature is lowered for fixed density, quantum corrections become more important. In the fermion case, the repulsion causes the gas to exert a positive pressure even at absolute zero. In the boson case, the attractions give rise to Bose–Einstein condensation. For such low temperatures and/or high densities, there is strong degeneracy, and the expansion used in this section is not valid. Section 2.4.2 explores the techniques used to study strong degeneracy.

2.4.2   Strong Degeneracy: Bose Gas

The average particle density for a Bose gas of zero-spin particles, as obtained from Eq. (2.83), is

$$\frac{\langle N \rangle}{V} = \frac{2\pi}{h^3}(2m)^{3/2} \int_0^\infty \frac{\epsilon^{1/2}\, d\epsilon}{\exp[\beta(\epsilon - \mu) - 1]} \tag{2.90}$$

where the factor $2S + 1 = 1$ for spinless bosons. The standard treatment (Pathria, 1996, Chapter 7; Huang, 1987, Chapter 12) of the Bose gas is as follows. The left-hand side of this equation is independent of temperature at a fixed number density, so the right-hand side should also be independent of temperature. For high temperatures, μ is large and negative, and its value changes with temperature in such a way that the right-hand side of the equation remains temperature independent. But as the temperature is lowered, μ reaches its limiting value 0 (recall that $\mu \le 0$ from Section 2.3.2), and the right-hand side reaches a limiting value given by

$$\frac{2\pi}{h^3}(2m)^{3/2} \int_0^\infty \frac{\epsilon^{1/2}\, d\epsilon}{\exp(\beta\epsilon) - 1} \tag{2.91}$$

at a certain temperature, say T_0. For $T = T_0$, one may write

$$\left(\frac{\langle N \rangle}{V}\right)_{T_0} = \frac{2\pi}{h^3}(2m)^{3/2} \int_0^\infty \frac{\epsilon^{1/2}\, d\epsilon}{\exp(\beta_0\epsilon) - 1} \tag{2.92}$$

where $\beta_0 = 1/k_B T_0$. The integral in Eq. (2.92) can be evaluated term by term by expanding the integrand in powers of $\exp(-\beta_0\epsilon)$, making the substitution $y = \epsilon^{1/2}$, and evaluating the resulting Gaussian integrals. Equation (2.92) then becomes

$$\left(\frac{\langle N \rangle}{V}\right)_{T_0} = \frac{1}{\lambda_0^3} \sum_{l=1}^\infty \frac{1}{l^{3/2}} \tag{2.93}$$

where λ_0 is the thermal wavelength at T_0. Here, the sum, which is known as the *zeta function* of argument $\frac{3}{2}$, is approximately equal to 2.612 (Pathria, 1996, Appendix D). The special temperature T_0 is given by

$$T_0 = \frac{h^2}{2\pi m k_B (2.612 v)^{2/3}} \tag{2.94}$$

which is the same order of magnitude as the degeneracy temperature, Eq. (2.74).

What happens as one lowers the temperature below T_0? Certainly, μ cannot increase above zero (recall that $\mu \le 0$). It must remain at the value 0. So it appears that the right-hand side of Eq. (2.91) is temperature dependent while the left-hand side is not. The standard way out of this contradiction, going back to Einstein (Einstein, 1925), is to say that an error was made in replacing the triple sum in Eq. (2.78) by an integral. This error was carried over in Eq. (2.90). One term of the triple sum, the one corresponding to the ground state $s = 0$, becomes as important as the rest of the integral below T_0, and, therefore, should be pulled out of the sum. This separate term is found to correspond to what is called the *Bose condensate*. Equation (2.90) is then replaced by

$$\frac{\langle N \rangle}{V} = \frac{\langle n_0 \rangle}{V} + \frac{2\pi V}{h^3}(2m)^{3/2} \int_0^\infty \frac{\epsilon^{1/2}\, d\epsilon}{\exp[\beta(\epsilon - \mu)] - 1} \qquad (2.95)$$

where Eq. (2.71) and the definition of absolute activity give

$$\langle n_0 \rangle = \frac{1}{\exp(-\beta\mu) - 1} = \frac{z}{1 - z} \qquad (2.96)$$

For $T \leq T_0$, one has

$$\frac{\langle N \rangle}{V} = \frac{\langle n_0 \rangle}{V} + \frac{2.612}{\lambda^3} = \frac{\langle n_0 \rangle}{V} + \frac{\langle N \rangle}{V}\left(\frac{T}{T_0}\right)^{3/2} \qquad (2.97)$$

where Eq. (2.94) has been used. Therefore, the temperature dependence of the condensate is given by

$$\langle n_0 \rangle = \langle N \rangle \left[1 - \left(\frac{T}{T_0}\right)^{3/2}\right] \qquad (2.98)$$

The condensate is equal to the total number of particles at $T = 0$, and the ground state empties as the temperature is raised, becoming unoccupied at temperatures $T \geq T_0$. This macroscopic occupancy of the ground state of a Bose system at low temperatures is called the *Bose–Einstein condensation*. The condensate has the same value 0 whether the temperature T_0 is approached from above or below. However, its derivative with respect to temperature is zero for $T = T_{0+}$ and infinity for $T = T_{0-}$. Mathematically, this is expressed by saying that the condensate is not an analytic function of temperature at $T = T_0$. Because if it were, all its derivatives would exist and would be continuous at that temperature.

The arbitrariness associated with the above derivation can be avoided by a more rigorous replacement of the triple sum in Eq. (2.70) by an integral. One technique, invented by Pathria and co-workers (Greenspoon and Pathria, 1974; Zasada and Pathria, 1976), involving the Poisson summation formula

(Schwarz, 1966), is extremely powerful not only in taking the thermodynamic limit in many-body systems but also in studying the approach to this limit under various boundary conditions. This technique has led to a variety of new results in the study of finite-size effects in the region near phase transitions in many exactly solvable models (e.g., Singh and Pathria, 1985). We give the outline of this method here; for details see the original papers.

The triple sum to be evaluated [Eq. (2.70)] is

$$\langle N \rangle = \sum_{\epsilon} \frac{1}{\exp(\alpha + \beta\epsilon) - 1} = \sum_{\epsilon} \sum_{j=1}^{\infty} \exp[-j(\alpha + \beta\epsilon)]$$

$$= \sum_{j=1}^{\infty} \exp(-j\alpha) \sum_{\{s_i\}=-\infty}^{\infty} \exp(-jws^2) \quad (2.99)$$

where $\alpha = -\beta\mu$ and

$$w = \frac{\beta h^2}{2mL^2} = \pi(\lambda/L)^2 \quad (2.100)$$

The Poisson summation formula is

$$\sum_{n=-\infty}^{\infty} F(n) = \sum_{q=-\infty}^{\infty} \widetilde{F}(q) \qquad \widetilde{F}(q) = \int_{-\infty}^{\infty} F(n) \exp(2\pi i q n)\, dn \quad (2.101)$$

The function $\widetilde{F}(q)$ is the Fourier transform of $F(n)$. It is seen that the term $\mathbf{q} = 0$ would have resulted if the sum on the left-hand side of the first part of Eq. (2.101) had been replaced by an integral. The $\mathbf{q} \neq 0$ terms represent the corrections. We shall show how the condensate appears naturally when these terms are investigated in detail.

In our case, Eq. (2.101) becomes

$$\sum_{n=-\infty}^{\infty} \exp(-jwn^2) - \sum_{q=-\infty}^{\infty} \left(\frac{\pi}{jw}\right)^{1/2} \exp\left(-\frac{\pi^2}{jw}q^2\right) \quad (2.102)$$

where the Fourier transform integral was evaluated by completing the square in the argument of the exponential and shifting the variable. By using this equation three times in Eq. (2.99), we get

$$\langle N \rangle = \frac{V}{\lambda^3} \sum_{\{q_i\}=-\infty}^{\infty} \sum_{j=1}^{\infty} \frac{\exp(-j\alpha)}{j^{3/2}} \exp\left(-\frac{\pi^2}{jw}q^2\right) \quad (2.103)$$

where Eq. (2.72) has been used. The $\mathbf{q} = 0$ term is given by

$$\frac{V}{\lambda^3} \sum_{j=1} \frac{\exp(-j\alpha)}{j^{3/2}} \tag{2.104}$$

For $T \leq T_0$, $\alpha = -\beta\mu = 0$, and the sum becomes a zeta function equal to 2.612, so that the $\mathbf{q} = 0$ term becomes

$$2.612 \frac{V}{\lambda^3} \tag{2.105}$$

This corresponds to the second term on the right-hand side in Eq. (2.97). The term for $\mathbf{q} \neq 0$ of Eq. (2.103) corresponds to the condensate in the limit $\alpha \rightarrow 0$. There is a rigorous way to take this limit. Here we give a quick derivation. The $\mathbf{q} \neq 0$ term of Eq. (2.103) is

$$\frac{V}{\lambda^3} \sum_{\{q_i\}=-\infty}^{\infty}{'} \sum_{j=1}^{\infty} \frac{\exp(-j\alpha)}{j^{3/2}} \exp\left(-\frac{\pi^2}{jw}q^2\right) \tag{2.106}$$

where the prime on the summation indicates the omission of the $\mathbf{q} \neq 0$ term. The triple sum over $\mathbf{q}$ is replaced by a triple integral, which turns out to be a triple Gaussian integral. Performing the integration, the term reduces to

$$\frac{V}{\lambda^3} \sum_{j=1}^{\infty} \frac{\exp(-j\alpha)}{j^{3/2}} \left(\frac{wj}{\pi}\right)^{3/2} = \sum_{j=1}^{\infty} \exp(-j\alpha) \tag{2.107}$$

where Eq. (2.100) has been used. The sum is a geometric series and yields

$$[\exp(\alpha) - 1]^{-1} = [\exp(-\beta\mu) - 1]^{-1} = \langle n_0 \rangle \tag{2.108}$$

using Eq. (2.96). By collecting terms, Eq. (2.103) reduces to

$$\langle N \rangle = 2.612 \frac{V}{\lambda^3} + \langle n_0 \rangle \tag{2.109}$$

which is identical to Eq. (2.97). A more rigorous analysis shows that corrections to this are of the order of $\exp(-L/\lambda)$ and therefore vanish in the thermodynamic limit.

All the thermodynamic functions can be calculated in the limits of weak and strong degeneracy by using the techniques outlined here. The nonanalyticity of the condensate at $T = T_0$ results in the nonanalyticity of all thermodynamic functions at $T = T_0$. For example, we show the heat capacity at constant volume in Figure 2.1. The curve for $T/T_0 < 1$ represents strong degeneracy and the curve for $T/T_0 > 1$ represents weak degeneracy.

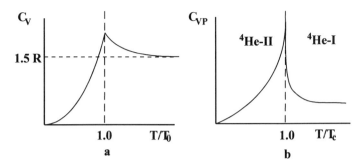

Figure 2.1 Molar heat capacity of (*a*) the ideal Bose gas and (*b*) the liquid ^{4}He under its own vapor pressure.

For low temperatures, $C_V \sim T^{3/2}$, and for high temperatures $C_V = 3R/2$, as expected. Shown on the right is the experimental heat capacity of liquid ^{4}He. Although the two figures are dissimilar, the λ-type transition of ^{4}He at the temperature T_c is thought by many to be an example of Bose–Einstein condensation. Needless to say, all this behavior is totally different from an ideal classical gas.

Recent observation of Bose–Einstein condensation (BEC) in ^{87}Rb atoms, confined in a magnetic trap (Anderson et al., 1995), has created quite a sensation. A readable account of what the "fuss" is all about is given by Kleppner in a reference frame article in *Physics Today* entitled *The Fuss about Bose–Einstein Condensation* (Kleppner, 1996). Although BEC is believed to occur in superfluids and superconductors, the observation in ^{87}Rb is the first time BEC has been seen in weakly interacting systems, where the theory is much easier. Therefore, a detailed comparison between experiment and theory can be made. Among practical applications, the BEC makes it possible to produce, in principle, a coherent beam of condensate atoms making an atomic laser. Such a laser will make it possible to increase the capability of an atomic interferometer a millionfold, which, in turn, will certainly lead to new discoveries in atomic physics. The only way to keep up with exciting new discoveries in BEC is to look at the *Physics News Update* on the *World Wide Web* (see Appendix A).

Bose–Einstein condensation is an example of critical phenomena, which are discussed in detail in Chapter 3.

2.4.3 Strong Degeneracy: Fermi Gas

The average particle density for a Fermi gas, as obtained from Eq. (2.83), is

$$\frac{\langle N \rangle}{V} = (2S + 1)\frac{2\pi}{h^3}(2m)^{3/2} \int_0^\infty \frac{\epsilon^{1/2}\, d\epsilon}{\exp[\beta(\epsilon - \mu) + 1]} \tag{2.110}$$

It is convenient to introduce the quantity

$$f(\epsilon) = \frac{1}{\exp[(\epsilon - \mu)/k_B T] + 1} \tag{2.111}$$

where $f(\epsilon)$ is the probability that a state with energy ϵ and a given spin component is occupied; it is simply $\langle n \rangle$, the average number of fermions in the state with energy ϵ and a given component of spin. The function $f(\epsilon)$ is often called the *Fermi function*. In terms of this function, Eq. (2.110) becomes

$$\frac{\langle N \rangle}{V} = (2S + 1)\frac{2\pi}{h^3}(2m)^{3/2} \int_0^\infty \epsilon^{1/2} f(\epsilon)\, d\epsilon \tag{2.112}$$

It is instructive to study the behavior of $f(\epsilon)$ at various temperatures as a function of ϵ. At $T = 0$ there are two different results depending on whether ϵ is less than or greater than μ. In the former case, $f(\epsilon) = 1$, whereas in the latter case it equals zero, as shown in Figure 2.2. This figure shows the Pauli principle at work. There can never be more than one particle in any orbital, characterized by ϵ and a given spin component. The special energy equal to the value of the chemical potential at $T = 0$ is called the *Fermi energy* and is labeled ϵ_F. As the temperature is raised, the Fermi distribution is changed as shown in Figure 2.2. The width of the slope is of the order of $k_B T$.

The temperature corresponding to the Fermi energy is $T_F = \epsilon_F/k_B$ and is called the *Fermi temperature*. To evaluate it, we calculate the integral in Eq. (2.112) at $T = 0$. Because the Fermi function is zero for $\epsilon > \epsilon_F$, the integral is zero in this range. For $0 < \epsilon < \epsilon_F$, the Fermi function is unity, and the integral is readily evaluated. We get

$$\frac{\langle N \rangle}{V} = (2S + 1)\frac{2\pi}{h^3}(2m)^{3/2}\frac{2}{3}\epsilon_F^{3/2} \tag{2.113}$$

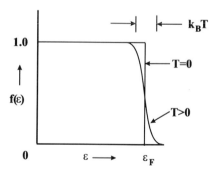

Figure 2.2 Fermi–Dirac distribution.

which gives

$$\epsilon_F = \frac{h^2}{2m} \left[\frac{3\langle N \rangle}{4\pi(2S + 1)V} \right]^{2/3} \tag{2.114}$$

Hence, the Fermi temperature is given by

$$T_F = \frac{h^2}{2mk_B} \left[\frac{3\langle N \rangle}{4\pi(2S + 1)V} \right]^{2/3} \tag{2.115}$$

which is the same order of magnitude as the degeneracy temperature [Eq. (2.74)]. The physical significance of this temperature is that above this temperature the gas is weakly degenerate, and below or near this temperature the gas is strongly degenerate.

In the same way, the total energy of the Fermi gas at $T = 0$, also called the zero-point energy, can be calculated. It is given by

$$E_0 = \frac{3}{5}\langle N \rangle \epsilon_F \tag{2.116}$$

Other thermodynamic quantities can be calculated similarly. We get

$$P_0 = \frac{2\langle N \rangle}{5V} \epsilon_F \tag{2.117}$$

$$A_0 = \frac{3}{5}\langle N \rangle \epsilon_F \tag{2.118}$$

$$G_0 = \langle N \rangle \epsilon_F \tag{2.119}$$

The entropy and the specific heat at $T = 0$ are found to be zero.

As the temperature is raised slightly above zero, the lowest orbitals are unaffected. This happens because the particles can go only to vacant orbitals, none of which are available near the ground orbital. The only orbitals affected are the ones near the Fermi energy, because there are vacant orbitals above this energy. Detailed calculations show that the only orbitals involved in determining the thermodynamic properties of a Fermi gas at low temperatures lie in a fractional width of the order of $k_B T/\epsilon_F$ around the Fermi energy. As a result, the specific heat of a Fermi gas is proportional to $k_B T/\epsilon_F$ at low temperatures. This behavior is completely different from an ideal classical gas for which the specific heat is a constant.

In order to evaluate thermodynamic functions for a degenerate Fermi–Dirac gas at a nonzero absolute temperature, one may calculate the grand

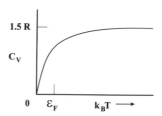

Figure 2.3   Molar heat capacity of a Fermi gas.

partition function given by Eq. (2.82). Standard methods are available for doing this (e.g., Reif, 1965, Section 9.17). The molar heat capacity C_V for a Fermi–Dirac gas is shown in Figure 2.3. At room temperature, $C_V \approx 10^{-2}R$ for electrons in metals. For $T \ll T_F = \epsilon_F/k$, the heat capacity curve is linear. Notice that, unlike the Bose–Einstein case, there is no break in the curve at the Fermi temperature.

Unlike the Bose gas, where all the particles occupy the lowest energy level as $T \rightarrow 0$, the Fermi particles occupy all the lowest energy states, $(2S + 1)$ particles per state. Above $T = 0$, the population of the occupied highest energy states falls below 1.0 as shown in Figure 2.2. This behavior is important for explaining quantum phenomena such as the behavior of electrons in metals and semiconductors.

2.4.4   Example: Black-Body Radiation

As an application of Bose–Einstein statistics, we consider the radiation emitted and absorbed by a black body. A black body, by definition, is one that emits radiation of all frequencies and absorbs all radiation that falls on it. According to quantum mechanics, the radiation consists of elementary particles called photons. We assume a photon field to be in thermal equilibrium with the walls of a black body. The black-body–photon system is isolated, so that the energy is constant. Because photons of different frequencies are being absorbed and emitted at constant energy, the number of photons is not constant. The criterion of thermodynamic equilibrium requires $\nu_a\mu + \nu_e\mu = 0$, where μ is the chemical potential of the photon and ν_a and ν_e are the stoichiometric coefficients for absorbed and emitted photons, respectively. Since $\nu_a \neq -\nu_e$, $\mu = 0$. In this case the Bose–Einstein distribution, given by the lower sign in Eq. (2.69), is

$$\langle n \rangle = \frac{1}{\exp(\epsilon/k_B T) - 1} = \frac{1}{\exp(\hbar\omega/k_B T) - 1} \tag{2.120}$$

where $\epsilon = \hbar\omega$. This equation was derived by Planck before the advent of the quantum theory and is called *Planck's distribution*. It is possible to change from a discrete to a continuous distribution of frequencies by multiplying Planck's distribution by the appropriate density of states. This leads to Planck's radiation formula, which describes the energy density (intensity) of black-body radiation (Pathria, 1996, Section 7.2).

2.4.5 Example: Electrons in Semiconductors

Semiconductors are elements such as boron, silicon, and germanium that lie between the metals and nonmetals in the periodic table. Also, they may be compounds of transition metal oxides such as TiO_2, ZnO, and Cu_2O. A semiconductor is characterized by an electrical conductivity in the range of 1–10^{-4} mho m^{-1} and a band gap in the range of 1–5 eV. On the left of Figure 2.4 is shown [Figure 2.4(a)] a schematic representation of the energy levels of a semiconductor. On the right [Figure 2.4(b)] is shown the Fermi distribution (cf. with Figure 2.2) for the electrons at room temperature. The Fermi energy ϵ_F is equal to $\epsilon_g/2$, where ϵ_g is the band-gap energy, which is the energy necessary for an electron to be excited from the top of the valence band to the bottom of the conduction band. As shown in Figure 2.4(b), the valence band is full and the conduction band is empty. No electrons are present in the gap between the valence and conduction bands. A high-temperature Fermi distribution is shown in Figure 2.4(c), where some electrons have moved from the valence to the conduction band. As electrons leave the valence band, their absence can be thought of as creating fictitious positively charged particles called holes.

We wish to determine $\langle N_c \rangle$, the number of conduction electrons in the semiconductor, under the condition that $\exp(\epsilon - \mu)/k_B T \gg 1$. These are

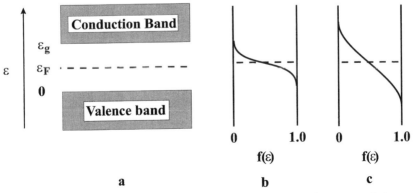

Figure 2.4   (*a*) Semiconductor energy levels. (*b*) Room temperature electron population. (*c*) High-temperature electron population.

electrons with energies in the range ϵ_g to ∞. The number $\langle N_c \rangle$ is given by

$$\langle N_c \rangle = C \int_{\epsilon_g}^{\infty} \frac{\epsilon^{1/2}\, d\epsilon}{\exp(\epsilon - \mu)/k_B T} \tag{2.121}$$

where $C = 4\pi(2m/h^2)^{3/2}V$ and we have used the fact that for electrons $2S + 1 = 2$. Specifically focusing on the energy above ϵ_g, we let $E = \epsilon - \epsilon_g$. Equation (2.121) becomes

$$\langle N_c \rangle = C \exp[(\mu - \epsilon_g)/k_B T] \int_0^{\infty} \exp(-E/k_B T)E^{1/2}\, dE \tag{2.122}$$

The integral in Eq. (2.122) has the value $[\pi/4(k_B T)^3]^{1/2}$. For an intrinsic semiconductor (one that has no electronic states in the band gap), $\epsilon_F = \epsilon_g/2$, and Eq. (2.122) becomes

$$\langle N_c \rangle = 2(2\pi m k_B T/h^2)^{3/2}V \exp(-\epsilon_g/2k_B T) \tag{2.123}$$

This equation requires that $\epsilon_g - \mu \gg k_B T$. Since $k_B T \approx 1/40$ eV at room temperature, this equation should be good for all intrinsic semiconductors. Doped semiconductors, which have electronic states in the band gap, require a more complicated analysis (Kittel, 1983, Chapter 8).

2.5   EXERCISES AND PROBLEMS

1. Evaluate $[\hat{x}, \hat{x}]$, $[\hat{p}, \hat{p}]$, and $[\hat{x}, \hat{p}]$.

2. Show that the eigenvalues of a Hermitian operator are real.

3. Use Eq. (2.21) to evaluate $d\hat{x}/dt$ for the harmonic oscillator.

4. Are the following systems in pure or mixed states?

Nitrogen molecules at STP.

Protons in resonance with a radio frequency (rf) field in a nuclear magnetic resonance (NMR) experiment.

Sodium atoms vibrating in the solid at 4 K.

Sodium atoms excited by an intense pulse of laser light.

5. Imagine that the mixed state described in Section 2.2.3 is sent through an inhomogeneous magnetic field aligned along the z direction. In this case, the state vectors are

$$|+z\rangle$$

and

$$|+x\rangle = \frac{1}{\sqrt{2}} \left(|+z\rangle + |-z\rangle\right)$$

where $|+z\rangle$ is an eigenstate and $|+x\rangle$ is a superposition state, and the basis states are $|k\rangle = |+z\rangle$ and $|-z\rangle$. Use Eqs. (2.7) and (2.30) to find the elements of the density matrix.

6. From the definition, Eq. (2.30), prove that the density matrix is Hermitian, that is,

$$\rho_{ij}(t) = \rho_{ji}^*(t)$$

7. Show that the matrix of Exercise 5 is Hermitian.

8. Determine the total number of states accessible to a system of 10 particles if the particles obey (a) Maxwell–Boltzmann statistics (assume distinguishable particles), (b) Bose–Einstein statistics, and (c) Fermi–Dirac statistics. Consider the following number of energy (eigen) states: 100, 10, 2, and 1.

9. Determine the thermal de Broglie wavelength for ^{4}He gas at Standard Temperature and Pressure (STP). Use Eq. (2.84) to determine $\exp(\mu/k_B T)$. What statistics does ^{4}He obey at STP?

10. Show that the specific heat of a weakly degenerate quantum gas is given by

$$C_V = \frac{3}{2} N k_B \left(1 \mp \frac{R_d}{2^{7/2}} + \cdots\right)$$

11. Calculate the first non-Gaussian correction to Maxwell's law of velocities in the weak degeneracy case.

12. For a Fermi–Dirac gas, evaluate the zero-point energy and the zero-point pressure. Calculate the Fermi energy for copper metal assuming one free electron per atom. The electron density of copper is 8.45×10^{22} cm^{-3}. What is the Fermi temperature of copper?

13. The relation between $g(n)$, the number of states per quantum number, and $g(k)$, the number of states per wave vector, is

$$g(n)\,dn = (Vk^2/2\pi^2)\,dk = g(k)\,dk$$

Starting from

$$dN_k = g(k)\langle n_\omega\rangle\,dk$$

use the relation $k = \omega/c$ for photons and multiply by 2 for the two directions of polarization to derive an expression for

$$dN_\omega = g(\omega)\langle n_\omega\rangle\,d\omega$$

the number of photons in the frequency interval between ω and $\omega + d\omega$. Then multiply by the energy per photon to get dE_ω, the total energy of photons in the frequency interval. The resulting equation is the Planck black-body radiation formula that gave birth to quantum mechanics.

14. Silicon has a band gap energy of 1.11 eV. Find the concentration of conduction electrons in pure silicon at $25°$ C.

PART II

EQUILIBRIUM STATISTICAL MECHANICS

3

PHASE TRANSITIONS AND CRITICAL PHENOMENA

3.1 INTRODUCTION

In Chapters 1 and 2, the fundamental assumptions and techniques of statistical mechanics were presented. With the use of these assumptions and techniques, any problem in a many-body system can be solved, in principle, with the following recipe:

1. Use suitable variables to write the Hamiltonian of the system.
2. Solve Hamilton's classical equations or Schrödinger's quantal equation for the allowed energies and microstates of the system.
3. Choose a suitable ensemble and calculate the partition function and the appropriate free energy. In most cases, it is enough to do so in the thermodynamic limit.
4. Use the free energy to calculate any other necessary information about the system.

Ideal classical and quantum systems were discussed using this recipe in Chapters 1 and 2. In the rest of the book, we concentrate on interacting many-body systems.

It is usually thought that the most profound steps are the first two steps above: the discovery of the laws of motion and the solution of the equations of motion. This seems to imply that once the fundamental laws of particle motion are discovered, the rest is either easy or just consists of applications. This point of view, dubbed the "constructionist hypothesis," was demolished in a brilliant and influential article entitled *More is Different* by P. W. Anderson (Anderson, 1972, pp. 393–396). [See also his more recent, *Theoretical Paradigms for the Sciences of Complexity* (in Anderson, 1994, p. 584).] His main point is that, even if the laws of motion of constituent particles and the *Theory of Everything* are known, much fundamental work remains to be done in the more "applied" condensed matter disciplines in order to understand the physical and chemical systems of interest to these disciplines. The reason is many-body systems have certain properties that emerge only if these systems are sufficiently large and complex.

A similar point was made by Michael Fisher (Fisher, 1983, p. 47). Even if we had a large and fast enough computer to calculate the specific heat and compressibility of a given liquid, we would have understood nothing. We must go beyond the calculations and look for universal behavior. Only then shall we find that certain properties of the system are fundamental to its understanding and certain others are just unessential details. Two prime examples of emergent properties are *broken symmetry* and *universal scaling behavior*. Broken symmetry is the phenomenon by which the macroscopic state of a system has a lower symmetry than the microscopic Hamiltonian. An example is the existence of crystalline solids, which have discrete translational symmetries, whereas the Hamiltonian of a system of individual molecules has full continuous translational symmetry. Universality means that certain essential properties of the system are independent of irrelevant details and depend only on certain basic parameters of the system. A famous example is the specific heat of solids at low temperatures. The specific heat at constant volume is a universal function of T/Θ_D, where Θ_D is the Debye temperature (Pathria, 1996, p. 176). It results because, at low temperatures, only the large wavelength modes are important, and the solid can be considered to be a continuum, thus washing out the details of the crystal structure.

In order to discover the universal and other emergent properties, one must study as many models as possible. Only from such studies can new behavior, which is absent in few-body systems, emerge. One way to characterize the essential behavior of large interacting systems is by the concept of *correlation length*. Assume that there is a particle at the origin in a fluid of a large interacting system. Now go in all directions and calculate the probability per unit volume of finding another particle at a given distance **r**. This conditional probability density is called the *pair correlation function* and is denoted by $g(\mathbf{r})$ (Pathria, 1996, p. 457). If the particles are totally independent from or uncorrelated with one another, as they are in a classical ideal gas, $g(\mathbf{r})$ is identically equal to 1. So the difference $g(\mathbf{r}) - 1$ is a measure of the correlations between the particles in a nonideal fluid. Because the interactions generally decay with increasing distance, this difference decreases to zero for large r. The characteristic distance beyond which only negligible correlations exist in the system is called the correlation length and is denoted by ξ. The correlation length becomes very large in a system near its critical point, which occurs for certain values of its thermodynamic parameters. For a fluid, these are the values of the critical temperature, pressure, and density. Other systems have other critical properties. It is because of their large correlation lengths that many systems behave identically in the vicinity of their critical points. In this chapter, the emergent phenomena of broken symmetry, scaling, and universality near critical points are discussed with the help of several models. The renormalization technique, which is the modern method of choice for the study of many-body systems, is also introduced.

In Chapter 4, liquids are studied by means of the exactly soluble Takahashi model and other analytical techniques. The computer-intensive approaches of molecular dynamics and Monte Carlo simulation are the subjects of Chapters 5 and 6, respectively. In both cases, the aim is to calculate thermodynamic averages. In the molecular dynamics simulation method, the averages are time averages taken over the simulated trajectories. In the Monte Carlo method, the averages are taken over stationary ensembles obtained by picking out random microstates. In Chapter 7, polymers are studied. These systems consist of monomers arranged randomly in space such that the monomers repel strongly at short distances, giving rise to the excluded-volume, or the self-avoiding, effect. In polymer systems, the correlation length is a function of the number of monomer units, which plays the role of the inverse of the distance from the critical temperature. Finally, heteropolymers, such as proteins, are studied by means of spin glass and hydrophobic-polar (HP) models.

3.2 BROKEN SYMMETRY AND CRITICAL PHENOMENA

3.2.1 Ising Model: Introduction

Almost all the essential ideas of this chapter can be explained with the help of the Ising model, which we shall use because of the simplicity of its formulation. For a history of the Ising model, see Brush (Brush, 1967). This model grew out of attempts to explain ferromagnetism, which results from the interaction among electron spins of the atoms comprising the ferromagnetic material. Ferromagnetism is the existence of spontaneous magnetization in a certain temperature range. Here, "spontaneous" means magnetization in the absence of any applied magnetic field. A more complete model, the Heisenberg model, is based on spin–spin coupling, where all three components of the spin operators interact. In the Ising model only the z components interact

(Ising, 1925). The three-dimensional Ising model is applicable to phase transitions in liquids, anisotropic magnets, and binary alloys. The Ising model consists of a lattice of N fixed points, each of which has attached to it a spin variable S_i. Associated with this spin variable is a spin operator $\widehat{S}_i$ and a spin eigenket $|s_i\rangle$, whose eigenvalue is $s_i = \pm 1$. Here, we are choosing units in which the spins are measured in units of $\hbar/2$. Following the recipe in Section 3.1, we first write the Hamiltonian of the system. The Hamiltonian in the Ising model in an external magnetic field H is

$$\widehat{\mathcal{H}} = -J \sum_{n.n.} \widehat{S}_i \widehat{S}_j - \mu_B H \sum_i \widehat{S}_i \tag{3.1}$$

where J is a spin-coupling parameter, taken to be greater than 0 for interaction between spins S_i and S_j^- located at the nearest-neighbor (*n.n.*) lattice sites, and μ_B is the Bohr magneton. Henceforth, we choose units such that $\mu_B = 1$. A given many-body microstate of the system can be represented by

$$|s_1, s_2, s_3, \ldots, s_N\rangle = |s_1\rangle |s_2\rangle |s_3\rangle \cdots |s_N\rangle \tag{3.2}$$

where the essential distinguishability of the lattice sites means that symmetrization or antisymmetrization is not necessary. Since each s_i can have two values, there are 2^N such microstates.

The two terms in the Hamiltonian have interesting symmetry properties. Assume that all the spins are flipped. If a given spin is up, it is flipped down, and if down, it is flipped up. The first term, which represents mutual interactions, is made up of pairs of spins and, consequently, is unchanged upon flipping. One says that this term has an up–down symmetry. Upon flipping, the second term, which includes the effects of the magnetic field, changes sign. Hence, this term is said to break the up–down symmetry of the Hamiltonian. The field is therefore called a *symmetry-breaking field*. This kind of symmetry breaking in a finite field is expected and is seen experimentally in the Zeeman effect, for example.

The canonical partition function is given by

$$Z_N(H, T) = \operatorname{tr} \exp(-\beta \widehat{\mathcal{H}}) \tag{3.3}$$

where $\beta = 1/k_B T$, as usual. It is seen that Z_N is an even function of the magnetic field H, since the trace includes the sum over both $s_i = +1$ and $s_i = -1$, so that terms involving both $+H$ and $-H$ are present. Therefore, the Helmholtz free energy

$$A_N(H,T) = -k_B T \ln Z_N(H,T) \tag{3.4}$$

is also an even function of H.

The magnetization or the magnetic moment per site of the Ising model is given by

$$M_N(H,T) = \frac{1}{N} \sum_i \langle S_i \rangle = -\frac{1}{N} \frac{\partial A_N}{\partial H} \tag{3.5}$$

where the angular brackets denote the ensemble average. Changing the sign of H will change the sign of the last term on the right-hand side of Eq. (3.5). This operation is equivalent to changing the signs of the spins, for fixed H, in the middle term. Thus, it is seen that

$$M_N(-H,T) = -M_N(H,T) \tag{3.6}$$

By using the symmetry properties of the Ising model, the following argument can be made. Since the partition function [Eq. (3.3)] is a finite sum of exponential functions, it should be an analytic function of H and T (i.e., continuous and differentiable). Also, since all terms in the sum are positive, Z_N should be strictly positive, implying that the free energy [Eq. (3.4)] is an analytic function. Therefore, its first derivative with respect to the field should be continuous, making the magnetization zero at zero field. So, in the absence of the field, there is no magnetization and ferromagnetism cannot exist. The flaw in this argument has to do with the thermodynamic limit and will be explained below.

3.2.2   Ising Model: Exact Solution in One Dimension

The partition function in Eq. (3.3) has not yet been calculated for a three-dimensional model, where the spins lie on a cubic lattice, for example. A solution for the two-dimensional square lattice was first obtained by Onsager in 1944 (Pathria, 1996, pp. 377–388). Here, continuing the recipe of Section 3.1, we show how the partition function of the linear, or the one-dimensional, model can be calculated exactly. The edge effects coming from the end spins S_1 and S_N can be eliminated by using periodic boundary conditions in which $S_{N+1} \equiv S_1$. The partition function [Eq. (3.3)] can be written as

$$Z_N = \sum_{S_1} \sum_{S_2} \cdots \sum_{S_N} \langle s_1, s_2, s_3, \ldots, s_N |$$

$$\exp\left\{\sum_{i=1}^{N}\left[K\widehat{S}_i\widehat{S}_{i+1} + \frac{1}{2}h(\widehat{S}_i + \widehat{S}_{i+1})\right]\right\} |s_1, s_2, s_3, \ldots, s_N\rangle$$

$$= \sum_{S_1} \sum_{S_2} \cdots \sum_{S_N} \prod_{i=1}^{N} \langle s_1, s_2, s_3, \ldots, s_N |$$

$$\exp\left[K\widehat{S}_i\widehat{S}_{i+1} + \frac{1}{2}h(\widehat{S}_i + \widehat{S}_{i+1})\right] |s_1, s_2, s_3, \ldots, s_N\rangle \tag{3.7}$$

where the notations $K = \beta J$ and $h = \beta H$ have been introduced and the field term has been written in a symmetric form for convenience. Use Eq. (3.2) to write Eq. (3.7) as

$$Z_N = \sum_{S_1} \sum_{S_2} \cdots \sum_{S_N} \langle s_1|\mathbf{T}_1|s_2\rangle\langle s_2|\mathbf{T}_2|s_3\rangle \cdots \langle s_N|\mathbf{T}_N|s_1\rangle \tag{3.8}$$

where the operator $\mathbf{T}_i$ has the matrix representation

$$\langle s_i|\mathbf{T}_i|s_{i+1}\rangle = \exp[Ks_is_{i+1} + \tfrac{1}{2}h(s_i + s_{i+1})] \tag{3.9}$$

Since $s_i = \pm 1$, all the matrices are identical and equal to

$$\mathbf{T} = \begin{pmatrix} \exp(K + h) & \exp(-K) \\ \exp(-K) & \exp(K - h) \end{pmatrix} \tag{3.10}$$

With this, Eq. (3.8) can be written as

$$Z_N = \sum_{S_i} \langle s_1|\mathbf{T}^N|s_1\rangle = \operatorname{tr}\mathbf{T}^N = \lambda_1^N + \lambda_2^N \tag{3.11}$$

where the closure condition [Eq. (2.11)] has been used for spins $S_2, S_3, \ldots, S_N$, and λ_1 and λ_2 are the two eigenvalues of the matrix (3.10). The eigenvalues of Eq. (3.10) are found to be

$$\lambda_{1,2} = \exp(K)\cosh(h) \pm \left[\exp(-2K) + \exp(2K)\sinh^2(h)\right]^{1/2} \tag{3.12}$$

All the thermodynamic quantities can now be obtained. In particular, the magnetization is obtained from Eqs. (3.4) and (3.5) and is given by

$$M_N = k_B T \left(\frac{\lambda_1^{N-1} \lambda_1' + \lambda_2^{N-1} \lambda_2'}{\lambda_1^N + \lambda_2^N} \right) \tag{3.13}$$

where the prime indicates the derivative with respect to the field H.

3.2.3   Thermodynamic Limit of the Linear Ising Model

We use Eq. (3.12) to reiterate the argument at the conclusion of Section 3.2.1. From Eq. (3.12), we take $\lambda_1 > \lambda_2 > 0$ for any finite temperature. The partition function [Eq. (3.11)] must then be positive. Hence, the free energy, which is essentially the log of the partition function, is an analytic function of H and T, because the logarithmic function is analytic when its argument is strictly positive. Therefore, the magnetization is an analytic function of H and T. As seen above, since the magnetization is an odd function of H, it vanishes for zero field at any temperature. There is no spontaneous magnetism and no ferromagnetism. This lack of magnetism can be seen more explicitly by expanding the eigenvalues for low temperatures and weak fields. For $H \to 0^+$, we get

$$\lambda_{1,2} \approx \exp(K)(1 \pm h) \tag{3.14}$$

By using this result in Eq. (3.13), it is found that the magnetization vanishes. The same is true if $H \to 0^-$.

Now let us take the thermodynamic limit. We shall see how the spontaneous magnetization develops in this limit. Since $\lambda_1 > \lambda_2$, we see that $\lambda_1^N \gg \lambda_2^N$ in the limit $N \to \infty$. Consequently, the partition function for large N is

$$Z_N \approx \lambda_1^N \tag{3.15}$$

and the free energy is

$$A_N \approx -N k_B T \ln \lambda_1 \tag{3.16}$$

from which the magnetization for large N is obtained using Eqs. (3.5) and (3.12):

$$M_N \approx \frac{\sinh(h)}{\left[\exp(-4K) + \sinh^2(h) \right]^{1/2}} \tag{3.17}$$

For any finite temperature, the magnetization vanishes when there is no field. But at zero temperature, $\exp(-4K)$ is zero and the magnetization is ± 1 for $H \to 0^\pm$. Thus, spontaneous magnetism, or ferromagnetism, can exist in the one-dimensional Ising model provided that the system is in the thermodynamic limit at $T = 0$. The flaw in the argument at the end of Section 3.2.1 was to consider finite systems, where the free energy is analytic. It is only in the thermodynamic limit that nonanalytic behavior can arise. This statement is expressed by the following equations:

$$\lim_{N \to \infty} \lim_{H \to 0^\pm} M_N = 0 \tag{3.18}$$

$$\lim_{H \to 0^\pm} \lim_{N \to \infty} M_N = M \tag{3.19}$$

For a discussion of the analytic properties of the free energy in magnets and fluids, see Griffiths (Griffiths, 1972, pp. 10–33).

3.2.4 Important Definitions

The point at which a function is nonanalytic is called a *singular point*. Note that the function does not have to go to infinity at a singular point. Singularity simply means that the function or one of its derivatives is discontinuous or infinite. For example, in the linear Ising model, the magnetization jumps discontinuously from $+1$ to -1 as the field values pass from positive to negative at $T = 0$. In other words, $M(H, T)$ or $A(H, T)$ is singular at the point $H = 0$ and $T = 0$ in the two-dimensional space of H and T. Apart from this singular point, the free energy is analytic everywhere. One says that the system exists in a single phase in the region in which the free energy is analytic. At the singular point, the system is considered to have split into two phases, one corresponding to $M = +1$ and the other to $M = -1$. This behavior is a simple example of a *bifurcation*. Bifurcations are important in understanding the behavior of nonlinear systems. We shall discuss them in detail in Chapter 12.

In the general case, such as the two-dimensional Ising model, it turns out that the region in which two ferromagnetic phases coexist is not restricted to $T = 0$ but extends over a range of temperatures, say from $T = 0$ to $T = T_c$. One says that this region is a region of coexistence of the two phases. For a given temperature less than T_c, as the field passes through zero, the system goes from one phase to another, indicating a phase transition from a state of positive to negative magnetization. The spontaneous magnetization $\pm M$ depends on temperature and vanishes at T_c. The two phases having magnetizations $+M$ and $-M$ become more and more similar as the temperature is raised. At T_c the two phases become identical, and above T_c there are no singularities, and the system exists in a single phase called the paramagnetic phase. The point at which the two phases become identical ($H = 0, T = T_c$) is called a *critical point*, and the phase transition that takes place at this point is called a *critical transition* or a *continuous phase transition*. (The old terminology of calling this a second-order phase transition point has been largely abandoned as it was found to be too narrow.) The transitions from positive to negative magnetization ($H = 0, T < T_c$) are called *discontinuous* or *first-order phase transitions*.

Let us look once again at the one-dimensional Ising model at $T = 0$, where the model is in its ground macrostate, which may be obtained by minimizing the energy. The following discussion is based on Goldenfeld (Goldenfeld, 1992, pp. 55–59). One argues as follows. In the absence of the field ($H = 0$), the energy is minimized by two microstates, one in which all spins are $+1$ or up and a second state in which all spins are -1 or down. The ensemble average requires that the ground macrostate be an equal mixture of these two microstates and thus have the up–down symmetry of the Hamiltonian. The magnetization should be zero in this ground macrostate. The system is continually making transitions from the spin-up microstate to the spin-down microstate owing to spontaneous fluctuations. In a finite system, the lifetime of these fluctuations is finite, and the time average equals the ensemble average, which gives a zero spontaneous magnetization. It makes no difference if the thermodynamic limit is now taken, as expressed in Eq. (3.18).

In the thermodynamic limit, however, the lifetime of fluctuations becomes infinite, so that the system is trapped in one of its microstates; which one it is depends on the initial conditions, and the time average gives a nonzero spontaneous magnetization. The ergodic hypothesis is not valid, and this phenomenon is said to be *ergodicity breaking*. In order to make the two averages come out equal, the following procedure is used. One keeps a small field on the sample, so that the time average is equal to the ensemble average, takes the thermodynamic limit, and then lets the field go to zero. A nonzero spontaneous ± 1 magnetization will be obtained, as required and expressed in Eq. (3.19). The sign of M depends on the initial H. The ground state does not have the up-down symmetry of the Hamiltonian, which is expressed by saying that the symmetry has been spontaneously broken (the word "spontaneously" implying the absence of the symmetry-breaking field H).

Often the notions of *order parameter* and *ordering field* are used to describe the phenomenon of broken symmetry. The order parameter is the quantity, such as the magnetization, that signals the development of spontaneously broken symmetry by becoming different from zero in the two-phase region. The word "order" here is used in two senses. First, when the order parameter is nonzero, the system shows order as opposed to the disorder seen in the single-phase region. The second sense is a type of command in that, owing to interactions, neighboring spins tend to line up, thus commanding the others to fall in place, so to speak. The ordering field is the field that couples to the order parameter in the Hamiltonian. In the magnetic case, it is the uniform magnetic field H. It may be noted that sometimes the order parameter is also called the *long-range order parameter*.

The symmetry-breaking phenomenon is very important and is responsible for crystalline solids, glasses, ferromagnetism, superconductivity, superfluidity of helium and of neutrons in neutron stars, anisotropic properties of liquid crystal displays and, perhaps, early universe phenomenology. The up–down symmetry is an example of a discrete symmetry. In nature, there are many examples of a continuous symmetry, such as the rotational symmetry of an isotropic ferromagnet or the gauge symmetry of electromagnetism. It turns out that, whenever a quantum mechanical Hamiltonian has continuous symmetry, the system automatically has certain low-energy excitations, called *Higgs bosons*, which are a consequence of Goldstone's theorem (Anderson, 1984). Familiar examples are photons, phonons, spin waves, and Higgs elementary particles.

We explained above how phase transitions and critical points arise, in principle, when the system develops singularities in the thermodynamic limit. Now, using the Ising model, we explain, in a physical way, why this happens. We start from a high temperature. The interaction term in the Hamiltonian may be neglected, and the spins are essentially independent and take on random configurations so as to maximize the total entropy S. As the temperature is lowered, the interactions will try to line up the neighboring spins parallel to each other, because this is how the energy E can be minimized. The aim, of course, is to minimize the free energy $A = E - TS$. These two influences, namely, the ordering effects due to interaction and the disordering effects due to entropy, compete at all temperatures. The energy becomes more important as the temperature is lowered. At $T = 0$, there is no entropy and the energy wins by lining up all the spins in the ground macrostate, giving rise to the full magnetization 1. If now the temperature is raised, this state may or may not be stable against fluctuations caused by the tendency of the spins to increase the entropy. If the ground state is unstable, the magnetization is destroyed as soon as the temperature rises above zero, which is what happens in the one-dimensional Ising model. In the two-dimensional case, however, a finite fraction of the spins can stay lined up to provide the system with a nonzero spontaneous magnetization even at a temperature above zero. This is the phenomenon of ferromagnetism. Of course, as the temperature rises, the fluctuations begin to dominate, thus reducing the tendency of the spins to line up and ultimately destroying the ferromagnetism at $T = T_c$.

3.3   MEAN-FIELD THEORIES

3.3.1   Weiss Mean-Field Theory of Ferromagnetism

Since most realistic models cannot be solved exactly, a host of approximate techniques, now called the mean-field theories, were developed. Here, we discuss the one for the Ising model [Eq. (3.1)]. To begin, let us suppose that $J = 0$. Then the Hamiltonian is that of noninteracting spins in a uniform field.

$$\mathcal{H} = -H \sum_i S_i \tag{3.20}$$

For simplicity, we now suspend the operator designations for $\mathcal{H}$ and S_i as well as the bracket notation for expectation values. The partition function for this Hamiltonian is

$$Z_N = \sum_{S_i} \exp\left(\sum_i \beta H S_i\right) \tag{3.21}$$

where the sum over the two values of each of the spins S_i can be performed independently to yield

$$Z_N = (2 \cosh \beta H)^N \tag{3.22}$$

Note that there is no factor of $1/N!$ because the lattice sites, being fixed, are distinguishable. The free energy and the magnetization in the thermodynamic limit can be obtained as above. We have

$$M = \tanh \beta H \tag{3.23}$$

Equation (3.23) is the equation of a paramagnet, to which all ferromagnetic systems approach at high enough temperature. As expected, in the absence of interactions, there is no phase transition except the one at $T = 0$.

Now let us rewrite a typical interaction term as

$$- J S_i S_j = -J S_i \langle S_j \rangle - J S_i (S_j - \langle S_j \rangle) \tag{3.24}$$

In a finite system the ensemble average $\langle S_j \rangle$ may be dependent on site, because some sites may be in the interior and others may be near the boundary. But in an infinite system, with no boundaries, all spins are equivalent. Therefore, $\langle S_j \rangle$ must be independent of site and thus equal to the magnetization per spin M.

Equation (3.24) can be written

$$- JS_iS_j = -JS_iM - JS_i(S_j - M) \tag{3.25}$$

The first term on the right-hand side represents the interaction of a spin with an average spin, and the second term represents the fluctuation around this interaction. So far, no approximation has been made. The central idea of the mean-field theory is to neglect the second term and consider only the first term, which looks like a spin in a magnetic field M.

The Hamiltonian in the mean-field approximation becomes a sum of single-site Hamiltonians, each one given by

$$\mathcal{H}_i = -\left(H + J\sum_{n.n.} M\right)S_i = -(H + qJM)S_i \tag{3.26}$$

where q is the number of nearest neighbors, equal to 2 for a one-dimensional lattice, to 4 for a planar square lattice, and to 6 for a cubic lattice or a planar triangular lattice. Equation (3.26) is exactly the same form as the Hamiltonian for a paramagnet in an effective field

$$H_{\text{eff}} = H + qJM \tag{3.27}$$

where the first term is the actual applied field and the second is called the Weiss mean field, or internal field, arising from the interactions among the spins. Therefore, the magnetization in the thermodynamic limit is given by

$$M = \tanh(\beta H + \beta qJM) \tag{3.28}$$

The appearance of M on both sides of the equation makes it an implicit equation. It also gives this approximation the name *self-consistent mean-field approximation*, because M has to be calculated from H_{eff}, which depends on M itself.

3.3.2   Weiss Theory: Critical Behavior

In the absence of the field H, Eq. (3.28) becomes

$$M = \tanh(\beta qJM) \tag{3.29}$$

At $T = 0$, the right-hand side of this equation is $+1$ or -1, depending on whether M is positive or negative, respectively. Consequently, there are two solutions, $M = \pm 1$, to this equation, as expected. Above $T = 0$, both solutions decrease in magnitude, coming closer till a temperature T_c is reached, when the two solutions become 0. To study the behavior of M for fixed T near T_c, expand the right-hand side of Eq. (3.29) for small M to get

$$M \approx \beta q J M - \tfrac{1}{3}(\beta q J M)^3 \tag{3.30}$$

which gives three solutions

$$M = 0 \qquad M = \pm \frac{\sqrt{3}}{(\beta q J)} \left(1 - \frac{1}{\beta q J}\right)^{1/2} \tag{3.31}$$

The solution $M = 0$ is applicable in the paramagnetic phase, $T > T_c$, whereas the other two become relevant in the ferromagnetic region. The temperature at which the other two solutions merge gives the critical point at $\beta_c = 1/qJ, T_c = qJ/k_B$. Inspection of Eq. (3.31) shows that for $T > T_c$ the two ferromagnetic solutions are imaginary and thus unphysical.

Expressing Eq. (3.31) in terms of the relative distance t from the critical point,

$$t = (T - T_c)/T_c \tag{3.32}$$

the substitution $\beta q J = T/T_c$ gives

$$M \approx \pm \sqrt{3}|t|^{1/2} \qquad H = 0, t < 0, |t| \ll 1 \tag{3.33}$$

where the restriction is made that $|t|$ be small. There is a singularity at $T = T_c$, because the derivative of M with respect to T is strictly zero above T_c and becomes infinite as T approaches T_c from below.

Equation (3.33) is our first equation in the critical region. The power $\tfrac{1}{2}$ is called a *critical exponent* and the amplitude $\sqrt{3}$ is called a *critical amplitude*. In the general case, the magnetization is assumed to go to zero as

$$M \approx B|t|^\beta \qquad H = 0, t < 0, |t| \ll 1 \tag{3.34}$$

where β (unrelated to $1/k_B T$) here stands for a critical exponent, and B is the critical amplitude. In general, the critical exponent can be calculated from the relation

$$\beta = \lim_{|t| \to 0} \frac{\ln M}{\ln |t|} \tag{3.35}$$

Since all the thermodynamic quantities are related, there are singularities in those too. Some definitions of critical exponents follow. The zero-field isothermal magnetic susceptibility is defined as

$$\chi = \lim_{H \to 0} \left(\frac{\partial M}{\partial H} \right)_T \tag{3.36}$$

and has the general critical equation

$$\chi/\chi_{id} \approx C^{\pm}|t|^{-\gamma} \qquad H = 0, |t| \ll 1 \tag{3.37}$$

where $\chi_{id} \equiv 1/k_B T$ is the susceptibility of the ideal paramagnet and the signs $\pm$ on the amplitude correspond to temperatures above and below T_c. It should be noted that we have used the same notation for the critical exponent, whether the field goes to 0 from the positive or the negative side and whether the critical point is approached from above or below T_c. The fact that the exponent is independent of the sign of the vanishing field is obvious from the evenness of χ with respect to the field. The independence of the exponent from the sign of t follows from well-established general arguments. Another exponent defines the behavior on the critical isotherm:

$$M \approx \text{sign}(H)D|H|^{1/\delta} \qquad T = T_c, H \to 0 \tag{3.38}$$

Finally, the zero-field specific heat behaves as

$$C_0/Nk_B \approx A^{\pm}|t|^{-\alpha} \qquad H = 0, |t| \ll 1 \tag{3.39}$$

It should be remembered that these definitions are applicable in the critical region, that is in the neighborhood of the critical point where $|H|/k_B T_c \ll 1$ and $|t| \ll 1$. Outside the critical region there are extra terms, which may involve other exponents, on the right-hand side of these defining equations.

The method of obtaining the exponent β was outlined above. To obtain the remaining critical exponents in the Weiss theory, we study Eq. (3.28) near the critical point. By expanding the right-hand side in a Taylor's series about $H = 0$, we get

$$M \approx \tanh(\beta q JM) + \beta H \text{sech}^2(\beta q JM) \tag{3.40}$$

Expanding Eq. (3.40) for small M and then taking $|t|$ to be small gives

$$H \approx q JM(t + M^2/3) \qquad |M| \ll 1, |H| \ll 1, |t| \ll 1 \tag{3.41}$$

From this equation it can be shown, by taking $M = 0$ above T_c and by using Eq. (3.33) below T_c, that

$$\gamma = 1 \qquad C^+ = 1, C^- = \tfrac{1}{2} \tag{3.42}$$

$$\delta = 3 \qquad D = (3/q J)^{1/3} \tag{3.43}$$

The specific heat can be found by noting that the energy is given by

$$E = -NH_{\text{eff}}M = -NM(H + qJM) \tag{3.44}$$

and the zero-field specific heat is $(\partial E/\partial T)_{H=0}$. For $T > T_c$, $M = 0$, so that the specific heat is zero. For $T < T_c$, as $T \to T_c$, the specific heat in zero-field approaches the value $3k_B$, so that the specific heat has a discontinuity at the critical temperature. By using Eq. (3.39), we get

$$\alpha = 0 \qquad A^+ = \text{undefined} \qquad A^- = 3 \tag{3.45}$$

3.3.3 Problems with the Weiss Approximation

The Weiss approximation is not exact, although it shows a phase transition and critical behavior. It explicitly neglects the fluctuations. For a treatment of fluctuations in the mean-field theory, see (Pathria, 1996, Sections 11.11–11.13). We now show that the results obtained in the Weiss approximation are wrong. Isotherms in the critical region are shown in Figure 3.1 obtained from Eq. (3.41). For $T > T_c$, M always increases with increasing H, or, in other words, the magnetic susceptibility χ is positive. This behavior is explained by the following argument. Consider a magnetic system described by an ensemble in which the magnetic field is allowed to fluctuate around

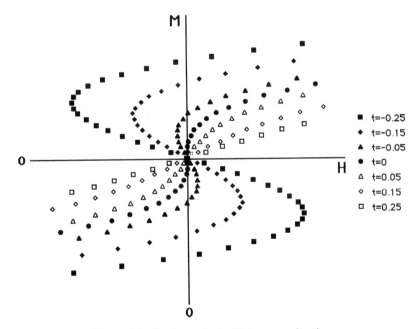

Figure 3.1 Isotherms in the Weiss approximation.

its mean value H. Let us divide the system into two subsystems A and B. Assume that in A the field fluctuates above its average value. There will then be a "flow" of magnetization (i.e., a flipping of spins) from A to B to equalize the field. Lowering the magnetization in A will lower the field in A because $\chi > 0$. On the other hand, for $T < T_c$, there are regions called loops, where M decreases as H increases, so that χ is negative. These regions are unphysical and cannot arise in a real system. The reasoning goes exactly as before. But in this case the lowering of magnetization in A, due to the flow, will increase the field because of negative χ. This behavior violates the second law of thermodynamics and is impossible.

What happens in a real system, as opposed to the Weiss mean-field theory, is the following. In the region where χ is negative, the real system does not remain homogeneous, but breaks up into two phases, which coexist. In each of the phases, χ is positive, thus avoiding the violation of the second law. The Weiss approximation can be "patched up" to yield the two phases with the help of a construction suggested by Maxwell. We shall discuss this construction in detail in Section 3.3.4. In the present case, the construction implies that the loops should be ignored and, when the isotherms reach zero field, the magnetization is assumed to jump discontinuously from a positive value to the equal negative value. This discontinuity was implicitly assumed above.

3.3.4   The van der Waals Theory

The basic mechanism for the gas–liquid phase transition is also provided by the argument in Section 3.2.4, which pitted the high-temperature entropy against the internal energy, which increased with the strength of the interaction field. For high temperatures, the potential energy of interaction between the gas molecules can be neglected, and the fluid behaves as an ideal gas. Thus, the molecules move independently of one another. As the temperature is lowered, the potential energy as well as the correlations between molecules become important. This ultimately leads to condensation. The potential energy has a repulsive hard-core part and an attractive long-range tail. Students of chemistry and physics are well aware of the heuristic kinetic theory derivation of the van der Waals equation as an equation of state of a nonideal gas based on mean-field ideas. It is often written in the form

$$P = \frac{nRT}{V - b} - \frac{n^2 a}{V^2} \qquad (3.46)$$

where n is the moles of sample, R is the gas constant, and a and b are the van der Waals constants representing the attractive interaction between molecules and the volume of the hard-sphere molecules, respectively. A derivation of this equation from statistical mechanics may be found in Pathria (Pathria, 1996, pp. 240–242). By rewriting Eq. (3.46) as an equation in the molar volume V_m,

$$V_m^3 - (b + RT/P)V_m^2 + aV_m/P - ab/P = 0 \qquad (3.47)$$

we see that it is a cubic equation. A cubic equation implies that, for a given P and T, there are three possible values of V_m. This behavior is very similar to the Weiss theory of the Ising model, where three values of M were found for a given H and T. In analogy, we expect only one solution to be physical above T_c and the other two to become physical below T_c. To identify the critical point, we must look for two roots that become identical. The only way for this behavior to occur is for the $P - V$ curves to have a maximum and a minimum that merge at a point of inflection. This point is given by

$$\left(\frac{\partial P}{\partial V_m}\right)_T = \left(\frac{\partial^2 P}{\partial V_m^2}\right)_T = 0 \qquad (3.48)$$

By inserting Eq. (3.46) into Eq. (3.48), the critical parameters are obtained. These are

$$P_c = a/27b^2 \qquad V_c = 3b \qquad T_c = 8a/27Rb \qquad (3.49)$$

To study critical behavior, we write Eq. (3.47) in terms of the reduced variables $p = (P - P_c)/P_c$, $v = (V - V_c)/V_c$, and $t = (T - T_c)/T_c$ and get

$$[p + 1 + 3/(v + 1)^2][3(v + 1) - 1] = 8(t + 1) \qquad (3.50)$$

or

$$p = [-3v^3 + 8t(v + 1)^2](3v^3 + 8v^2 + 7v + 2)^{-1} \qquad (3.51)$$

It is convenient to expand p in a Taylor's series in powers of v and t about $v = 0$ and $t = 0$:

$$p = -\tfrac{3}{2}v^3 + 4t - 6vt + 9v^2t + \cdots \qquad (3.52)$$

We shall see later that $|v|$ is of the order of $|t|^{1/2}$, so that we can neglect higher order terms on the right-hand side and write

$$p \approx -\tfrac{3}{2}v^3 + 4t - 6vt + 9v^2t \qquad (3.53)$$

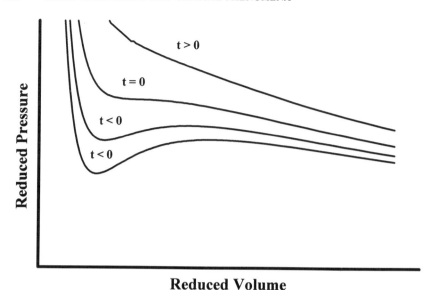

Reduced Volume

Figure 3.2   Isotherms of the van der Waals equation.

Equation (3.53) is a cubic equation just like the equation in the Weiss theory. The isotherms resulting from this equation, shown in Figure 3.2, are very similar to those of Figure 3.1 if the figure is rotated by 90°.

Once again the unphysical loops are present, indicating the approximate nature of the theory. The loops can be eliminated in favor of the physical two-phase region as follows. For a one-component system at mechanical and thermal equilibrium,

$$d\mu = \frac{1}{n}(-S\,dT + V\,dP) \tag{3.54}$$

If this expression is integrated between the liquid and gas phases at constant temperature, we have

$$\mu_g - \mu_l = \int_l^g (V/n)\,dP \tag{3.55}$$

For material equilibrium, $\mu_g = \mu_l$, so that the integral in Eq. (3.55) vanishes, and the equation reduces to

$$\int_l^g V\,dP = 0 \tag{3.56}$$

Because $P_g = P_l$ at mechanical equilibrium, a horizontal line must be drawn such that the areas between the isotherm and the equation of state sum to zero according to Eq. (3.56). Equation (3.56) is then known as *Maxwell's equal area rule* or the *Maxwell construction*. The phases at the ends of the horizontal line are in equilibrium, and the unphysical situation in between is ignored.

From Eq. (3.53) all the critical exponents can be obtained. On the coexistence curve, $p_l = p_g$ and $t_l = t_g = t$ so that

$$0 = p_l - p_g$$
$$\approx -(\tfrac{3}{2})(v_l^3 - v_g^3) - 6t(v_l - v_g) + 9t(v_l^2 - v_g^2) \tag{3.57}$$

Very near the critical point, using Eq. (3.56), one may show that the Maxwell construction implies that $v_g \approx -v_l$. Equation (3.57) then yields the three solutions

$$v_l \approx 0 \qquad v_l \approx \pm(-4t)^{1/2} \tag{3.58}$$

For $t > 0$ the first solution is physical, and for $t < 0$ the only physical solution is

$$v_l \approx -(4|t|)^{1/2} \tag{3.59}$$

which gives

$$v_g \approx (4|t|)^{1/2} \tag{3.60}$$

Since the difference between v_g and v_l goes to zero as the two phases become identical, the order parameter is taken as $v_g - v_l$. Also, since pressure naturally couples to the volume, p is taken as the ordering field. So there is an analogy between the magnetic and the fluid models:

$$M \Longleftrightarrow v_g - v_l \quad H \Longleftrightarrow p \tag{3.61}$$

The analogue of magnetic susceptibility is the isothermal compressibility κ_T defined as

$$\kappa_T = -\frac{1}{V}\left(\frac{\partial V}{\partial P}\right)_T \tag{3.62}$$

By using this analogy, all the exponents and the amplitudes can be calculated as before. The results are (Pathria, 1996, Section 11.2)

$$\beta = \tfrac{1}{2} \qquad B = 4 \tag{3.63}$$

$$\gamma = 1 \qquad C^+ = \tfrac{1}{6} \qquad C^- = \tfrac{1}{12} \tag{3.64}$$

$$\delta = \tfrac{1}{3} \qquad D = (\tfrac{2}{3})^{1/3} \tag{3.65}$$

$$\alpha = 0 \qquad A^+ = \text{undefined} \qquad A^- = \tfrac{9}{2} \tag{3.66}$$

Similar to the Weiss and van der Waals theories, many other schemes have been proposed for other phase transitions in, for example, ferroelectrics, binary alloys, and binary fluids. Landau proposed a more general approach encompassing all these theories, an approach to which we now turn.

3.3.5   Landau Theory

The Landau theory of phase transitions is a phenomenological theory based upon very general and plausible looking assumptions (Landau and Lifshitz, 1970, Chapter 14). We illustrate Landau's method for the case of a Hamiltonian having an up–down symmetry. Let the order parameter, such as the magnetization, be ψ and the ordering field, such as the magnetic field, be ζ. Assume that near the critical point a suitable free energy can be expanded in a Taylor's series around $\psi = 0$. The free energy is then given by

$$G(T, \psi) = G_0 - \psi\zeta + A\psi^2 + B\psi^4 \tag{3.67}$$

where the coefficients of odd powers of ψ vanish because of the up–down symmetry, which is broken by the $-\psi\zeta$ term. We have neglected higher order terms, which are not needed to discuss a critical point. Equation (3.67) is assumed to be an estimate for the free energy, which is to be minimized with respect to ψ to obtain the "true" free energy. The coefficients A and B are assumed to have the temperature dependence

$$A \approx A'(T - T_c) \qquad B \approx B' \tag{3.68}$$

For large ψ, G is large and decreases as ψ decreases. A minimum can be found from the equation

$$\left[\frac{\partial G(T, \psi)}{\partial \psi}\right]_T = -\zeta + 2A'(T - T_c)\psi + 4B'\psi^3 = 0 \tag{3.69}$$

with the constraint that

$$\left[\frac{\partial^2 G(T, \psi)}{\partial \psi^2}\right]_T = 2A'(T - T_c) + 12B'\psi^2 > 0 \tag{3.70}$$

TABLE 3.1   Experimental Data on Critical Exponents

Critical Exponents	Magnetic Systems	Gas–Liquid Systems	Binary Fluid Mixtures	Ferroelectric Systems	Superfluid ^{4}He	Mean-Field Theory
α	0.0–0.2	0.1–0.2	0.05–0.15		-0.026	0
β	0.30–0.36	0.32–0.35	0.30–0.34	0.33–0.34		0.5
γ	1.2–1.4	1.2–1.3	1.2–1.4	1.0 ± 0.2	Inaccessible	1
δ	4.2–4.8	4.6–5.0	4.0–5.0		Inaccessible	3

Equation (3.69) is a cubic equation, just like the equations in the Weiss and van der Waals theories. It can be treated in exactly the same way to yield the critical exponents near the critical point $\zeta = 0$ and $T = T_c$. In fact, by making a correspondence with the Weiss equation, the results can be written by inspection. We leave the calculation as an exercise for the reader.

Landau theory may be criticized on the same grounds as other mean-field theories. The main point to the presentation of the Landau theory is to emphasize the concept of the order parameter and to show that all critical phenomena occurring in many diverse systems are similar. For a summary of experimental results on exponents, see Table 3.1. [For a more extensive list of data with references, see Pathria (Pathria, 1996, Table 11.1).] Two statements can be made about these data. First, all the experimental exponents have similar values. Second, these values are different from the mean-field theory results. Accordingly, the task of any theory of the critical region is twofold. The first task is to understand the "universality" of critical exponents over a variety of systems with manifestly different Hamiltonians that pose different, yet difficult many-body problems. The second task is to calculate the exponents and thus find the ways they are related to the properties of the underlying Hamiltonians.

Many exact and other analytical studies were done starting from the 1940s, and these led to the concepts of scaling and universality and to the discovery of the technique known as the renormalization group. The rest of this chapter is devoted to the elucidation of these concepts.

3.4   SCALING AND UNIVERSALITY

3.4.1   Scaling in Mean-Field Theories

Scaling can be presented in two ways, graphical and analytical. In the graphical way, one looks at the isotherms in Figure 3.1 and notices the following. The isotherms for $t > 0$ and $t < 0$ are qualitatively different, because the latter must have a spontaneous magnetization. But all the isotherms for $t > 0$ are qualitatively similar, as are those for $t < 0$. Would it not be nice to present the data in a suitable way so that we do not have so much of it? By trial and error one may find that if, instead of M versus H, a plot of $M/|t|^{1/2}$ versus $H/qJ|t|^{3/2}$ is made, all the data collapse onto two curves as shown in Figure 3.3. This collapsing of the data is called scaling.

Analytically, one may notice that Eq. (3.41) can be rewritten in the scaled form

$$\frac{H}{qJ|t|^{3/2}} \approx \frac{M^3}{3|t|^{3/2}} \left(1 + \frac{3t}{M^2} \right) \tag{3.71}$$

The scaling is usually expressed by saying that the magnetization obeys the *scaling relation*

$$M \approx \sqrt{3}|t|^{1/2} \mathcal{M}(x) \qquad x = H/qJ|t|^{3/2} \tag{3.72}$$

where x is called the *scaled magnetic field* and $\mathcal{M}$ is called the *scaling function*. In the present case, the scaling function is given by the implicit relation

$$x = \sqrt{3}\mathcal{M}[\text{sign}(t) + \mathcal{M}^2] \tag{3.73}$$

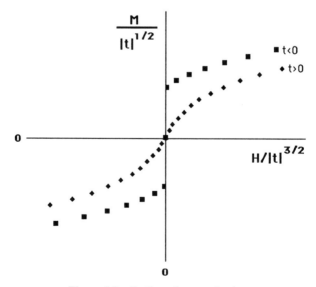

Figure 3.3   Scaling of magnetization.

In the same way, it can be seen that the van der Waals equation (3.53) also satisfies a scaling relation. The scaling function $\mathcal{V}$ is given by

$$x = \tfrac{3}{2} \mathcal{V}[4\,\mathrm{sign}(t) + \mathcal{V}^2] \qquad (3.74)$$

where

$$x = (p - 4t)/|t|^{3/2} \qquad v \approx |t|^{1/2}\mathcal{V} \qquad (3.75)$$

Comparing the two scaling functions of the Weiss and van der Waals theories, one can see that they are very similar. Likewise, one can express the order parameter in the Landau theory in a scaling form.

3.4.2 Scaling Beyond the Mean-Field Theory

To speculate further, one might say that a real physical system (as contrasted with a Weiss approximate one) can satisfy a similar scaling hypothesis with its own exponents (Widom, 1965). We illustrate this hypothesis with the help of the magnetization order parameter. The procedure is quite general. We postulate

$$M \approx B|t|^{\beta} \mathcal{M}^{\pm}(x) \qquad x = AH|t|^{\Delta} \qquad (3.76)$$

where the scaling function $\mathcal{M}^{\pm}$ may not be the same as the one in the Weiss theory, just as the exponent β may not be equal to 0.5. The superscript $\pm$ denotes $t > 0$ and $t < 0$, respectively. The exponent Δ, which is equal to $\tfrac{3}{2}$ in the mean-field theory, is an unknown at this stage in the general theory. The scaling hypothesis is supposed to be valid for $|M| \ll 1, |H| \ll 1, |t| \ll 1$ and $-\infty < x < \infty$. From this hypothesis follow many important conclusions, which show the utility of the hypothesis. We now discuss some of these conclusions.

We study the scaling hypothesis as the critical point is approached from different directions. For $H = 0$ and $t \to 0$ from above $(T > T_c)$, we must have $M = 0$, so that $\mathcal{M}^{+}(0) = 0$. On the other hand, on approaching the critical point from below $(T < T_c)$, Eq. (3.34) must be satisfied, so that $\mathcal{M}^{-}(0) = 1$. To obtain Eq. (3.38), we put $t = 0$ or $x = \infty$. How should $\mathcal{M}(x)$ behave for $x \to \infty$? In order to cancel the power of $|t|$ in Eq. (3.76), we must assume that

$$\mathcal{M}^{\pm} \approx \mathcal{M}(\infty)x^{\beta/\Delta} \qquad (3.77)$$

When substituted in Eq. (3.76), we get

$$M \approx B\mathcal{M}(\infty)H^{\beta/\Delta} \tag{3.78}$$

By comparing Eq. (3.78) with Eq. (3.38), we find that we must have

$$\Delta = \beta\delta \qquad D = BA^{1/\delta}\mathcal{M}(\infty) \tag{3.79}$$

Thus, we find the exponent Δ in terms of known exponents.

To obtain the behavior of the susceptibility, we differentiate Eq. (3.76) with respect to the field to get

$$\chi/\chi_{id} \approx k_{\mathrm{B}}T_cAB\mathcal{M}^{\pm\prime}(x)|t|^{\beta-\beta\delta} \tag{3.80}$$

where the prime means differentiation of a function with respect to its own argument. If we assume that in zero field $\mathcal{M}^{\pm\prime}(0)$ is finite and compare this equation to Eq. (3.37), we get

$$\gamma = \beta(\delta - 1) \tag{3.81}$$

and

$$C^{\pm} = k_{\mathrm{B}}T_cAB\mathcal{M}^{\pm\prime}(0) \tag{3.82}$$

The relation (3.81) is one of the many *scaling laws*, and it tells us that γ, the exponent responsible for the behavior of the zero-field susceptibility, is not an independent exponent but depends entirely on the values of the order parameter exponent β and the critical isotherm exponent δ.

The scaling law (3.81) is satisfied by the mean-field theories, since $\gamma = \frac{1}{2}(3 - 1) = 1$. What is even more important is that it is found to be satisfied by all the known exactly solvable models. For a recent survey of exactly solvable models, see Baxter (Baxter, 1989). Furthermore, it is satisfied also by all the known experiments on all the known critical systems within experimental error. See the reviews by Fisher and Kadanoff et al. (Fisher, 1967; Kadanoff et al., 1967). Historically, this result was the first indication that the scaling hypothesis is not an idle generalization of the mean-field theory but a very potent one. This result was one of the first universal features that people were looking for in critical phenomena.

Even more importantly, experimental data from many diverse magnetic and fluid systems were found to obey the scaling hypothesis, and it was possible to find the scaling function by suitable plotting (Stanley, 1971, Figs. 11.4 and 11.5; Vicentini-Missoni, 1971). This finding sent the theorists on a search for the origin of the scaling laws. At the same time, by studying many exactly soluble models and experimental systems, the phenomenon of *universality* was discovered. We turn to it now.

3.4.3 Universality

Scaling laws imply that not all critical exponents are independent. It turns out there are only two independent exponents. So, if one knows any two exponents of a system and the scaling hypothesis holds, all other exponents, relating to the behavior of any thermodynamic quantity along any path to the critical point, may be obtained using scaling laws. One more independent exponent is needed when correlation functions are studied. The independent exponent may be chosen as the exponent for the divergence of the correlation length ξ. The exponent is defined by

$$\xi \sim |t|^{-\nu} \tag{3.83}$$

The question of what features determine the exponents is answered by the concept of universality. It says that the exponents depend on a very few essential properties of the system and not on any others. An example is seen in the mean-field theories. The exponent β is 0.5, because it is assumed that the first leading power of the order parameter in the expansion of the free energy [Eq. (3.67)] is 2 and not any other. The exponent does not depend on the critical temperature or any other coefficients such as A' or B' of the Landau theory. From various studies, it has been found that the exponents depend on the symmetry of the Hamiltonian, on the dimensionality of the system, and on the range of interactions (Fisher, 1974). We discuss each of these in turn now.

3.4.4 Symmetry

Critical phenomena occur when the symmetry of a Hamiltonian is spontaneously broken. The symmetry of the Hamiltonian can be expressed typically in terms of the number of components n, which the order parameter has. The standard cases are the scalar case ($n = 1$), the planar vector case ($n = 2$), and the three-dimensional vector case ($n = 3$). In these cases, the Hamiltonian has the up–down, planar rotational, and three-dimensional rotational symmetry, respectively. Examples of the first case are the Ising model, the lattice gas model, and various fluid models. Examples of the second case are the planar ferromagnets, the superfluid, and the superconductivity models. The latter two models fall into this category, because the order parameter is a complex number. Consequently, for the Hamiltonian to be real, it must be unchanged under rotations in the complex plane. The case ($n = 3$) occurs in the Heisenberg model of ferromagnetism. The values of the exponents are dependent on n, but they are the same for a given value of n. Thus, the same exponent values apply for a planar ferromagnet and the superfluid transition in helium, but they are different from those of the Ising model.

Sometimes intermediate cases arise. For example, consider the following Hamiltonian for an anisotropic magnet

$$\widehat{\mathcal{H}} = -J_1 \sum_{n.n.} \widehat{S}_i^x \widehat{S}_j^x - J_2 \sum_{n.n.} \widehat{S}_i^y \widehat{S}_j^y \tag{3.84}$$

where J_1 and J_2 are two positive coupling constants representing interactions between the x and the y components of a pair of neighboring spins. Nominally, $n = 2$, because the spin vector has two components. But, unless $J_1 = J_2$, the Hamiltonian does not have the planar rotational symmetry. In general, one of the directions dominates, and the exponents are those of the Ising model! Only in the case of equality are the exponents those of the planar model.

3.4.5   Dimensionality

The dimensionality denoted by d is the number of topological dimensions to which the thermodynamic limit has been taken. It does not matter if the system is a continuum (like a fluid) or a lattice (like a crystal). The values of the critical exponents do not depend on the value of the lattice constant. We already know that d plays an important role in phase transitions. For example, the Ising model does not have a critical point at a nonzero T for $d = 1$. Only for $d > 1$ does it have a nonzero T_c. Universality says that the critical behavior of a one-dimensional Ising model and of a one-dimensional model having a finite thickness in the other two directions are the same. The latter does not have a critical point at a nonzero T either. It also implies that the critical exponents of a planar square lattice Ising model are exactly the same as that of a thick plate made of a finite stack of such models. Of course, the critical temperature of the stack will depend on the thickness of the stack, but the values of the critical exponents will not.

A natural question arises as to what happens if the thickness of the stack becomes very large, making it a case of d almost equal to 3. This question is answered by the theory of *crossover scaling*, which says the following. For the planar Ising model of finite but large length L in the third dimension, all the quantities depend on the ratio L/ξ, where ξ is the correlation length of the truly three-dimensional system. Since ξ depends on temperature, the following *crossover phenomenon* will be seen. When the system is far from the critical temperature $T_c(L)$ of the $d = 2$ system, the system will behave as if it is three dimensional. The effective exponents will correspond to $d = 3$.

But on approaching $T_c(L)$, the exponents will cross over to the $d = 2$ case, thus revealing the true behavior of the system, since, after all, it is really infinite in only two dimensions.

The phenomenon of crossover also helps to answer the following question, which sometimes arises about the thermodynamic limit (see Robertson, 1993, p. 152). All the real physical systems are finite, even the universe is finite; there are only a finite number of protons in the universe. So, how come we have to take the limit $N \to \infty$ to see sharp phase transitions? One can hold water in a finite three-dimensional pot, boil it, and take it to the critical point experimentally, even though it is not at the thermodynamic limit. The answer lies in the crossover phenomenon. The number of molecules in a pot of water is so large ($\approx 10^{23}$) that, within experimental error, the pot behaves effectively like a system of $d = 3$, a system to which the thermodynamic limit has been applied in all three directions. If we could really get close enough to the critical point, we would see effects arising from the true finite nature of the water sample. The accuracy in temperature to see the finiteness turns out to be of the order of 10^{11}, which is unattainable by present experiments. That is why we need to take the thermodynamic limit. The pot of water behaves like a system in the thermodynamic limit, so we treat it as one.

The dimensionality above which there is a nonzero T_c is called the *lower critical dimension*. In the Ising model it is equal to 1, whereas it is equal to 2 in the Heisenberg model. On studying higher dimensional models, it is found that for $d > 4$, the exponents of these models become the same as in the mean-field approximation. This dimension 4 is called the *upper critical dimension*. For more discussion, see Pathria (Pathria, 1996, pp. 441–448).

3.4.6   Range of Interactions

The third parameter to influence critical exponents involves the range of interactions. Universality says that all models with short-range interactions have the same behavior. Short-ranged systems are the ones for which either the interactions have a sharp cutoff, as in the Ising model, or the interaction parameter decays faster than r^{-d} at large distances. Thus, interactions that include next-nearest neighbors or even farther neighbors will not change the critical behavior of the Ising model, as long as the interactions do not extend over all possible spins.

Long-ranged interaction models typically have interaction parameters decaying at least as slowly as $r^{-(d+\sigma)}$, where $\sigma(> 0)$ is the range parameter. It turns out that for $\sigma > 2$, these models behave as short-ranged models near the critical point. The exponents for long-ranged models depend on σ, in general, and are different from those of short-ranged models.

The consequences of universality are immense. They allow us to divide all the known systems into *universality classes* based on d, n, and σ. Once the exponents are known as functions of these three parameters, the behavior of the systems is known. A very large number of diverse systems undergoing critical phenomena in their own complex many-body ways are thus reduced to simple bare-essential models, which have the behavior of their universality classes. That is a profound achievement by any measure.

Two questions still remain. Why is there such a thing as universality, and how do we calculate the exponents and scaling functions? The answer to the first question lies in the fact that the correlation length, which dictates the range of correlations between interacting particles, grows to infinity in the thermodynamic limit at the critical point. Hence, the short-distance features of the system are washed out, and only qualitative features remain, such as the symmetry of the Hamiltonian (n), the number of dimensions in which the system extends to infinity (d), and the range of interactions (σ).

To answer the second question, we note that there are many ways to calculate the critical behavior of a given model system. Very few models can be solved exactly, so most of the methods are approximate. A survey may be found in Baxter (Baxter, 1989). The most successful method to explain scaling and universality on the one hand and yield practical methods for calculating the properties of the system on the other is the *renormalization group* (RG). This method is so astounding that one of the leading authorities of the field, Michael Fisher, has said that the renormalization group "is as general and vigorous as the partition function approach to statistical mechanics was" [quoted in Anderson (Anderson, 1994, p. 305)]. We now turn to the renormalization group.

3.5 RENORMALIZATION GROUP

3.5.1 The Ising Model and the Kadanoff Construction

For a qualitative introduction to RG theory see Wilson (Wilson, 1979), and for examples of one-dimensional Ising systems treated exactly by the RG see Pathria (Pathria, 1996, Section 13.4). In a typical system in physics or chemistry, the correlations between the particles are short-ranged, generally reflecting the range of interactions. However, near the critical point the correlations extend over the whole system, and consequently we are forced to treat the entire system, not just small parts of it. We cannot treat the system as just one or two particles in the average potential of all the other particles as we did in mean-field theory. A way to extend the treatment to the entire system was suggested by Kadanoff (Kadanoff, 1966).

Consider a two-dimensional Ising model Hamiltonian weighted by $\beta(\equiv 1/k_B T)$:

$$\mathcal{H} = -h \sum_i S_i + K_1 \sum_{i,j} S_i S_j + K_2 \sum_{i,j,k} S_i S_j S_k + \cdots \tag{3.85}$$

where $h = \beta H, K_1 = -\beta J, K_2$ is a three-spin coupling constant, and the dots represent all other higher order couplings. We have to include all couplings from the outset, because they will be "generated" by the process of the RG. A given spin has an indirect correlation with a distant spin, even if the given physical interactions are only between nearest neighbors. The Kadanoff construction consists in carrying out a block-spin transformation by reducing the length scale by a factor b such that

$$1 < b \ll \xi \tag{3.86}$$

where ξ is the correlation length of the spins. This transformation is illustrated in Figure 3.4.

In this figure, we take a 8×8 lattice containing 64 lattice blocks (squares in two dimensions), each block one length unit on a side. Imagine a spin ± 1 located at the center of each lattice block, and assume cyclic boundary conditions. Now choose $b = 2$. The new length scale is one unit per two original units, which reduces the original length to 0.5 in new length-scale units. If the basic length unit is a_1, we have $a_2 = a_1/b$. Repeating this, we have $a_3 = a_1/b^2$, $a_4 = a_1/b^3$, and so on, where a_1 is the basic length

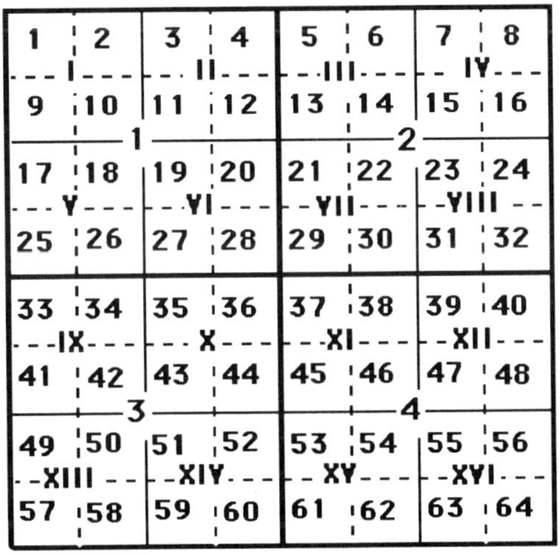

Figure 3.4   The block construction.

in the new length scale and a_2, a_3, a_4, and so on are the original lengths after rescaling. In Figure 3.4, the original blocks are labeled 1–64. After one rescaling, the blocks are labeled I–XVI, and after two rescalings the blocks are labeled 1–4. The entire figure is the one block remaining after three rescalings. Likewise, the new correlation lengths are $\xi_2 = \xi_1/b$, and so on. In this way, the correlation length has been reduced by a factor of $b > 1$. If the transformation is repeated, the new correlation length will be even smaller. By doing this many times, the correlation length in the system can be decreased to the order of the lattice constant. At that stage, the system can be dealt with by ordinary methods, which work well when the correlation length is of the order of a lattice constant.

Let N be the total number of spins ($= 64$); after one rescaling each block contains b^2 ($= 4$) spins. The Kadanoff construction replaces the b^2 block spins with one spin, the assumption being that the spins are very strongly correlated locally. There are now a total number of N/b^2 ($64/4 = 16$) spins. It is necessary that the spin of each block be determined. There are several ways of doing this. One popular way is to use the "majority rule," which sets the spin of block J, namely, S_J, equal to ± 1 depending on the sign of $\sum_b S_i$, where the subscript b indicates a sum over the spins in one block. If $\sum_b S_i$ turns out to be 0, we let S_J alternate between $+1$ and -1 with equal probability.

3.5.2   The Ising Hamiltonian and the RG

A renormalization group transformation R_b is one that changes the coupling constants in the Hamiltonian upon a rescaling of the spins S_i by a factor b to a new set of spins S_J, normalized such that $S_J = \pm 1$ for all J. The operator R_b operates directly on the coupling constants

$$K_b = R_b K \tag{3.87}$$

and, consequently, on the Hamiltonian

$$\mathcal{H}_b(\{K_b\}) = R_b \mathcal{H}(\{K\}) = \mathcal{H}(\{R_b K\}) \tag{3.88}$$

where $\{K\}$ stands for the set of coupling constants $K_1, K_2, \ldots$ and $\{K_b\}$ is the new set of coupling constants generated by the transformation. In group theory, a set of operations is a *group* if (1) the associative law is valid [i.e., $(AB)C \equiv A(BC)$], (2) the operations contain the identity, (3) the operations are closed [i.e., if $AB = C$, then C must be a member of the group], and (4) the operations contain the inverse. The nature of the RG transformations is such that they proceed in only one direction; consequently, there is no inverse operation. The lack of an inverse transformation means that the RG is not a true group; it is more properly called a *semigroup*. The RG transformation must be such that the canonical partition function is left unchanged:

$$Z_b(\{K_b\}, N/b^2) = Z(\{K\}, N) \tag{3.89}$$

(see Goldenfeld, 1992, p. 239). The new partition function $Z_h(\{K_h\}, N/b^2)$ must have the same form as the old one, but the values of the coupling constants are changed.

For the Ising model, the renormalization group transformation consists of the following steps:

1. Coarse graining of the Ising lattice into blocks by a scaling factor b.
2. Use of the majority rule to assign block spins S_J.
3. Evaluation of R_b.

It is the third step that presents a difficulty. The approach used to find R_b depends on how close to criticality the initial state is. The renormalization procedure generates a new set of coupling constants $\{K_b\}$ associated with the Hamiltonian [Eq. (3.85)] and, in general, the new Hamiltonian is further from criticality than the original Hamiltonian, since the correlation length has been reduced. In order to better understand this behavior of the coupling constants, we consider a "phase diagram" in a hyperspace whose axes are the coupling constants K_1, K_2, and so on. We consider the "motion" of the system under RG transformations in this hyperspace as shown in Figure 3.5.

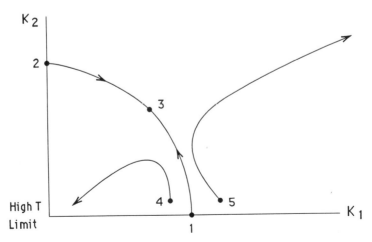

Figure 3.5 Flow diagram for Ising model coupling constants under RG transformation.

For convenience, we consider only K_1 and K_2. If desired, K_1 can be further divided into nearest-neighbor, next-nearest-neighbor, and so on parts, each with its own axis, but details of this sort need not concern us.

For $K_2 = 0$ and a fixed coupling constant J_1, point 1 represents the critical temperature where $K_1 = K_1^*$. The designation K_1^* indicates what is called a fixed point. Under RG transformation

$$K_1^* = R_b K_1^* \qquad (3.90)$$

so that at T_c the coupling constant K_1 is fixed. At $T > T_c$, Eq. (3.87) gives $K_{1b} < K_1$ and the coupling constant moves away from the critical point toward the origin, which is the high-temperature limit. At $T < T_c$, $K_{1b} > K_1$ and the coupling constant moves toward the right away from K_1^*. In this manner, rescaling moves the Hamiltonian away from criticality. The same behavior occurs when $K_1 = 0$ instead of K_2. In this case, point 2 is the fixed point.

In the general case, where neither K_1 nor K_2 are zero, we assume there is a fixed point, labeled point 3 in the figure. For a system near criticality, such as point 4, it is not immediately clear what trajectory the system will take under a RG transformation. However, calculations have shown that the system will first begin to move toward the fixed point (point 3) and then will proceed toward the high-temperature limit as shown in the figure. Likewise, starting from point 5 the system will first move toward the fixed point and then will move away toward the low-temperature limit. One might guess that there is a region between points 4 and 5 where the system will move neither toward the high- nor low-temperature limits but will proceed to the fixed point. This produces a curve that runs between points 1 and 2 via point 3. Any initial points along this curve will always proceed to the fixed point under a RG transformation. In general, this region in the coupling-constant hyperspace is called a *critical hypersurface* (also called a *critical domain* or a *critical manifold*). All points on the critical hypersurface have an infinite correlation length.

Even if only nearest-neighbor spins are coupled, upon renormalization each block will have interblock as well as intrablock couplings. This frustrates any attempt at an exact determination of R_b. Various schemes, such as perturbation techniques, have been devised to solve this problem. If the system is very close to criticality (i.e., is in the immediate vicinity of the fixed point), it is possible to consider the Monte Carlo renormalization group, which is discussed in Chapter 6.

A record of continuing research in phase transitions and critical phenomena can be found in the series edited by Domb and Green and by Domb and Lebowitz (Domb and Green, 1972–1983; Domb and Lebowitz, 1984–).

3.6   EXERCISES AND PROBLEMS

1. Derive Eq. (3.17).

2. Derive Eq. (3.45).

3. Derive Eq. (3.49).

4. Derive Eqs. (3.63)–(3.66).

5. Show that the exponents α, β, γ, and δ of the Landau theory are identical to the Weiss mean-field theory.

6. Show that the order parameter of the Landau theory obeys a scaling relation.

7. Study the scaling function $\mathcal{M}(x)$ of the Weiss mean-field theory and show that the critical amplitudes D and $C^{\pm}$ obtained from Eqs. (3.79) and (3.82) are the same as given in Eqs. (3.43) and (3.42), respectively.

4

THE LIQUID STATE

4.1 INTRODUCTION

Compared to a gas or a solid, the partition function of a liquid is difficult to evaluate. The problem resides in the intermolecular potential energy, which involves many-body interactions. In gases, where kinetic energy is dominant, one starts with an ideal gas with no potential energy, a problem that can be solved exactly. Then one tries to include the effects of potential energy in a perturbative manner. In solids, where potential energy is dominant, one starts with a perfect crystal and exploits its periodicity in order to study its properties exactly at absolute zero. Then one tries to include the effects of kinetic energy in a sensible way. Unfortunately, for liquids, both kinetic and potential energies are equally dominant. Hence, the theory of liquids lacks a convenient starting point.

Still, as shown in Chapter 3, the behavior of all liquids near their critical points is so similar that their universal critical properties are described by the Ising model. Outside of the critical region, however, liquids may have different properties. For example, water has a density maximum at about 4°C. To explain such specific properties, different methods are needed.

This chapter formulates several such techniques for monatomic liquids. An important concept is the radial distribution function, in terms of which all the thermodynamic properties can be expressed. After introducing and discussing this concept, we illustrate it with the exactly soluble Takahashi model. The chapter concludes with an example of the important Ornstein–Zernike theory as applied to light scattering near critical points.

116

4.2 DISTRIBUTION FUNCTIONS

4.2.1 Configurational Partition Function

We consider a classical liquid containing N particles, each of mass m. The Hamiltonian is

$$\mathcal{H} = \sum_i \frac{\mathbf{p}_i \cdot \mathbf{p}_i}{2m} + \Phi(\mathbf{r}) \qquad (4.1)$$

where $\mathbf{p}_i$ is the momentum of the ith particle, and $\Phi(\mathbf{r})$ is the total potential energy of interaction between the centers of masses of all the particles. Here, $\mathbf{r}$ stands for the vector positions of all the particles. The canonical partition function Z is given by Eq. (1.89),

$$Z = \frac{1}{N! h^{3N}} \int \exp[-\beta \mathcal{H}(\mathbf{p}, \mathbf{r})] \, dp \, dr \qquad (4.2)$$

where $dp \, dr$ stands for $\Pi_i \, dp_{ix} \, dp_{iy} \, dp_{iz} \, dx_i \, dy_i \, dz_i$. The $3N$ integrals over the momentum are Gaussian integrals, which are readily evaluated. After integration over the momenta, we get

$$Z = \frac{1}{N! \lambda^3} \int \exp[-\beta \Phi(\mathbf{r})] \, dr$$

$$= \frac{Q_N}{N! \lambda^3} \qquad (4.3)$$

where λ represents the thermal wavelength, defined in Eq. (2.72). The function Q_N is called the *configurational partition function* (recall configuration space). It depends on the total potential energy, which can be any kind of many-body potential energy between the particles. In practice, it is common to make the *binary*, or the *two-body approximation*, where it is assumed that the total potential is made entirely of interactions between pairs of particles. In that case,

$$\Phi(\mathbf{r}) = \sum_{i<j} \phi(r_{ij}) \qquad (4.4)$$

where the two-body potential depends on the mutual distance, $r_{ij} = |\mathbf{r}_i - \mathbf{r}_j|$, between the particle centers. For studies of nonideal or almost perfect gases, it is possible to expand Q_N in a Taylor's series in powers of $\Phi(\mathbf{r})/k_B T$ and to relate the terms in this expansion to the virial coefficients. Unfortunately, a Taylor's series expansion does not work for liquids, because it is not possible to truncate the series for various degrees of interaction. In a liquid, the interaction is a multibody one, and any series expansion is bound to fail.

4.2.2   Reduced Distribution Function

In analogy to the canonical distribution function [Eq. (1.72)], which depends on both momentum and position coordinates, we define a distribution function for position only,

$$\rho^{(N)}(\mathbf{r}_1, \mathbf{r}_2, \ldots, \mathbf{r}_N) = \frac{1}{Q_N} \exp[-\beta\Phi(\mathbf{r})] \tag{4.5}$$

This distribution is the probability for finding the particle labeled 1 at $\mathbf{r}_1$, the particle labeled 2 at $\mathbf{r}_2$, and so on. Although a given particle interacts with all the other particles in the system, as a practical matter most of the correlations are short ranged and involve the given particle and a small number, $n - 1$, of neighboring particles, where $n \ll N$. We therefore consider a *reduced (marginal) distribution function* obtained by integrating over the unimportant coordinates.

$$\rho_N^{(n)}(\mathbf{r}_1, \mathbf{r}_2, \ldots, \mathbf{r}_n) = \int \rho^{(N)}(\mathbf{r}_1, \mathbf{r}_2, \ldots, \mathbf{r}_n, \mathbf{r}_{n+1}, \ldots, \mathbf{r}_N)\, dr_{n+1} \cdots dr_N$$

$$= \frac{1}{Q_N} \int \exp[-\beta\Phi(\mathbf{r})]\, dr_{n+1} \cdots dr_N \tag{4.6}$$

This distribution function is for n particles and is normalized to unity, since the integral over the positions of the n particles on the left-hand side of Eq. (4.6) results in an integral over the positions of all the particles in the numerator on the right-hand side, which is just the denominator Q_N. Since the particles are distinguishable, any interchange of particles does not change the macrostate, but it does contribute to the degeneracy of the configuration. This degeneracy arises because any n of the N particles can occupy the n positions. This degeneracy number can be found by asking how n positions can be distributed among N distinguishable particles. The answer is $N!/(N - n)!$. We can now correct for the indistinguishability of the particles by dividing Q_N by $N!/(N - n)!$. A new distribution function for indistinguishable particles is defined as

$$\rho^{(n)}(\mathbf{r}_1, \mathbf{r}_2, \ldots, \mathbf{r}_n) = \frac{N!}{(N - n)!} \rho_N^{(n)}(\mathbf{r}_1, \mathbf{r}_2, \ldots, \mathbf{r}_n) \tag{4.7}$$

and is normalized so that its integral is equal to $N!/(N - n)!$. The distribution function [Eq. (4.7)] represents the probability density for one (not a particular one) particle at $\mathbf{r}_1$, another at $\mathbf{r}_2$, and so on, irrespective of their velocities and of the positions of the other $N - n$ particles.

In the case of noninteracting systems or under the circumstance that the potential energy can be neglected, as in the case of interacting systems when all the interparticle separations are very large, many simplifications take place. The configurational partition function then becomes a volume integral and is equal to V^N [see Eq. (4.3)]. The right-hand side of Eq. (4.6) becomes $V^{N-n}/V^N = 1/V^n$. The reduced distribution function defined in Eq. (4.7) becomes $N!/[(N-n)!V^n]$, which is equal to ρ^n in the thermodynamic limit, where ρ is the particle density. It has been suggested that a dimensionless reduced distribution function may be defined in the interacting case as

$$g^{(n)}(\mathbf{r}_1, \mathbf{r}_2, \ldots, \mathbf{r}_n) = \rho^{(n)}(\mathbf{r}_1, \mathbf{r}_2, \ldots, \mathbf{r}_n)/\rho^n \qquad (4.8)$$

It reduces in the noninteracting, or the large separation case, to unity, so that its deviations from unity give us some information about interactions, which are usually significant for short distances.

4.2.3   Radial Distribution Function

For $n = 1$, the reduced distribution function [Eq. (4.6)] is $\rho^{(1)}(\mathbf{r}_1)$, which has to be a constant, C, because the system is uniform. Since the integral over dr_1 of a one-particle reduced distribution function is simply the total number of particles, N, and the integral of a constant over dr_1 is CV, the constant is N/V. We get

$$\rho^{(1)}(\mathbf{r}_1) = N/V = \rho \qquad (4.9)$$

where, as above, ρ is the particle density. For values of n greater than 1, the reduced distribution functions depend on the potential and are difficult to evaluate.

The function $g^{(2)}(\mathbf{r}_1, \mathbf{r}_2)$ for a two-body interaction has assumed especial importance in statistical theories; it is called the *radial distribution function*, or *pair distribution function*, and for spherical molecules and a homogeneous system is written simply as $g(r)$, where $r = |\mathbf{r}_2 - \mathbf{r}_1|$. From Eq. (4.8), we can write

$$\rho^{(1)}(\mathbf{r}_1)g(r) = \rho^{(2)}(\mathbf{r}_1, \mathbf{r}_2)/\rho^{(1)}(\mathbf{r}_1) \qquad (4.10)$$

The left-hand side of Eq. (4.10) is recognized as a conditional probability density [see Eq. (1.42)]:

$$\rho^{(1)}(\mathbf{r}_1)g(r) \equiv \rho(\mathbf{r}_2|\mathbf{r}_1) \qquad (4.11)$$

The conditional probability $\rho(\mathbf{r}_2|\mathbf{r}_1)\, dr_1\, dr_2$ is the probability of finding a molecule between $\mathbf{r}_2$ and $\mathbf{r}_2 + d\mathbf{r}_2$ conditioned upon the occurrence of a molecule between $\mathbf{r}_1$ and $\mathbf{r}_1 + d\mathbf{r}_1$. Therefore, the function $g(r)$ weighted by $\rho^{(1)}$ is a conditional probability density. As we shall see, $g(r)$ is important because (1) all thermodynamic functions can be written in terms of it in the case of pairwise interactions and (2) it can be experimentally determined by X-ray or neutron scattering studies.

The radial distribution function is normalized such that

$$\int_{V_1} \int_{V_2} g^{(2)}(\mathbf{r}_1, \mathbf{r}_2)\, dr_1\, dr_2 = \frac{N(N-1)}{\rho^2} \tag{4.12}$$

which, on using the homogeneity of the liquid and the central nature of the potential, can be simplified to

$$\int_V g(r) 4\pi r^2\, dr = V \frac{N-1}{N} \tag{4.13}$$

This result has the following physical interpretation. With respect to a central molecule, $\rho g(r) 4\pi r^2\, dr$ is the number of molecules in a thin spherical shell of thickness dr at a distance r about the central molecule. Thus,

$$\rho \int_V 4\pi r^2 g(r)\, dr = N - 1 \tag{4.14}$$

We note that $g(r) \rightarrow 0$ as r becomes very small, since no two molecules can occupy the same space, and $g(r) \rightarrow 1$ as $r \rightarrow \infty$, since there is certainly another molecule at $r = \infty$.

In order to evaluate $g(r)$, it is necessary to know the pair potential $\phi(r)$ between two molecules. Typical potentials are the hard-sphere potential with the hard-sphere diameter d,

$$\phi(r) = \begin{cases} \infty & r < d \\ 0 & r \geq d \end{cases} \tag{4.15}$$

the square-well potential with well depth ϵ

$$\phi(r) = \begin{cases} \infty & r < d \\ -\epsilon & d < r < d_1 \\ 0 & r \geq d_1 \end{cases} \tag{4.16}$$

and the Lennard–Jones potential

$$\phi(r) = 4\epsilon \left[\left(\frac{\sigma}{r}\right)^{12} - \left(\frac{\sigma}{r}\right)^6 \right] \tag{4.17}$$

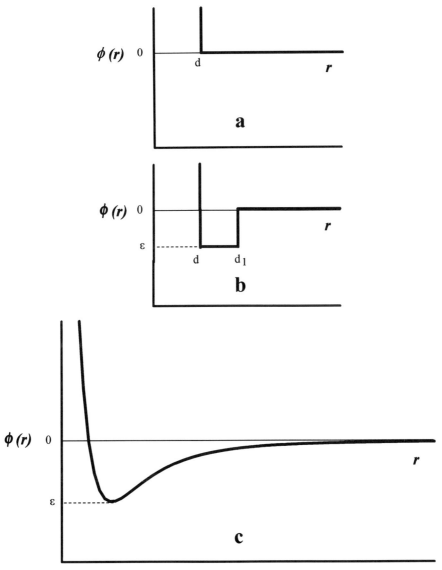

Figure 4.1   (*a*) The hard-sphere potential. (*b*) The square-well potential. (*c*) The Lennard–Jones potential.

where σ is the intermolecular separation for which the potential energy vanishes. These potential functions are shown in Figure 4.1. We shall use some of these potentials later to study the properties of one-dimensional liquids.

The radial distribution functions for the hard-sphere and Lennard–Jones potentials are shown in Figures 4.2 and 4.3, respectively. Inspection of these figures reveals the similarity between the two functions. Because of this

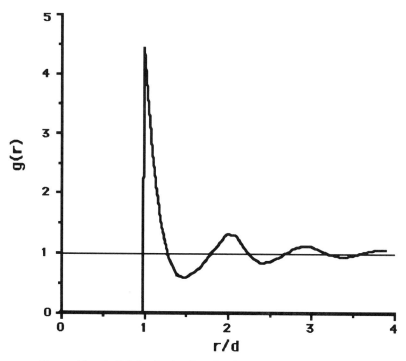

Figure 4.2 Radial distribution function for the hard-sphere potential.

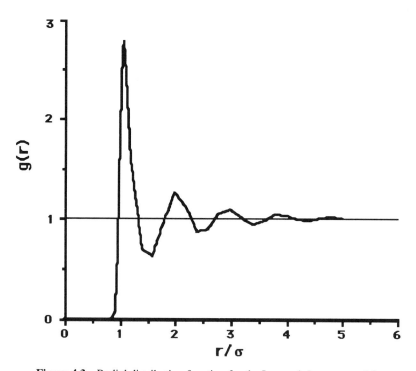

Figure 4.3 Radial distribution function for the Lennard–Jones potential.

similarity, the choice of an interparticle potential for the calculation of $g(r)$ is not crucial, and the simple hard-sphere potential is often chosen to begin with.

4.3 THERMODYNAMIC FUNCTIONS

We now show that the thermodynamic functions can be expressed as a function of $g(r)$. By using Eqs. (4.3) and (1.88), the internal energy can be written as

$$
\begin{aligned}
E &= k_B T^2 \left(\frac{\partial \ln Z}{\partial T} \right)_{N,V} \\
&= \frac{3}{2} N k_B T + k_B T^2 \left(\frac{\partial \ln Q_N}{\partial T} \right)_{N,V} \\
&= \frac{3}{2} N k_B T + E_N
\end{aligned}
\tag{4.18}
$$

where E_N is called the *configurational energy*. By evaluating E_N in terms of two-body (pairwise additive) interactions (e.g., McQuarrie, 1976, p. 261),

$$
E_N = \frac{1}{2} N \rho \int_0^\infty \phi(r) g(r) 4 \pi r^2 \, dr
\tag{4.19}
$$

We see that, if the intermolecular potential is known, the radial distribution function can be evaluated and the internal energy calculated.

The pressure can be written in terms of the configurational partition function as

$$
P = k_B T \left(\frac{\partial \ln Q_N}{\partial V} \right)_{N,T}
\tag{4.20}
$$

The differentiation here is somewhat more involved than in Eq. (4.18) (see McQuarrie, 1976, p. 262). The final expression is

$$
P = k_B T \rho - \frac{1}{6} \rho^2 \int_0^\infty r \frac{d\phi(r)}{dr} g(r) 4 \pi r^2 \, dr
\tag{4.21}
$$

This equation is sometimes called the "pressure equation." In the case of a low-density fluid or a gas, this equation is called the "virial equation," and $w(r) = r \, d\phi(r)/dr$ is called the "intermolecular virial function." A series expansion of $g(r)$ as a power series in ρ (the density) gives the well-known virial equation of state.

The chemical potential of the liquid can be calculated similarly. In this case, we follow McQuarrie (McQuarrie, 1976, pp. 263–264) and write

$$\mu = A(N, V, T) - A(N - 1, V, T) \tag{4.22}$$

where $A = -k_B T \ln Z$. It follows that

$$\mu = -k_B T \ln(Q_N/Q_{N-1}) + k_B T \ln N - k_B T \ln(2\pi m k_B T/h^2)^{3/2} \tag{4.23}$$

In Eq. (4.23), Q_N/Q_{N-1} can be evaluated by a method used by Kirkwood (1935). This method involves incorporating a coupling parameter ξ into the intermolecular potential. The coupling parameter couples Molecule 1 with all other molecules in the system. By varying ξ from 0 to 1, it is possible to go from no coupling to complete coupling. When the coupling parameter is incorporated, the intermolecular potential becomes

$$\Phi(\mathbf{r}_1, \mathbf{r}_2, \ldots, \mathbf{r}_N, \xi) = \sum_{j=2}^{N} \xi\phi(r_{1j}) + \frac{1}{2} \sum_{i \neq j=2}^{N} \phi(r_{ij}) \tag{4.24}$$

In order to evaluate Q_N, we take $\xi = 1$, and for Q_{N-1}, we take $\xi = 0$. Insertion of Eq. (4.24) into the expression for Q_N [see Eq. (4.3)] shows that $Q_N/Q_{N-1} = VQ_N(\xi = 1)/Q_N(\xi = 0)$. Rewriting $\ln[Q_N(\xi = 1)/Q_N(\xi = 0)]$ as $\int_0^1 (\partial \ln Q_N/\partial\xi)\,d\xi$ gives

$$\mu = k_B T \ln[\rho(2\pi m k_B T/h^2)^3] + \rho \int_0^1 \int_0^\infty \phi(r)g(r, \xi')4\pi r^2 \, dr \, d\xi' \tag{4.25}$$

Having found expressions for E, P, and μ, all the other thermodynamic functions can be calculated for a given $\phi(r)$ provided that $g(r, \xi)$ can be evaluated.

By considering the distribution functions in the grand canonical ensemble, one can show that the integral of the radial distribution function is related to the isothermal compressibility, $\kappa_T = -(\partial V/\partial P)_T/V$.

$$\rho \int [g(r) - 1]\,d^3r = \rho k_B T \kappa_T - 1 \tag{4.26}$$

This equation is called the *compressibility equation*. For a derivation, see Pathria (Pathria, 1996, Section 14.2).

4.4   THE KIRKWOOD EQUATION

4.4.1   An Exact Equation for $g(r)$

From Eqs. (4.5)–(4.7) and with the use of the coupling parameter, the reduced distribution function of Eq. (4.8) can be written as

$$g^{(n)}(\mathbf{r}_1, \mathbf{r}_2, \ldots, \mathbf{r}_n, \xi) = \frac{N!}{\rho^n Q_N(\xi)(N-n)!} \int \exp[-\beta\Phi(\xi)]\, dr_{n+1} \cdots dr_N$$

(4.27)

The difficulty with the direct evaluation of this equation is that integrals must be evaluated over a large number of molecules, especially if we take $n = 2$. The Kirkwood method to circumvent this problem is to first differentiate Eq. (4.27) with respect to ξ. Then integration from 0 to ξ, letting $n = 2$, gives (see Hill, 1987, Section 32)

$$-k_B T \ln g^{(2)}(\mathbf{r}_1, \mathbf{r}_2, \xi) = \xi\phi(r_{12})$$
$$+ \rho \int_0^\xi \int_V \phi(r_{13}) \left[\frac{g^{(3)}(\mathbf{r}_1, \mathbf{r}_2, \mathbf{r}_3, \xi')}{g^{(2)}(\mathbf{r}_1, \mathbf{r}_2, \xi')} - g^{(2)}(\mathbf{r}_1, \mathbf{r}_3, \xi') \right] dr_3 d\xi' \qquad (4.28)$$

This introduces a hierarchy of coupled equations giving $g^{(2)}$ in terms of $g^{(3)}$, $g^{(3)}$ in terms of $g^{(4)}$, and so on. This hierarchy of equations is called the BBGKY hierarchy (after Born, Bogoliubov, Green, Kirkwood, Yvon). Ideally, we would like to truncate this hierarchy at some low term. However, this will introduce large errors, especially if long-range forces are important as, for example, in ionic systems. Consequently, an approximation must be made in which all terms are kept somehow. This approximation takes the form of mean-field theory.

4.4.2   Potential of Mean Force

Because it is difficult to know the details of the intermolecular potential in condensed systems, a simplification consists in using some sort of averaged potential. Such a potential $w^{(n)}(\mathbf{r}_1, \mathbf{r}_2, \ldots, \mathbf{r}_n)$ can be defined by

$$\exp[-\beta w^{(n)}(\mathbf{r}_1, \mathbf{r}_2, \ldots, \mathbf{r}_n)] = g^{(n)}(\mathbf{r}_1, \mathbf{r}_2, \ldots, \mathbf{r}_n) \qquad (4.29)$$

By differentiating the log of both sides of Eq. (4.29) with respect to the position of the jth molecule, there is obtained

$$-\nabla_j w^{(n)}(\mathbf{r}_1, \mathbf{r}_2, \ldots, \mathbf{r}_n) = \frac{\int_V \exp[-\beta\Phi(\mathbf{r})]\{-\nabla_j\Phi(\mathbf{r})\}\, dr_{n+1} \cdots dr_N}{\int_V \exp[-\beta\Phi(\mathbf{r})]\, dr_{n+1} \cdots dr_N}$$

$$(4.30)$$

where $-\nabla_j\Phi(\mathbf{r})$ is the force acting on the jth molecule of the set 1 through n. Therefore, $-\nabla_j w^{(n)}(\mathbf{r}_1, \mathbf{r}_2, \ldots, \mathbf{r}_n)$ is a mean force, and $w^{(n)}(\mathbf{r}_1, \mathbf{r}_2, \ldots, \mathbf{r}_n)$ is a kind of mean potential, called the *potential of mean force*. Thus, $w^{(2)}(\mathbf{r}_1, \mathbf{r}_2)$ is the potential between two molecules at a distance $r = |\mathbf{r}_2 - \mathbf{r}_1|$ in the average field of the other $N - 2$ molecules.

4.4.3 Superposition Approximation

Equation (4.28) can now be written as

$$-k_B T \ln \exp[-\beta w^{(2)}(\mathbf{r}_1, \mathbf{r}_2, \xi)]$$

$$= \xi\phi(r_{12}) + \rho \int_0^\xi \int_V \phi(r_{13})$$

$$\times \left(\frac{\exp[-\beta w^{(3)}(\mathbf{r}_1, \mathbf{r}_2, \mathbf{r}_3, \xi')]}{\exp[-\beta w^{(2)}(\mathbf{r}_1, \mathbf{r}_2, \xi')]} - \exp[-\beta w^{(2)}(\mathbf{r}_1, \mathbf{r}_3, \xi')] \right) dr_3\, d\xi'$$

$$(4.31)$$

This equation still does not permit a truncation. In analogy to the very good approximation of pairwise additivity of potential energy functions, $\Phi(\mathbf{r}) = \sum \phi(r_{ij})$, we assume that $w^{(3)}(\mathbf{r}_1, \mathbf{r}_2, \mathbf{r}_3)$ can be written as

$$w^{(3)}(\mathbf{r}_1, \mathbf{r}_2, \mathbf{r}_3) \approx w^{(2)}(\mathbf{r}_1, \mathbf{r}_2) + w^{(2)}(\mathbf{r}_1, \mathbf{r}_3) + w^{(2)}(\mathbf{r}_2, \mathbf{r}_3) \tag{4.32}$$

From this it follows that

$$g^{(3)}(\mathbf{r}_1, \mathbf{r}_2, \mathbf{r}_3) \approx g^{(2)}(\mathbf{r}_1, \mathbf{r}_2) + g^{(2)}(\mathbf{r}_1, \mathbf{r}_3) + g^{(2)}(\mathbf{r}_2, \mathbf{r}_3) \tag{4.33}$$

This equation is called the *superposition approximation*. By its use, Eq. (4.28) can be written

$$-k_B T \ln g^{(2)}(\mathbf{r}_{12}, \xi)$$

$$= \xi\phi(r_{12}) + \rho \int_0^\xi \int_V \phi(r_{13}) g^{(2)}(\mathbf{r}_{13}, \xi')[g^{(2)}(\mathbf{r}_{23}) - 1]\, dr_3\, d\xi' \tag{4.34}$$

and the equation is truncated at $g^{(2)}(r)$. Equation (4.34) is called the *Kirkwood equation*.

4.5   TAKAHASHI ONE-DIMENSIONAL FLUID MODELS

4.5.1   Hard-Sphere Potential

In this section, we shall use the exactly soluble one-dimensional fluid models to illustrate the ideas of the previous sections. Consider a line of length L of rigid hard-sphere fluid particles of diameter d arranged along the x axis. Let their coordinates be $x_1, x_2, x_3, \ldots, x_N$. Because of the hard-sphere potential, the particles can never penetrate one another. Therefore, when moving, they can never change their order. So, we can write $0 \leq x_1 \leq x_2 \leq x_3 \leq \cdots \leq x_N \leq L$. This condition holds at all temperatures and pressures and is the reason the one-dimensional models can be solved exactly. This condition is easily violated in higher dimensions, because the particles in higher dimensions, not being confined to a line, can move around one another, thus changing their order.

Two hard spheres, labeled i and j, interact in accordance with the pair potential

$$\phi(|x_{ij}|) = \begin{cases} \infty & |x_{ij}| < d \\ v(|x_{ij}| - d) & d < |x_{ij}| < 2d \\ 0 & |x_{ij}| \geq 2d \end{cases} \tag{4.35}$$

where $x_{ij} = x_i - x_j$ and $v(|x_{ij}| - d)$ is any suitable potential. Because the potential vanishes for $x_{ij} \geq 2d$, the interactions exist only between the adjacent particles. This model is called the *Takahashi model* and was solved exactly by him (Takahashi, 1942).

We shall start with the purely hard-sphere case, which was solved exactly by Tonks (Tonks, 1936). In this case, $v(|x|) = 0$. The particle motion is simple: They collide elastically when they touch, and move freely otherwise. Still, as we shall see, they are not independent, and they have correlations that are different from the ideal gas fluid. We shall use the isothermal-isobaric ensemble, introduced in Problem 10 in Chapter 1. The partition function is given by

$$Z(N, P, T) = \frac{1}{N! h^N} \int_{L^*}^{\infty} dL \int \exp[-\beta(PL + \mathcal{H})] \, dp \, dx \tag{4.36}$$

where $L^* = (N - 1)d$ is the lowest possible length of the fluid system when all the particles just touch. The quantity P is the "pressure." In the one-dimensional case, it is just the tension at the two ends. The "volume" of the system is the total length L. The N momentum integrals can be evaluated to give

$$Z(N, P, T) = \frac{1}{N! \Lambda^N} \int_{L^*}^{\infty} dL \exp(-\beta PL) Q_N(L, T) \tag{4.37}$$

where $Q_N(L, T)$, the configurational partition function for a given L and T, is

$$Q_N(L, T) = \int_0^L dx_N \cdots \int_0^L dx_i \cdots \int_0^L dx_1 \exp\left[-\beta \sum_{i=1}^{N-1} \phi(|x_{i+1} - x_i|)\right] \tag{4.38}$$

To illustrate the procedure now to be followed, let us calculate $Q_N(L, T)$ for the case of two particles. We have

$$Q_2(L, T) = \int_0^L dx_2 \int_0^L dx_1 \exp[-\beta \phi(|x_2 - x_1|)] \tag{4.39}$$

The two-dimensional integral is over the square shown in Figure 4.4. The square is divided into two regions by the diagonal, where x_1 is equal to x_2. By splitting the integral over x_1 into two parts, we get

$$Q_2(L, T) = \int_0^L dx_2 \int_{x_2}^L dx_1 \exp[-\beta \phi(|x_2 - x_1|)]$$
$$+ \int_0^L dx_2 \int_0^{x_2} dx_1 \exp[-\beta \phi(|x_2 - x_1|)] \tag{4.40}$$

The first integral is over the upper region and the second one is over the lower region. We focus on the first integral and show that it is equal to the second. In the first integral in Eq. (4.40), x_2 is held fixed first, integration over x_1 is done from x_2 to L, and then x_2 is integrated from 0 to L. Interchanging the order of integration in the first integral does not change its value, but it changes the integral to

$$\int_0^L dx_1 \int_0^{x_1} dx_2 \exp[-\beta \phi(|x_2 - x_1|)] \tag{4.41}$$

Now interchange the names of the dummy integration variables. This procedure changes this integral to the second integral in Eq. (4.40). Note that the

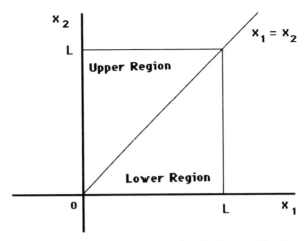

Figure 4.4   The region of integration for the case $N = 2$.

potential energy is symmetric in x_1 and x_2 and, thus, is unchanged by this procedure. We have proved that the double integral in Eq. (4.39) is twice the integral over the lower region:

$$Q_2(L, T) = 2 \int_0^L dx_2 \int_0^{x_2} dx_1 \exp[-\beta\phi(|x_2 - x_1|)] \qquad (4.42)$$

Now we use the crucial property of the hard-sphere potential. Since the particles can never come closer than d, the upper limit of the x_1 integral must be $x_2 - d$ and the lower limit of the x_2 integral must be d. Therefore, Eq. (4.42) becomes

$$Q_2(L, T) = 2 \int_d^L dx_2 \int_0^{x_2 - d} dx_1 \qquad (4.43)$$

where we have eliminated the exponential factor, because, in the indicated range of variables, $\phi(|x_2 - x_1|) \equiv 0$. The integration in Eq. (4.43) can be performed by the substitutions

$$y_1 = x_1 \qquad y_2 = x_2 - d \qquad (4.44)$$

which reduce Eq. (4.43) to

$$Q_2(L, T) = 2 \int_0^{L-d} dy_2 \int_0^{y_2} dy_1 \qquad (4.45)$$

Now,

$$Q_2(L, T) = (L - d)^2 \tag{4.46}$$

The same procedure can be used for the general case. Instead of two regions, we have $N!$ regions into which the N-dimensional region of integration is divided. The equation corresponding to Eq. (4.42) is

$$Q_N(L, T) = N! \int_0^L dx_N \int_0^{x_N} dx_{N-1} \exp[-\beta(\phi(|x_N - x_{N-1}|)]$$
$$\cdots \int_0^{x_2} dx_1 \exp[-\beta\phi(|x_2 - x_1|)] \tag{4.47}$$

The substitutions required are

$$y_j = x_j - (j - 1)d \qquad 1 \le j \le N \tag{4.48}$$

and the result is

$$Q_N(L, T) = [L - (N - 1)d]^N \tag{4.49}$$

By using this result in Eq. (4.37), we get

$$Z = \frac{1}{N! \lambda^N} \int_{L^*}^{\infty} \exp(-\beta PL)(L - L^*)^N dL \tag{4.50}$$

With the substitution $y = \beta P(L - L^*)$, this equation becomes

$$Z = \frac{1}{N! \lambda^N (\beta P)^{N+1}} \exp(-\beta PL^*) \int_0^{\infty} \exp(-y)y^N \, dy \tag{4.51}$$

The dimensionless integral

$$I_N = \int_0^{\infty} \exp(-y)y^N \, dy \tag{4.52}$$

can be evaluated as follows. Integrating by parts shows that the following recurrence relation holds:

$$I_N = NI_{N-1} \tag{4.53}$$

Therefore, we may write

$$I_N = NI_{N-1} = N(N-1)I_{N-2} = \cdots = N(N-1)(N-2)\cdots 2.1 = N! \quad (4.54)$$

where we have used the fact that $I_0 = 1$. Hence, the isothermal-isobaric partition function is given by

$$Z = \frac{1}{\lambda^N(\beta P)^{N+1}} \exp(-\beta P L^*) \quad (4.55)$$

From this, the Gibbs free energy is obtained by using

$$G(N,P,T) = -k_B T \ln Z \quad (4.56)$$

from which all the quantities in the thermodynamic limit may be obtained. For example, the equation of state can be obtained from $L = (\partial G/\partial P)_T$ for $N \gg 1$ and is given by

$$P = \frac{\rho k_B T}{1 - d\rho} \quad (4.57)$$

where $\rho = N/L$ is the particle density. The equation of state reduces to the ideal gas equation for low densities, as expected. The zero temperature represents a singularity of the equation of state at which the highest possible density $1/d$ is obtained. At the highest density all the particles touch one another with no gaps (a crystalline state). As soon as the temperature rises above zero, one gets a liquid with no more singularities in the equation of state. One says that the solid existing at zero T melts and remains melted thereafter.

4.5.2 Other Potentials

The partition function of the Takahashi model, whose potential is given by Eq. (4.35), can be solved along similar lines (Lieb and Mattis, 1966). We omit the details. The Gibbs free energy in the thermodynamic limit is given by

$$G(P,T) = Nk_B T \ln[\lambda/\kappa(P,T)] \quad (4.58)$$

where

$$\kappa(P,T) = \int_d^\infty \exp[-\beta P x - \beta\phi(x)]\,dx \quad (4.59)$$

To calculate the thermodynamic properties, one simply chooses a suitable potential, evaluates the κ integral and the Gibbs free energy.

Recently, the Takahashi model was employed to explore the density maximum in water at about $4°C$ (Cho et al., 1996). The basic idea is the following. Water has many solid phases called ices. They are all similar in their nearest-neighbor crystalline structures. But one of them called ice-II has a second-neighbor structure that is different from the ordinary ice, called ice-I. The second neighbors in ice-II are closer to a given water molecule than in ice-I. This makes ice-II denser than ice-I. Various studies show that both ice-I-like and ice-II-like structures may be present in liquid water. On increasing the temperature, the fraction of the ice-II-like denser structure grows, offsetting the normal expansion of water. The increased fraction of ice-II gives rise to the density maximum. To model this behavior of water, the potential of two wells separated by a barrier was used in the Takahashi model by Cho et al. (1996). The calculated density as a function of temperature is strikingly similar to the density of real water. This model also explains the well-known experimental fact that, with increasing pressure, the density maximum shifts to lower temperatures, ultimately disappearing completely. For more complete discussion of this and a general review of water, see Robinson et al. (Robinson et al., 1996).

4.5.3 Radial Distribution Function

The radial distribution function of the hard-sphere model was obtained by Zernike and Prins (Zernike and Prins, 1927), well before the calculation of the partition function of the model. All the reduced distribution functions of the Takahashi model were obtained later (Salsburg et al., 1953). Here we follow the Zernike–Prins approach as discussed by Frenkel (Frenkel, 1946, pp. 125–129).

We first discuss the ideal fluid. In this case, we already know that $g(x) = 1$. Let us derive this result in a way that will be instructive and useful in the calculation of $g(x)$ of the hard-sphere liquid. Consider a segment of length x on the straight line of length L with periodic boundary conditions. Because of ideality, the system is uniform and the probability of finding a given particle in this segment is x/L independent of the location of the segment and the other particles. We shall assume the segment to start from the origin, where one particle is assumed to be located with probability one. This segment may contain $0, 1, 2, \ldots, N$ particles. The probability that the segment contains exactly n particles out of N is equal to the probability that a random walker is n steps away from the origin with one-step probability x/L. This occurs because putting n particles in the segment is like taking n independent steps.

The result is given in Eq. (1.1).

$$P_N(n) = \frac{N!}{n!(N-n)!} \left(\frac{x}{Na}\right)^n \left(1 - \frac{x}{Na}\right)^{N-n} \tag{4.60}$$

where we have introduced the mean interparticle distance $a = L/N$. It is possible to carry out the calculation for a finite N, but we shall focus on the thermodynamic limit. In this limit, $N, L \to \infty$ in such a way that a is fixed. The probability of finding a given particle in the segment x becomes very small. These are precisely the conditions under which the binomial distribution, given in Eq. (4.60), reduces to the Poisson distribution.

$$P(n) = \frac{x^n}{n! a^n} \exp\left(-\frac{x}{a}\right) \tag{4.61}$$

By using the results of Chapter 1, we can say that the average number of particles in the segment of length x is x/a and the standard deviation is $\sqrt{x/a}$. Also, for a large number of particles n in the segment, the distribution becomes Gaussian,

$$\frac{1}{\sqrt{2\pi}} \exp\left(-\frac{z^2}{2}\right) \tag{4.62}$$

where $z = (n - x/a)/\sqrt{x/a}$.

We now focus on the small segment δx between x and $x + \delta x$. The probability S_k, say, that the kth particle is found in the small segment is the difference of two probabilities, one of which is the probability that this particle is found at a distance greater than x and the other is that it is found at a distance greater than $x + \delta x$. The probability that the kth particle is indeed at a distance greater than x is obtained by summing Eq. (4.61) over n from 0 to $k - 1$. Consequently, S_k is

$$S_k = \sum_{n=0}^{k-1} \frac{x^n}{n! a^n} \exp\left(-\frac{x}{a}\right) - \sum_{n=0}^{k-1} \frac{(x + \delta x)^n}{n! a^n} \exp[-(x + \delta x)/a] \tag{4.63}$$

Expanding Eq. (4.63) for small δx to lowest order gives

$$S_k \approx \delta x \sum_{n=0}^{k-1} \frac{x^n}{n! a^n} \exp\left(-\frac{x}{a}\right) \left(\frac{1}{a} - \frac{n}{x}\right) \tag{4.64}$$

All the terms in the sum in Eq. (4.64) cancel in pairs except one. We get

$$S_k(x) \approx \frac{\delta x}{a} P(k) \tag{4.65}$$

The total number of particles in the range x and $x + \delta x$ is given by summing Eq. (4.65). The total number of particles in the range x and $x + \delta x$ is also given by $\rho g(x)\delta x$, by definition. Therefore,

$$\rho g(x)\delta x = \sum_{k=0}^{\infty} S_k(x) \approx \sum_{k=0}^{\infty} \frac{\delta x}{a} P(k) \tag{4.66}$$

As δx approaches zero, Eq. (4.66) gives

$$g(x) = \sum_{k=0}^{\infty} \frac{1}{a\rho} P(k) \tag{4.67}$$

or

$$g(x) = 1 \tag{4.68}$$

as expected, where we have used the fact that the probabilities $P(k)$ sum to unity and $\rho = 1/a$ in one dimension.

The calculation for the hard-sphere liquid proceeds exactly along the same lines. There are just two differences, both due to the excluded-volume effect of the hard spheres. The probability that a given particle is found in a segment of length x is not x/L anymore. The entire length of the segment is no longer available to the hard spheres. Only the length $x - d(n + 1)$, which is the sum of the gaps between the nth and the $(n + 1)$th particle (not counting the reference particle at the origin), is available. For example, between the first two particles ($n = 0$) only $x - d$ is available. By the same token, out of the total length L of the system, only the length $Na - Nd$ is available. If we use the binomial distribution of the random walk and take the thermodynamic limit, we find that Eq. (4.61) is replaced by

$$P^h(n) = \frac{x_0^n}{n!a_0^n} \exp\left(-\frac{x_0}{a_0}\right) \tag{4.69}$$

where we have defined, for convenience,

$$x_0 = x - d(n + 1) \qquad a_0 = a - d \tag{4.70}$$

Clearly, Eq. (4.69) is valid only for $x \geq nd$, because the minimum length of a possible segment is nd. We must define $P^h(n) = 0$ for $x < nd$. Once again, for a large number of particles in the segment, the distribution becomes Gaussian.

We have taken into account the effects of the hard-sphere interaction. When the particles do not touch, they cannot interact, and so they are independent. Hence, the rest of the calculation proceeds exactly as in the ideal case up to Eq. (4.65), which is replaced by

$$S_k(x) \approx \frac{\delta x}{a_0} P^h(k) \tag{4.71}$$

To obtain $g(x)$ we cannot sum $P^h(k)$ to infinity, because the segment under discussion can have only a finite number of particles. The sum should be up to a maximum k, denoted by k^*. The maximum number of particles in a segment of length x is the largest integer not exceeding x/d. For example, for $x = 3.14d$, the largest number of particles in the segment is 3 and the sum should go up to $k^* = 3$. With these considerations, $g(x)$ is found to be

$$g(x) = \frac{a}{a_0} \sum_{k=0}^{k^*} \frac{[x - d(k+1)]^k}{k! a_0^k} \exp\left[\frac{-\{x - d(k+1)\}}{a_0} \right] \tag{4.72}$$

where k^* is the largest integer not exceeding x/d. This expression is valid for $x > d$ only. For $x < d$, we must have $g(x) = 1$, by definition. As an example, for $x = 2.14d$ we get

$$g(x) = \frac{a}{a_0} \exp\left[\frac{-(x-d)}{a_0} \right] + \frac{a(x-2d)}{a_0^2} \exp\left[\frac{-(x-2d)}{a_0} \right]$$
$$+ \frac{a(x-3d)^2}{2a_0^3} \exp\left[\frac{-(x-3d)}{a_0} \right] \tag{4.73}$$

In the limit of the ideal gas, $d \rightarrow 0$ and all the terms in the sum in Eq. (4.72) contribute, adding up to unity. We get $g(x) = 1$, as expected.

At absolute zero, we have a perfect crystal so that $a = d$, or, $a_0 = 0$. In this case, all the terms in the sum in Eq. (4.72) become delta functions. As the temperature rises, a becomes greater than d, the delta functions broaden, and their heights decrease as x increases. In the extreme high temperature limit, all structure is washed out, and a line of unit height is obtained.

4.6   ORNSTEIN–ZERNIKE EQUATION

4.6.1   The Direct Correlation Function

The Kirkwood equation for the radial distribution function suffers from the superposition approximation. Other approaches have derived integral equations that do not depend on the superposition approximation. One of these is the exact approach due to Ornstein and Zernike (Ornstein and Zernike, 1914).

We saw above that $g(r) = 1$ for an ideal fluid. The difference, $g(r) - 1$, must therefore represent the nonideal part of the pair distribution function for a nonideal fluid. For an ideal fluid, there are no correlations between particles. There are correlations between particles of a nonideal fluid, so that the difference, $g(r) - 1$, is called a *pair correlation function, h(r)*.

$$h(r) = g(r) - 1 \tag{4.74}$$

It is to be noted that $h(r) \rightarrow -1$ as $r \rightarrow 0$ and $h(r) \rightarrow 0$ as $r \rightarrow \infty$. This behavior is shown in Figure 4.5 for the Lennard–Jones potential. In this figure,

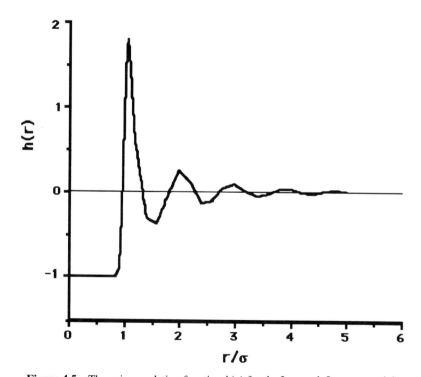

Figure 4.5   The pair correlation function $h(r)$ for the Lennard–Jones potential.

the structure in $h(r)$ is interpreted as short-range correlations and a long-range tail, which damps out, showing increasing structural disorder.

Ornstein and Zernike proposed dividing $h(r)$ into direct and indirect parts:

$$h(\mathbf{r}_{12}) = c(\mathbf{r}_{12}) + \rho \int c(\mathbf{r}_{13})h(\mathbf{r}_{23})\, dr_3 \tag{4.75}$$

where $\mathbf{r}_{12} = \mathbf{r}_2 - \mathbf{r}_1$. This equation is called the *Ornstein–Zernike equation*. The first term on the right-hand side is called the *direct correlation function*, and the second term is the indirect part. The direct part is supposedly shorter ranged and more amenable to calculation than is the entire function $h(\mathbf{r}_{12})$. The indirect part is what is known as a convolution of correlation functions; it supposedly represents the long-range, many-body effects. By virtue of Eq. (4.26), $h(r)$ is related to the compressibility,

$$\rho k_B T \kappa_T = \rho \int h(r)\, dr + 1 \tag{4.76}$$

The evaluation of the Ornstein–Zernike equation is as difficult as the evaluation of the Kirkwood equation. However, by ignoring the indirect part, only $c(r)$ need be calculated. There are graphical solutions for $c(r)$ (McQuarrie, 1976, Section 13-6), but these are beyond the scope of this discussion. It is interesting to note that the Fourier transform of $h(r)$ is directly related to the structure factor of X-ray and neutron diffraction and of light scattering. The relation between a correlation function and the structure factor is illustrated next by considering critical opalescence.

4.6.2 Light Scattering Near Critical Points

According to light-scattering theory (Berne and Pecora, 1976), the intensity of scattered radiation for a fixed wave vector $\mathbf{k}$ (e.g., 90° scattering) and frequency ω is proportional to the Fourier transform of the autocorrelation function of the dielectric constant fluctuations $\delta\epsilon(\mathbf{k}, t)$ (see Section 1.6.2 for a discussion of fluctuations):

$$\tilde{I}(\mathbf{k}, \omega) \sim \int_{-\infty}^{\infty} \langle \delta\epsilon(\mathbf{k}, 0)\delta\epsilon(\mathbf{k}, t)\rangle \exp(-i\omega t)\, dt \tag{4.77}$$

where the Fourier transform is denoted by a tilde ($\sim$). Because $\delta\epsilon = \epsilon - \epsilon_0$ is a function of the number density ρ, a Taylor's series expansion of ϵ about the average density ρ_0 gives

$$\epsilon = \epsilon_0 + (\rho - \rho_0) \left(\frac{\partial \epsilon}{\partial \rho} \right)_0 + \text{higher order terms} \qquad (4.78)$$

Neglecting the higher order terms,

$$\widetilde{I}(\mathbf{k}, \omega) \sim \int_{-\infty}^{\infty} \left(\frac{\partial \epsilon}{\partial \rho} \right)_0^2 \langle \delta\rho(\mathbf{k}, 0)\delta\rho(\mathbf{k}, t) \rangle \exp(-i\omega t)\, dt$$

$$\sim \left(\frac{\partial \epsilon}{\partial \rho} \right)_0^2 \widetilde{S}(\mathbf{k}, \omega) \qquad (4.79)$$

where

$$\widetilde{S}(\mathbf{k}, \omega) = \left(\frac{1}{2\pi} \right) \int_{-\infty}^{\infty} \langle \delta\rho(\mathbf{k}, 0)\delta\rho(\mathbf{k}, t) \rangle \exp(-i\omega t)\, dt \qquad (4.80)$$

is called the *dynamic-structure factor*. The integrated dynamic-structure factor is equal to the *static-structure factor* $\widetilde{S}(\mathbf{k})$.

$$\widetilde{S}(\mathbf{k}) = \int_{-\infty}^{\infty} \widetilde{S}(\mathbf{k}, \omega)\, d\omega = \langle [\delta\rho(\mathbf{k})]^2 \rangle \qquad (4.81)$$

4.6.3   Ornstein–Zernike Approach

For light scattering near the critical point of a homogeneous fluid phase (see Chapter 3 for a discussion of critical point phenomena), critical opalescence is observed, indicating the existence of long-range spatial correlations among the molecules. The long-range correlations lead to the isothermal compressibility becoming infinite at the critical point. For this to happen, $h(r)$ must go to zero slowly, and as a result it is dominated by its long-range tail, which says that the difference between $h(r)$ with long-range correlations near the critical point and $h(r)$ without these correlations away from the critical point must be, to a good approximation, the long-range tail on $h(r)$. Observations such as this prompted Ornstein and Zernike to propose their equation [Eq. (4.75)].

Now we make connection with the light-scattering theory of Section 4.6.2. Because the wave vector of visible light is very small (the wavelength of light is large compared to molecular dimensions), we let $\mathbf{k} \to 0$. It now follows that

$$\lim_{\mathbf{k} \to 0} \widetilde{S}(\mathbf{k}) = \langle (\delta\rho(\mathbf{k}))^2 \rangle = \langle (\delta N)^2 \rangle = \langle N^2 \rangle - \langle N \rangle^2 \qquad (4.82)$$

since

$$\lim_{\mathbf{k}\to 0} \delta\rho(\mathbf{k}) = \lim_{\mathbf{k}\to 0} \int \delta\rho(\mathbf{r})\exp(i\mathbf{k}\cdot\mathbf{r})\,d^3r = \int \delta\rho(\mathbf{r})\,d^3r = \delta N \quad (4.83)$$

From Eq. (1.99),

$$\lim_{\mathbf{k}\to 0} \widetilde{S}(\mathbf{k}) = \langle N\rangle^2 k_B T \kappa_T / V$$

$$= V\rho^2 k_B T \kappa_T \quad (4.84)$$

Therefore, the static structure factor is proportional to the isothermal compressibility. From Eq. (4.76),

$$\lim_{\mathbf{k}\to 0} \widetilde{S}(\mathbf{k}) = \lim_{\mathbf{k}\to 0} V\rho[\rho\widetilde{h}(\mathbf{k}) + 1] \quad (4.85)$$

where $\widetilde{h}(\mathbf{k}) = \int h(r)\exp(i\mathbf{k}\cdot\mathbf{r})\,dr$ is the Fourier transform of $h(r)$.

It can be shown (Stanley, 1971, Section 7.4.2; Pathria, 1996, Section 11.11) that, in the mean-field approximation to the Ornstein–Zernike theory,

$$h(r) \sim \frac{\exp(-r/\xi)}{r} \quad (4.86)$$

for $r \gg d$, where d is the range of interactions and ξ is the correlation length. The Fourier transform of $h(r)$ can be shown (Stanley, 1971, Section 7.4.2) to be

$$\widetilde{h}(\mathbf{k}) \sim \frac{1}{k^2 + \xi^{-2}} \quad (4.87)$$

Near the critical temperature the correlation length diverges as

$$\xi \sim |t|^{-1/2} \quad (4.88)$$

Equations (4.87) and (4.88) show that $\widetilde{h}(\mathbf{0})$ diverges as $1/|t|$ near the critical point. That is why the compressibility becomes huge, giving rise to large fluctuations in the dielectric constant, which results in critical opalescence. As a qualitative explanation, this is correct, but numerically there are problems with the mean-field approximation. In reality, the correlation length diverges with an exponent ν which is usually far from 0.5. In fluids, it is closer to 0.3. It is to explain such factual discrepancies that more modern theories like the renormalization group are needed. These are discussed in Chapter 3.

4.7 CONCLUDING REMARKS

Other approaches to the calculation of the radial distribution function along the lines of the Ornstein–Zernike equation give similar equations called the Percus–Yevick and hypernetted-chain equations. These will not be discussed and are mentioned only for reference. Comparisons of these integral equations to experimental data are given by McQuarrie (McQuarrie, 1976, Section 13-9). In any case, being mean-field theories, none of them work in the critical region, although they are very useful far from the critical point.

This chapter has focused on the radial distribution function of liquids, because the concept of a radial distribution function is important in any discussion of liquids, and this function can be measured experimentally. Although the calculation of the radial distribution function for real liquids is very difficult, the formalism of this chapter is important for understanding current theories of liquids. For example, the separation of short- and long-range forces suggested by the Ornstein–Zernike equation is an important concept.

We have not discussed perturbation theories of liquids. Some success has been achieved by Chandler, Weeks, and Andersen using a hard-sphere repulsive force coupled with a Lennard–Jones attractive force to calculate many properties of simple liquids including $\tilde{h}(\mathbf{k})$. The reader is referred to McQuarrie (McQuarrie, 1976, Chapter 14) for more details and references.

The study of polyatomic liquids is an active area of research that requires the inclusion of the rotational and vibrational degrees of freedom in the distribution and correlation functions. This, in turn, requires an extension of the results of this chapter, as well as approximations other than the Percus–Yevick and hypernetted-chain equations. A discussion can be found in Hansen and McDonald (Hansen and McDonald, 1986, Chapter 12).

We have illustrated the rather abstract concepts of this chapter with the help of the exactly soluble Takahashi model. Unfortunately, the techniques used in the model cannot be generalized to higher dimensions, where other methods must be employed.

4.8 EXERCISES AND PROBLEMS

1. Show that Eq. (4.3) follows from Eq. (4.2).

2. Show for any function $f(\mathbf{r}_{12})$ that $\int f(\mathbf{r}_{12})\,dr_1\,dr_2 = V \int 4\pi r^2 f(r)\,dr$, where $\mathbf{r}_{12} = \mathbf{r}_2 - \mathbf{r}_1$.

3. Show dimensionally that $\rho g(r)4\pi r^2\,dr$ is the number of molecules in a thin spherical shell. What is the significance of $g(r)$?

4. The following data are for the radial distribution function using a Lennard–Jones potential, where $r^* = r/\sigma$. The reduced density is $\rho^* = \sigma^3 \rho = 0.880$.

r^*	$g(r^*)$	r^*	$g(r^*)$	r^*	$g(r^*)$	r^*	$g(r^*)$
0.84	0.000	1.24	1.286	1.64	0.663	2.04	1.268
0.88	0.001	1.28	1.058	1.68	0.720	2.08	1.257
0.92	0.086	1.32	0.891	1.72	0.790	2.12	1.223
0.96	0.688	1.36	0.781	1.76	0.873	2.16	1.168
1.00	1.871	1.40	0.699	1.80	0.960	2.20	1.103
1.04	2.701	1.44	0.650	1.84	1.040	2.24	1.044
1.08	2.785	1.48	0.616	1.88	1.113	2.28	0.981
1.12	2.453	1.52	0.606	1.92	1.180	2.32	0.936
1.16	2.003	1.56	0.612	1.96	1.221	2.36	0.896
1.20	1.596	1.60	0.631	2.00	1.250	2.40	0.869

a. Using these data, determine the number of molecules in the first coordination sphere.

b. Determine the number of molecules in the second coordination sphere.

c. For close-packed hard spheres, what would be the results of items a and b in two dimensions?

5. a. Starting from Eq. (4.21), expand $g(r)$ in a Taylors's series in $\rho = 0$. Use the well-known expression for the second virial coefficient [McQuarrie, 1976, Eq. (12-25)],

$$B_2(T) = -2\pi \int [\exp(-\beta\phi) - 1] r^2 \, dr \qquad (4.89)$$

to do an integration by parts in order to identify $g_0(r)$, the first term in the Taylor's series expansion of $g(r)$.

b. Evaluate the second virial coefficient for a hard-sphere potential.

6. Show that $\kappa = (\partial\rho/\partial P)_T/\rho$ agrees with the definition,

$$\kappa = -(\partial V/\partial P)_T/V$$

7. Derive Eq. (4.69).

5

MOLECULAR DYNAMICS METHODS

5.1 INTRODUCTION

By continuing with our study of liquids, we now turn to computer-simulation methods. The focus in this chapter is on molecular dynamics (MD) techniques, where classical equations of motion are integrated numerically to generate trajectories. In Chapter 6, we shall consider the Monte Carlo (MC) technique, where stochastic methods are used to sample phase space. In both cases, the physical quantities are obtained by taking averages over the calculated set of microstates.

5.2 EARLY ACHIEVEMENTS

We present a brief summary of the early successes in the computer simulation of liquids. For more details, the reprint volume, *Simulation of Liquids and Solids* (Ciccotti et al., 1987) should be consulted. The molecular dynamics method was introduced by Alder and Wainwright (Alder and Wainwright, 1957–1964; Alder and Wainwright, 1962) to study a system of particles interacting with a hard-sphere and a square-well potential. They established many important results in these pioneering studies. They found that the Boltzmann *H* function of a system prepared in a nonequilibrium state decays in time almost monotonically and relaxes to the equilibrium value in a small number of collisions. Alder and Wainwright showed that 100 particles are sufficient to describe the bulk hard-sphere system in the fluid region. But perhaps the most influential finding was that the hard-sphere fluid undergoes a first-order phase transition into a solid at a certain temperature. This transition had been predicted earlier by Kirkwood (Kirkwood, 1951, p. 67) and was thought to be controversial. The MD calculation essentially settled the matter.

Alder and Wainwright continued their studies (Alder and Wainwright, 1970) and made another discovery counter to intuition. They found that the velocity autocorrelation function of a hard-sphere fluid decays to zero as $t^{-d/2}$ for long times, where d is the dimensionality of the system. This caused a sensation, because until that time it was believed that such correlation functions decayed to zero exponentially fast. The reasoning was as follows. According to the molecular-chaos hypothesis of Boltzmann (see Chapter 8), each molecular collision is independent of the previous one. This theory implies that the processes involved in establishing equilibrium are memory-free or Markovian (see Chapter 9). This assumption immediately leads to a short correlation time and an exponential decay by Doob's theorem (see Section 10.4.2). The work of Alder and Wainwright (Alder and Wainwright, 1970) showed that these notions are not quite correct, in that collisions do influence one another over long times leading to a very slow power-law decay of correlation functions. This finding had a profound effect on the kinetic theory of dense fluids and led to its total reformulation.

The next important step was taken by Rahman (Rahman, 1964), who is rightly considered the father of MD. In this work, he studied liquid argon atoms interacting with a Lennard–Jones potential. He calculated the radial correlation function, velocity autocorrelation function, and the triplet correlation function $g^{(3)}(\mathbf{r}_1, \mathbf{r}_2, \mathbf{r}_3)$. Among other findings in this work, the following two are extremely important. The first is that the radial distribution function of a Lennard–Jones fluid is qualitatively very similar to that of a hard-sphere fluid. This finding implies that a good starting point for the theory of liquids is the hard-sphere fluid (see Section 4.2.3). The idea that the repulsive part of a potential, such as the Lennard–Jones potential, can be used as a "known" reference potential on which the attractive part is treated as a perturbation is the main idea behind the work of Weeks et al. (Weeks et al., 1971). See also the work of Barker and Henderson (Barker and Henderson, 1976). The second finding of the Rahman work was that the superposition approximation gives very poor triplet correlation functions. Previously, much time and effort had been devoted to this approximation, which had been found to be valid for the Takahashi model. In 1984, a special meeting was held at the Argonne National Laboratory to commemorate the twentieth anniversary of the 1964 paper of Rahman.

All the above MD calculations were for monatomic fluids. The first significant "molecular" dynamics study was performed by Rahman and Stillinger on water (Rahman and Stillinger, 1971). This paper was important because the authors took a rather complicated model of water and showed how the various properties of water can be reproduced with a computer simulation. This work pointed out the utility of simulations versus analytical techniques, for which the model had to be simplified in some way so as to make it tractable. The Rahman and Stillinger work proved that there was no limitation on the models used to tackle complex many-body problems. Thus, the hidden potential of the simulation process was openly exposed for all kinds of scientists working in the theory of liquids. As far as water is concerned, it may be stated that Rahman and Stillinger comprehensively addressed every single aspect of the molecular dynamics simulation of water. A history of further advances and important results obtained in MD may be found in Ciccotti and Hoover (Ciccotti and Hoover, 1986).

The essentials of the MC method were laid down by Metropolis et al. (Metropolis et al., 1953) in the first computer simulation of liquids using the MANIAC computer at Los Alamos in 1953. All important concepts of *importance sampling* and periodic boundary conditions were introduced in this work. Shortly thereafter, Wood and Jacobson (Wood and Jacobson, 1957) published the results of the first definitive study of the hard-sphere model using MC. This paper was published just one page before the Alder and Wainwright MD work mentioned above, and there is good agreement in the results from the two approaches. The fluid–solid phase transition was found by Wood some time later (Wood, 1968, 1970). A Lennard–Jones potential was used in a MC study in 1957 (Wood and Parker, 1957) to obtain $P - V - T$ data on argon for comparison with experimental values. Two important conventions called the *spherical cutoff* and the *minimum image* convention, which reduce the number of molecules required for a force calculation, were introduced by Metropolis et al. (Metropolis et al., 1953) and Wood and Parker (Wood and Parker, 1957), respectively. Liquid water was studied in 1969 by Barker and Watts (Barker and Watts, 1969) by using a potential based on data from the gas phase. Since then, the MC approach has been used to study complex many-body systems in order to obtain a great deal of information on realistic models, which could not have been reliably treated in any other way (see Binder, 1995). A very promising development is the amalgamation of the RG and the MC techniques to generate the Monte Carlo renormalization group (MCRG) method. Monte Carlo methods will be discussed in detail in Chapter 6.

5.3 ANALYTIC METHODS VERSUS SIMULATION

Ideally, one would like to have an analytic solution to a realistic model in the thermodynamic limit. The advantages of the availability of such a model are tremendous. After Onsager obtained his celebrated solution of the Ising model on a square lattice, the world of statistical mechanics was never the same. First of all, the solution laid to rest many doubts expressed by many famous scientists as to whether the formal machinery of statistical mechanics was really capable of explaining phase transitions. Second, the analytical properties of free energy and correlations in the Onsager solution, which laid the groundwork of innumerable future advances that are still continuing, provided invaluable insights into the problem of critical phenomena. Unfortunately, such exact solutions are difficult to come by. In the absence of analytical solutions to theoretical models, the comparison of models with the results of experiments is full of doubt. One always wonders whether the agreement between the two, if any, is just a lucky coincidence. In the case of disagreement, doubts may be placed on the approximation method involved, the adequacy of the model itself, and, as a last resort, on the experimental results.

Computer simulation provides a way out of these problems. The Hamiltonians and the equations of motion of almost all many-body systems of interest are presumably known. These can be solved numerically, and physical quantities can be obtained, in principle, by using the MD or the MC techniques. However, there are two main concerns that one has about these techniques. First, computer simulations deal with systems of finite volume, and second, they do this for a finite time. At the present time, one can study a few thousand molecules for a few thousand picoseconds. According to Hoover, "Our computers are now about one billion times quicker than humans, but they are still a billion times too slow to simulate the real-time dynamics of a cube of gas only one millionth of a meter on a side!" (Hoover, 1984). Even though computers are faster now than when this statement was made, the gist of the statement is still valid. Thus, a simulator's dream machine is described by K. G. Wilson: "A standard of reasonable speed is the ability to do a 3-D simulation, in color, at the speed of a movie, so the human can follow what's happening. A reasonable estimate is a factor of a million over a CRAY 1" (Wilson, 1985).

We describe how one deals, in practice, with the finiteness of a computer simulation in space and time. A finite number of molecules is not a big problem. The determining factor is the ratio of the length of the system to the correlation length, L/ξ. In ordinary liquids, away from the critical region, the correlation length is not too large, so that the finiteness of L does not pose a problem. In fact, it was found in the early stages of MD by Alder and Wainwright that 100 particles with periodic boundary conditions are enough to model a bulk hard-sphere system in the fluid phase. One regime where the finiteness of the systems can cause severe problems is in the vicinity of the critical region. Luckily, special methods, such as finite-size scaling (Pathria, 1996, Section 13.5) and the MCRG have been invented to deal with this singular regime.

The second concern in any simulation is whether enough phase space has been sampled in the allotted simulation time. Normally, it is seen that the various correlation and relaxation times are small, and choosing a total simulation time to be greater than these times is sufficient. Unfortunately, in many cases no estimate of such relaxation times is available, since that is the quantity to be determined by simulation. In other cases, there are many energy barriers in the multidimensional phase space, and the system may stay in one metastable state for a long time. In such a case, the system averages will appear to be time independent, giving the impression that equilibrium has been reached. Many techniques and criteria have been developed to judge if equilibrium has been truly reached. For example, radial distribution functions or order parameters oftentimes provide clues. In the special case of activated barrier crossing problems, where the reaction rate is very slow, special techniques may be required (Berne, 1985). In the critical regime, where critical slowing down occurs along with large spatial correlations, yet another set of special procedures is employed (Swendsen et al., 1995).

5.4 A TYPICAL MD SIMULATION

5.4.1 Start of the MD Simulation

In the rest of this chapter, we take the reader through a typical simulation of a monatomic system of particles interacting with a Lennard–Jones potential in order to lay bare the essentials of the molecular dynamics. We shall describe the microcanonical or the *NVE* (fixed number of particles, volume, and energy) method. For other potentials and ensembles, see Allen and Tildesley (Allen and Tildesley, 1989). For more details on the MD method, see the book by Haile (Haile, 1992). We choose to study $N = 864$ argon atoms of mass $m = 39.95 \times 1.6605 \times 10^{-24}$ g at a temperature of 94.4 K and density $mN/V = 1.374\,\mathrm{g\,cm^{-3}}$. These values give a system volume $V = 4.171 \times 10^{-20}\mathrm{cm^3}$, where the edge of the cubic simulation box is $L = 3.468 \times 10^{-7}\mathrm{cm}$. The particles are assumed to be points interacting with the Lennard–Jones potential having parameters $\sigma = 3.4 \times 10^{-8}\mathrm{cm}$ and $\epsilon = 1.657 \times 10^{-21}\mathrm{J}$, which is all the input that is needed at this stage.

Next, we introduce dimensionless or reduced units. We measure energies in units of ϵ, distances in units of σ, and mass in units of m. The various quantities in reduced units, denoted by an asterisk, are shown in Table 5.1. Reduced units are used so that the equations will not involve the physical parameters ϵ, σ, and m. Thus, once the simulation results are available for a given N-particle Lennard–Jones system at volume V^*, they apply to any Lennard–Jones system with the same reduced parameters. In other words, all Lennard–Jones fluids behave identically at the same reduced temperature, pressure, and density.

TABLE 5.1 Reduced Units for Lennard–Jones System

Quantity	Symbol	Reduced Units
Time	t	$t^* = t\sqrt{\epsilon/m\sigma^2}$
Velocity	$\mathbf{v}$	$\mathbf{v}^* = \mathbf{v}\sqrt{m/\epsilon}$
Acceleration	$\mathbf{a}$	$\mathbf{a}^* = \mathbf{a}m\sigma/\epsilon$
Force	$\mathbf{f}$	$\mathbf{f}^* = \mathbf{f}\sigma/\epsilon$
Pressure	P	$P^* - P\sigma^3/\epsilon$
Box length	L	$L^* = L/\sigma$
Temperature	T	$T^* = k_B T/\epsilon$

This statement is an expression of the law of corresponding states and is a universal property. Hence, by using reduced units, we are simulating all Lennard–Jones systems at once.

In order to use the molecular dynamics simulation, the Hamiltonian and the equations of motion (Hamilton's, Lagrange's, or Newton's) of the system must be known. We first determine the force on the ith particle. The acceleration of the ith particle is given by

$$\mathbf{a}_i = \mathbf{f}_i/m \tag{5.1}$$

where $\mathbf{f}_i$ is the total force acting on the ith particle due to all other $N-1$ particles. In reduced units, Eq. (5.1) becomes

$$\mathbf{a}_i^* = \mathbf{f}_i^* \tag{5.2}$$

The force between two particles located at $\mathbf{r}_i^*$ and $\mathbf{r}_j^*$, as obtained from the Lennard–Jones potential in reduced units, is

$$\phi^*(r^*) = 4\left[\left(\frac{1}{r^*}\right)^{12} - \left(\frac{1}{r^*}\right)^6\right] \tag{5.3}$$

where $r^* = r/\sigma$, and $\mathbf{r}^* = \mathbf{r}_i^* - \mathbf{r}_j^*$. The force on particle i due to particle j is obtained from the equation

$$\mathbf{f}_i^* = -\nabla_i \phi^*(r^*) \tag{5.4}$$

where the subscript on the del operator indicates differentiation with respect to the components of the vector $\mathbf{r}_i^*$. The force is

$$\mathbf{f}_i^* = 24\left(\frac{2}{r^{*14}} - \frac{1}{r^{*8}}\right)\mathbf{r}^* \tag{5.5}$$

It is seen that the force is attractive for $r^* > 2^{1/6}$ and repulsive for $r^* < 2^{1/6}$, where the potential minimum is at $r^* = 2^{1/6}$.

Next, we must decide on the initial positions and velocities of the particles. Since the final equilibrium state is independent of the initial conditions, at least in principle, we first position the particles randomly in the box. To conserve computer time, some thought must be given as to how the approach to equilibrium can be hastened. With this in mind, the particles are given random velocity components from a Maxwell–Boltzmann distribution for the temperature T^*. In our case $T^* = 94.4\,\text{K}/\epsilon = 0.787$. Also, to insure that the center of mass of the whole system is at rest at the beginning of the simulation, we require that, for each particle having a velocity $\mathbf{v}^*$, there is another one of velocity $-\mathbf{v}^*$. This requirement guarantees that the system is initially at a temperature T^* and minimizes the time necessary to reach thermal equilibrium.

The Maxwell–Boltzmann distribution of the x component of the velocities in reduced units is given by (see Section 1.6.3)

$$f(v_x^*)\,dv_x^* = \left(\frac{1}{2\pi T^*}\right)^{1/2} \exp[-v_x^{*2}/(2T^*)]\,dv_x^* \tag{5.6}$$

with similar equations for the other two components. Equation (5.6) is a Gaussian distribution with zero mean and variance T^*. Random number generators, which are now available as standard subroutines on most computers, provide us with a uniform distribution of random numbers in the interval $(0, 1)$. For the two random numbers U_1 and U_2, it can be verified (Box and Mueller, 1958) that the two numbers

$$v_1 = \sqrt{T^*}(-2\ln U_1)^{1/2}\cos(2\pi U_2)$$
$$v_2 = \sqrt{T^*}(-2\ln U_1)^{1/2}\sin(2\pi U_2) \tag{5.7}$$

are two desired random components of velocity. This procedure can be used to assign initial velocity components to all the particles.

Placing the particles at random positions, as above, turns out to be disastrous for the following reason. Some particles will be placed too close to each other giving rise to very large repulsive forces, which will give rise to unusually large accelerations, forcing the particles to move rapidly away from each other. To circumvent this, it is customary to start the particles in a closed-packed lattice configuration, such as the face centered cubic (fcc) lattice. In this scheme, particles are placed at each of the eight vertices and at the centers of each of the six faces of a unit cube. Then the unit cube is repeated to generate the entire system of particles in what is called a simulation box. The number of particles in the system is then $4 \times \text{integer}^3 = N$, which explains why it is common to use $N = 256, 500, 864, \ldots$. This placement ensures that the particles are evenly distributed, thus avoiding strong repulsions at the start of the simulation.

The origin of the system is chosen at one edge of the simulation box. This fixes the coordinates of all the particles at the start. We call this the edge coordinate system. In order to apply periodic boundary conditions, it is convenient to shift the origin to the center of the box. We call this the center coordinate system. If the coordinates of a given point are (x_e, y_e, z_e) and (x_c, y_c, z_c) with respect to the edge and the center coordinate systems, respectively, one may transform between the two systems using relations of the type

$$x_c = x_e - x_{co} \tag{5.8}$$

where (x_{co}, y_{co}, z_{co}) are the coordinates of the origin of the center coordinate system with respect to the edge coordinate system. Coming back to the edge coordinate system, the whole fcc lattice may be generated by using primitive unit vectors, which are given by

$$\mathbf{e}_1 = (\mathbf{j} + \mathbf{k})/2 \qquad \mathbf{e}_2 = (\mathbf{k} + \mathbf{i})/2 \qquad \mathbf{e}_3 = (\mathbf{i} + \mathbf{j})/2 \tag{5.9}$$

in terms of the unit vectors of the edge coordinate system. Any lattice point of the fcc lattice is given by

$$\mathbf{l} = n_1 \mathbf{e}_1 + n_2 \mathbf{e}_2 + n_3 \mathbf{e}_3 \tag{5.10}$$

where n_1, n_2, n_3 are integers, in general. For the edge coordinate system, they are positive integers or zero.

5.4.2 Predictor–Corrector Method

Knowing the initial positions $\mathbf{r}_i(0)$ and velocities $\mathbf{v}_i(0)$ of each of the particles, the subsequent positions and velocities can be calculated by integrating Eq. (5.2). The integration is done numerically by using a finite time step. In principle, the various quantities after time δt, the step length, can be obtained from a Taylor's series expansion:

$$\mathbf{r}_i(\delta t) = \mathbf{r}_i(0) + \delta t \mathbf{v}_i(0) + \left(\tfrac{1}{2}\right)(\delta t)^2 \mathbf{a}_i(0) + \cdots \tag{5.11}$$

$$\mathbf{v}_i(\delta t) = \mathbf{v}_i(0) + \delta t \mathbf{a}_i(0) + \cdots \tag{5.12}$$

$$\mathbf{a}(\delta t) = \mathbf{a}_i(0) + \cdots \tag{5.13}$$

Everything on the right-hand side of these equations is known, because, using Eq. (5.1), the accelerations at $t = 0$ can be calculated from a knowledge of the initial forces, which depend on the initial positions. The calculated quantities here are called the predicted quantities.

Within the given approximation, the accelerations at δt are identical to the ones at $t = 0$. This result cannot be correct, since the particles have moved to new positions. This result can be corrected as follows. Calculate the correct accelerations $\mathbf{a}_i' = \mathbf{a}_i + \Delta_i$ using the new positions and use the new corrected accelerations to obtain the corrected quantities as follows:

$$\mathbf{r}_i'(\delta t) = \mathbf{r}_i(\delta t) + c_0 \Delta_i \tag{5.14}$$

$$\mathbf{v}_i'(\delta t) = \mathbf{v}_i(\delta t) + c_1 \Delta_i \tag{5.15}$$

$$\mathbf{a}_i'(\delta t) = \mathbf{a}_i(\delta t) + c_2 \Delta_i \tag{5.16}$$

This method is called the *predictor–corrector* algorithm, and the values of the constants c_i depend on the order of the differential equation and the order of the terms kept on the right-hand side of these equations. In our case, $c_2 = 1$. Tables of these coefficients may be found in Appendix E of Allen and Tildesley (Allen and Tildesley, 1989, Appendix E).

One predictor and one corrector together make one step of the simulation. In principle, one may iterate the corrector to improve the accuracy and stability of the algorithm. In practice, this is not done for the following reason. Application of each corrector step requires the evaluation of forces on each of the particles. The calculation of forces is the most time-consuming part of the whole simulation process. Therefore, the evaluation of forces is kept to a minimum. In fact, many tricks are used to reduce the number of force evaluations.

To see why there are so many forces present in a molecular dynamics simulation, just consider that each particle interacts with all the others, so that $N(N - 1)/2$ force evaluations are needed. For a 864-particle system, this number is about 3.7×10^5. Furthermore, to minimize the effects of a finite simulation box, one must also introduce periodic boundary conditions. The idea of periodic boundary conditions is to replace the finite box by an infinite box in such a way that the central box is repeated to infinity in all directions. Each of the repeated boxes, called *image boxes*, is identical to the central box. Hence, whatever the particles in the central simulation box do is followed by the particles in the identical image boxes surrounding the central box. In principle, we have to keep track of an infinite number of particles. We also have to include the interactions of a given particle with all the image particles. The whole problem gives an infinite number of two-body forces that must be calculated. Obviously, the number of two-body forces must be reduced to a manageable number.

In practice, one argues that the forces from distant particles are small and constant, the largest effects coming from short range. Therefore, one introduces an arbitrary cutoff distance r_c^* into the problem. The forces are evaluated between a given particle and all its neighbors lying within a sphere of radius r_c^* surrounding it. The results for thermodynamic averages can be corrected for the missing part of the potential by using the exact results such as Eqs. (4.18), (4.21), and (4.23), where the radial correlation function may be assumed to be unity in the region beyond the cutoff. For examples of such corrections, see Allen and Tildesley (Allen and Tildesley, 1989, p. 65). The spherical cutoff convention reduces the number of force evaluations, since only the volume $4\pi r_c^{*3}/3$ is being considered instead of the full volume, L^{*3}, of the cubic box. The cutoff distance is typically set to $L^*/2$ or less.

What happens if a particle leaves the central simulation box? Because of periodic boundary conditions, another particle enters the central box from the opposite side so that the number of particles is conserved in each image box. In practice, if a particle has left the central box, a suitable multiple of the length L^* is subtracted or added in order to bring it back into the simulation box. It is rarely necessary to add or subtract L^* more than once, because the time steps are usually so small that the particles do not go very far in each step. Once again, it is argued that distant particles can be neglected, and their effect can be taken into account in an average way. With this in mind, in addition to the spherical cutoff convention mentioned above, a minimum image convention is also adopted. According to this convention, only the nearest images of a given particle must be considered in the evaluation of forces. The two conventions taken together drastically reduce the number of force evaluations required in each simulation step.

Rather than calculate the distances between all pairs of particles to see if they are greater or smaller than the cutoff, the following *neighbor list* scheme, originally due to Verlet (Verlet, 1967), is used. A sphere of radius slightly larger than the cutoff distance is drawn. It is slightly larger so that the particles crossing the cutoff sphere can be properly accounted for. A list of particles in the sphere is maintained, and only the interactions between the given particle and its neighbors are considered. Of course, as time goes by, particles may enter and leave the sphere. Therefore, the list of neighbors has to be renewed at suitable intervals. The effort needed to do this is much less than that needed to check every single pair of particles.

5.4.3 Equilibration

A molecular dynamics simulation consists of stepping the system through its microstates. Ideally, one would like to keep the step size as small as possible, since, in the limit of zero step size, the difference equations reduce to the exact differential equations. The error introduced by keeping a finite step size is called the *truncation error*. Another type of error is introduced, because the computers deal with only integers and not real numbers, although, depending on the precision used (single, double, quadruple), a real number may be represented quite accurately. The error introduced in this way is called the *roundoff error*. One would like to keep the roundoff error to a minimum by using higher precision, but that is time consuming. As expected, the truncation error dominates at large time steps, and the roundoff error becomes important at short time steps. Consequently, a certain intermediate time step has to be used. Typical time steps are about one-hundredth to one-thousandth of t^*, which is the characteristic time of the system.

While the system is evolving, it is necessary to monitor its total energy. The total energy is strictly conserved in an exact solution of Newton's equations, but the energy usually drifts because of various errors in the simulation. A check on the potential energy of the system can tell us if the particles are stuck in certain potential minima. The kinetic energy of the particles should be monitored to see if they are repelling one another so strongly that the temperature rises. In such a case, the system must be cooled by lowering or scaling down the velocities of all the particles.

In the end, after a sufficient number of steps, it is presumed that equilibrium has been reached. A necessary condition is that the various macroscopic quantities should be stable and fluctuate about their mean values. But this is not a sufficient condition, because the particles may be stuck in a small region of phase space. Worse, they may be in a solid phase, where they are performing small oscillations. There are many ways of distinguishing between a solid or liquid system. We mention two. One is the calculation of the positional-order parameter

$$\frac{1}{N} \sum_i \cos(\mathbf{k}^* \cdot \mathbf{r}_i^*) \tag{5.17}$$

where $\mathbf{k}^*$ is a reciprocal lattice vector. For the fcc lattice, one may choose $\mathbf{k}^* = (2\pi l^*)(-1, 1, -1)$ where $l^* = (4V^*/N)^{1/3}$ is the size of the unit cell. It can be shown that the order parameter is unity for a solid, whereas it fluctuates around zero with amplitude of the order of $N^{-1/2}$ in a liquid (see Verlet, 1967). Another is the calculation of the mean-square displacement defined as

$$\frac{1}{N} \sum_i \langle [\mathbf{r}_i^* - \mathbf{r}_i(0)]^2 \rangle \tag{5.18}$$

a quantity that is proportional to the time and the diffusion constant [see Eq. (9.22)] in liquids.

Once equilibrium has been reached, one may make a production run to calculate all the quantities of interest as follows. The kinetic energy, potential energy, and total energy can be obtained by averaging the expressions for these quantities over the trajectory. The temperature can be obtained from the equipartition theorem

$$\sum_i \langle \mathbf{p}_i^{*2} \rangle = N_d T^*$$
(5.19)

where N_d is the number of degrees of freedom. In our case, N_d is equal to $3N - 3$, since the three momentum components of the center of mass are fixed at zero. An instantaneous kinetic temperature, $\mathcal{T}$, may be defined by the relation

$$\sum_i \mathbf{p}_i^{*2} = N_d \mathcal{T}^*$$
(5.20)

which can be monitored as needed. Similarly, an instantaneous pressure, $\mathcal{P}$, may be defined (Cheung, 1977) as

$$\mathcal{P}^* = \frac{N\mathcal{T}^*}{V^*} - \frac{1}{3V^*} \sum_i \sum_{j>i} r^* \frac{d\phi^*}{dr^*}$$
(5.21)

which also can be monitored throughout the simulation. Its average over the trajectory gives the thermodynamic pressure. Fluctuations of various quantities give the expansion coefficients, specific heats, and the compressibilities. Detailed discussions of such relations can be found in Allen and Tildesley (Allen and Tildesley, 1989, Section 2.5). As an example, the specific heat at constant volume and the isothermal compressibility can be found from Eqs. (1.98) and (1.99), respectively. Correlation functions can be found using their definitions. For example, the radial distribution function may be found by a simple counting of the average number of particles in various spherical shells surrounding a given particle.

There are certain quantities that do not depend on the trajectory alone and, hence, cannot be calculated this way. A famous example is the entropy, which depends on the total number of microstates in the system. A MD or MC calculation never goes through all the microstates, only through a finite subset. So, entropy and the various free energies can never be calculated directly. However, many indirect methods are available. One method simply is to integrate the specific heat to obtain the entropy up to a constant. For example,

$$S^* = \int_{T_0^*}^{T^*} \frac{C_V^*}{T^*} dT^* + S_0^*$$
(5.22)

where S_0^* is the value of the reduced entropy at the reference temperature T_0^*. This reference temperature may be chosen in the regime where the properties of the system are known accurately by other means. Accordingly, the reference state may be at a very low or very high temperature, which still requires that the simulation be done over a range of temperatures and is very time consuming. Of course, once the entropy is known, the various free energies can be found by using thermodynamic relations. An important method called *umbrella sampling*, which avoids these problems, will be discussed in Chapter 6.

A FORTRAN program to simulate a one-dimensional Lennard–Jones fluid is given in Appendix D.6.

5.5   OTHER TECHNIQUES

More complicated systems may be simulated by generalizing these basic ideas. Molecular systems are usually dealt with by considering them to be made of interacting sites. Rotational equations of motion can be included using the principles of rigid-body motion. By assuming the sites to interact with intramolecular forces, vibrations can be studied. By associating suitable charges with the sites, polarizable molecules, such as water, can be studied. Of course, long-range interactions, such as the Coulomb potential, bring their own problems. Two important methods of dealing with them are the *Ewald summation* method and the *reaction field* method. In both methods, the range of forces is divided into two parts. In the first method, the point charges are replaced by continuous distributions and the difference is estimated by going into reciprocal space. In the second method, the forces are summed exactly up to a cutoff distance and the effect of the rest of the system is estimated by calculating the reaction field introduced by Onsager in his studies of dielectric systems.

Other ensembles are studied by modifying the equations of motion. For example, the constant NVT canonical ensemble is obtained as follows. An extra degree of freedom corresponding to the bath is introduced and the equations of motion modified in such a way that the system of interest exchanges energy with the bath degree of freedom, and the temperature remains fixed (see Nosé, 1984). Alternatively, equations of motion can be modified to incorporate a Lagrange multiplier so that the instantaneous temperature of the system at each step of the simulation can be kept constant (Evans and Morriss, 1983). Similar techniques can be used for simulating the grand canonical ensemble.

In principle, nonequilibrium quantities, such as the diffusion constant and viscosity, can be obtained from equilibrium averages of time correlation functions. Such relations are common in linear transport theory (see Chapter 10). But there is a problem. Steady-state quantities, such as the diffusion constant, depend on the long-time behavior of the correlation functions. Because of systematic and statistical errors that enter any computer simulation, calculated quantities are not accurate for long simulation times. For this reason new methods, called *Nonequilibrium Molecular Dynamics*, have been devised to deal with such systems. In these methods, one prepares the system in a state far from equilibrium by introducing an external field and watches the system behavior with time. The response of the system to these external constraints in a steady state yields the values of the nonequilibrium quantities. For examples, see Allen and Tildesley (Allen and Tildesley, 1989, Chapter 8). Nonequilibrium molecular dynamics has allowed studies of phenomena like shear flow and heat flow.

5.6   EXERCISES AND PROBLEMS

1. Calculate the reduced isothermal compressibility, the reduced specific heat at constant volume, the reduced volume expansion, and the reduced diffusion constant in terms of the unreduced quantities and the parameters of the Lennard–Jones potential.

2. Show that the quantities v_1 and v_2 in Eq. (5.7) satisfy the distribution in Eq. (5.6).

3. Write a computer program to solve the one-dimensional harmonic oscillator problem with the classical Hamiltonian

$$\mathcal{H} = \frac{p^2}{2m} + \frac{1}{2}m\omega^2 x^2$$

by solving the relevant differential equation, with the boundary conditions $x = 0$ at $t = 0$. Use reduced units and the predictor–corrector method. Plot the numerical solution and the exact solution on the same graph for comparison.

4. Calculate the number of force evaluations needed for a 864-particle system if the spherical cutoff distance is one-half the length of the simulation box.

6

MONTE CARLO METHODS

6.1 INTRODUCTION

In Chapter 5, we provided a brief comparison of analytical versus numerical techniques in studying liquids. We also provided a brief early history of the two main computational methods for such studies. In this chapter, the reader is introduced to the powerful Monte Carlo method for the generation of microstates by the use of stochastic methods such as the Markovian chain. In particular, the Monte Carlo method is applied to the Ising model. Finally, the important Monte Carlo renormalization group is introduced. Two FORTRAN programs demonstrate Monte Carlo calculations.

6.2 SIMPLE MONTE CARLO

6.2.1 Random Experiments

When casting a die, it is assumed that each face has an equal probability, $\frac{1}{6}$, of appearing. This assumption can be checked by performing a random experiment. A *random experiment* is one whose possible outcomes are known but for which the outcome of any given experiment is uncertain. Thus, a die can be cast many times and the results tabulated. Eventually, of course, as the number of trials becomes large the probabilities become obvious. Games of chance are good examples of random experiments. Perhaps the most famous gaming establishment is the casino at Monte Carlo. Basically, the games at Monte Carlo generate random numbers. The generation of random numbers is useful for performing mathematical operations, especially when analytical evaluation is not possible and complete numerical analysis is impractical. The solution of numerical problems by the use of random numbers is known as the Monte Carlo (MC) method.

6.2.2 Numerical Integration

The use of random numbers to solve mathematical problems can be illustrated by the evaluation of an integral. Consider the integral I of a function of x.

$$I = \int_{x_0}^{x_n} f(x)\, dx \tag{6.1}$$

The integral can be calculated numerically by performing the following summation

$$I \approx \frac{1}{n}(x_n - x_0) \sum_{i=0}^{n-1} f(x_i) \tag{6.2}$$

where the range of x extends from x_0 to x_n. Monte Carlo integration consists of the generation of random values of x_i from a random number generator. For each random value of x_i, a numerical value is calculated for $f(x_i)$. The integral is then calculated according to

$$I \approx \frac{1}{N}(x_n - x_0) \sum_{i}^{N} f(x_i) \tag{6.3}$$

for N random numbers, where $N \gg n$. For ordinary integrals, the MC method is a poor one. For example, consider the integral

$$\int_{-4}^{4} \frac{dx}{1 + x^2} \tag{6.4}$$

This integral is readily evaluated analytically to give a value of $2 \tan^{-1} 4 \approx 2.6516$. Use of Eq. (6.2) using equally spaced values of x_i ($n = 32$) gives a value of 2.6514. The MC method using the FORTRAN program MCINT in Appendix D.10 gives values in the range 2.64–2.65 for $N = 10{,}000$. Clearly, for such integrals there is no advantage to using MC integration, which gives at best three- or four-figure accuracy. However, for multiple integrals involving complicated regions of integration, where the usual numerical integration techniques are awkward or require an inordinate amount of computer time,

MC integration is the only suitable method. The advantage of MC integration is in the speed of random number generators, which are now available to microcomputers as well as to main frame computers. In the case of liquids, apart from the molecular dynamics method, MC is the only available choice for evaluating the configurational integrals (or the multidimensional averages) that occur in the theory of liquids (see Chapter 4).

6.3   MARKOVIAN CHAINS

6.3.1   Importance Sampling and the Metropolis Method

In evaluating the integral in Eq. (6.4) above, the random number sampling is what is known as uniform random sampling. This type of sampling is discussed in connection with Brownian motion (see Chapter 9). In the case of configurational averages, *importance sampling* is necessary (Metropolis et al., 1953). Importance sampling has its origins in the work of Metropolis et al., who describe a modified MC integration over a configuration space using the MANIAC computer at Los Alamos. The object is to calculate the equilibrium average values of physical observables for a system composed of particles (spheres), between which one can assume an intermolecular potential (see Chapter 4). In particular, only configurational averages are considered, the effect of kinetic energy being considered at the end of the calculation.

If G is a physical observable, the configurational average is

$$\langle G \rangle = \int G(\mathbf{r})\rho^{(N)}(\mathbf{r})\,d\mathbf{r} \tag{6.5}$$

where $\mathbf{r}$ represents a point in configuration space, and the distribution function $\rho^{(N)}(\mathbf{r})$ is defined in Section 4.2. In the MC method, the configuration space is discretized, each point representing a microstate of the system. In the canonical ensemble,

$$\langle G \rangle \approx \frac{1}{Q_N} \sum_{i=1}^{M} G(i)\exp[-\beta\Phi(i)] \tag{6.6}$$

$$Q_N \approx \sum_{i=1}^{M} \exp[-\beta\Phi(i)] \tag{6.7}$$

where Φ is the potential energy, Q_N is the configurational partition function, and the sum is over all states M. Of course $\beta = 1/k_{\mathrm{B}}T$ as usual. The simple MC method places the N particles randomly in a cube (square for two dimensions) with periodic boundary conditions. Then the potential energy $\Phi(1)$ is calculated for this initial configuration (state). Next, the Boltzmann factor $\exp[-\beta\Phi(1)]$ is evaluated, giving the first term in Q_N. The configuration is again randomized and the calculation repeated. This process is continued till Q_N is obtained to the desired accuracy. The same process is repeated for G. However, this method, called *uniform sampling*, which weights all microstates equally, is not the best method for condensed systems, because a very large number of states must be summed over, thereby requiring a large expenditure of computing time. Furthermore, a large percentage of the states chosen are high-energy configurations, which contribute little to the average of many observables (e.g., internal energy) but which still require the evaluation of an exponential function, adding greatly to the computing cost with very little benefit in accuracy.

Metropolis et al. (1953) solved this problem by first choosing only those configurations with a high Boltzmann-factor probability and then weighting them evenly. This solution is an application of the general method known as importance sampling. The MC method with importance sampling typically uses a *Markovian chain* for the purpose of sampling the microstates. It can be shown that under suitable conditions this method asymptotically generates canonical ensemble averages (Wood, 1968, 1970).

6.3.2 Transition Probability Matrix

A Markovian chain is a discrete Markovian process. Markovian processes are more fully discussed in Chapter 9. For now, we point out that the Markovian property of a discrete random process, such as a random walk, means that the probability of the jth event occurring, conditioned on a set of past events, depends only on the probability of occurrence of the $j-1$th event. We consider an event to be a transition between states of the system. Let the system have accessible to it M states. Initially, the probability of finding the system in the ith state is $p_i^{(0)}$. The probability of occurrence of each state will change when the system undergoes a one-step transition according to the equation

$$p_j^{(1)} = \sum_i P_{ji} p_i^{(0)} \tag{6.8}$$

where the transition probabilities P_{ji} are the elements of the transition probability matrix $\mathbf{P}$, which is the matrix that characterizes the Markovian chain. For a two-step transition,

$$p_j^{(2)} = \sum_k P_{jk} p_k^{(1)} = \sum_{k,i} P_{jk} P_{ki} p_i^{(0)} = \sum_i [\mathbf{P}^2]_{ji} p_i^{(0)} \tag{6.9}$$

where $\mathbf{P}^2$ stands for the square of the matrix $\mathbf{P}$, as usual. In general, for n steps

$$p_j^{(n)} = \sum_i [\mathbf{P}^n]_{ji} p_i^{(0)} \tag{6.10}$$

If the Markovian chain is *ergodic* (every state is visited in the limit $n \to \infty$) and *irreducible* (any state can be reached eventually from any other state), the final state is independent of the initial state and

$$p_j^{(\infty)} = \lim_{n \to \infty} \sum_i [\mathbf{P}^n]_{ji} p_i^{(0)} \tag{6.11}$$

This final state is said to be a *limiting state*.

6.3.3 Example: The Random Walk

Consider a three-step, one-dimensional random walk. Let p and q be the probabilities of steps to the right and left, respectively. For a random walker beginning at the origin, Eq. (6.10) is

$$
\begin{pmatrix} p_{-3}^{(3)} \\ p_{-2}^{(3)} \\ p_{-1}^{(3)} \\ p_0^{(3)} \\ p_1^{(3)} \\ p_2^{(3)} \\ p_3^{(3)} \end{pmatrix}
=
\begin{pmatrix}
0 & q & 0 & 0 & 0 & 0 & 0 \\
p & 0 & q & 0 & 0 & 0 & 0 \\
0 & p & 0 & q & 0 & 0 & 0 \\
0 & 0 & p & 0 & q & 0 & 0 \\
0 & 0 & 0 & p & 0 & q & 0 \\
0 & 0 & 0 & 0 & p & 0 & q \\
0 & 0 & 0 & 0 & 0 & p & 0
\end{pmatrix}^3
\begin{pmatrix} 0 \\ 0 \\ 0 \\ 1 \\ 0 \\ 0 \\ 0 \end{pmatrix}
=
\begin{pmatrix} q^3 \\ 0 \\ 3pq^2 \\ 0 \\ 3p^2q \\ 0 \\ p^3 \end{pmatrix}
\tag{6.12}
$$

If $p = q = \frac{1}{2}$, the last column of Eq. (6.12) becomes ($\frac{1}{8}$, 0, $\frac{3}{8}$, 0, $\frac{3}{8}$, 0, $\frac{1}{8}$) in agreement with the binomial distribution of random walk theory (see Section 1.2.1). For a large number of steps, where $p = q = \frac{1}{2}$, the distribution is symmetric, and the limiting distribution can be shown to be Gaussian.

The random walk Markovian chain is said to have period 2, because it takes at least two steps to return to the initial state. Much of Markovian chain theory is concerned with *aperiodic chains*, in which the initial state never recurs. The interested reader is referred to textbooks on probability theory and stochastic analysis (e.g., Feller, 1968, Vol. I, Chapters XV, XVI). In the remainder of this chapter, all Markovian chains will be assumed to be aperiodic.

6.4   EXAMPLE OF METROPOLIS MONTE CARLO

6.4.1   The Ising Model

The method of Metropolis et al. will be applied to the Ising model, which was introduced in Chapter 3. Here, we discuss the two-dimensional lattice, which exhibits a phase transition from a system of disordered spins at high temperatures to a ferromagnetic system of ordered spins. This problem has been solved exactly by Lars Onsager. Because the two-dimensional model has been solved exactly, it is a good system on which to test our MC method.

The Hamiltonian in the Ising model is

$$\widehat{\mathcal{H}} = -J \sum_{n.n.} \widehat{S}_i \widehat{S}_j \tag{6.13}$$

where J is a spin-coupling parameter, taken to be greater than 0 for interaction between spins S_i located at the nearest neighbor ($n.n.$) lattice sites. The energy of the Ising model for a given configuration of spins is

$$E = -J \sum_{n.n.} s_i s_j \tag{6.14}$$

where s_i and s_j are eigenvalues of the spin operator with values $+1$ or -1, and, as before, the spins are measured in units of $\hbar/2$. If all spins are in the same direction (ferromagnetic ground state), $E = -2NJ$ for a square lattice with periodic boundary conditions, where N is the number of lattice sites. If all spins alternate directions (antiferromagnetic ground state), $E = +2NJ$.

The calculation of ensemble averages and the study of phase transitions using the Ising model is really nothing but a combinatorial problem. In particular, if N_{ud} is the number of neighbors with spins in opposite directions, we see that the expression for the sum in Eq. (6.14) can be written

$$\sum_{n.n.} s_i s_j = 2(N - N_{ud}) \tag{6.15}$$

Therefore, the energy for a given configuration is

$$E = -2J(N - N_{ud}) \tag{6.16}$$

and the average energy per spin is

$$\frac{\langle E \rangle}{N} = -2J \left(1 - \frac{\langle N_{ud} \rangle}{N} \right) \tag{6.17}$$

which depends on the average number of antiparallel pairs. Obviously, this reduces to $-2J$ or $+2J$ if the system is in the ferromagnetic or the antiferromagnetic ground state, respectively.

In order to illustrate the application of the MC technique, we shall show how to calculate the average energy per spin in the Ising model. For this purpose, it is convenient to introduce the quantity $f_c = N_{ud}/2N$, which is the fraction of spin pairs that are antiparallel for a given configuration of spins. As can be verified readily using Eq. (6.15),

$$f_c = \frac{1}{4N} \sum_{n.n.} (1 - s_i s_j) \tag{6.18}$$

The average of f is defined as

$$\langle f \rangle = \frac{1}{2^N} \sum_c f_c \tag{6.19}$$

where the sum is over all configurations of spins obtained from the canonical distribution and f_c is the value of f for a given configuration denoted by c. The total number of configurations in Eq. (6.19) is 2^N, since each spin variable can have two values, $+1$ or -1. By using Eqs. (6.17) and (6.19), the energy per spin can be calculated from

$$\frac{\langle E \rangle}{N} = -2J(1 - 2\langle f \rangle) \tag{6.20}$$

6.4.2 Metropolis Monte Carlo Applied to the Ising Model

The Metropolis method consists of the following steps:

1. For a given temperature T, set up N particles on a square grid and number the particles (sites).
2. Set up the initial configuration by arbitrarily allotting spins to the particles, which can be done randomly or the totally parallel or antiparallel configuration can be chosen.
3. Count the number of antiparallel spins N_{ud} using periodic boundary conditions.
4. Calculate the configurational energy E_1 from Eq. (6.17).
5. Reverse one of the spins. The site can be selected randomly or chosen to be site 1. For the sake of concreteness, we assume ordered site selection.
6. Repeat Step 3 and calculate E_2.

7. Calculate $\Delta E = E_2 - E_1$ and the Boltzmann factor $\exp(-\beta\Delta E)$, which represents the relative probability of the occurrence of the two states in the physical system.

8. If $\Delta E \leq 0$, the spin is left reversed. If $\Delta E > 0$, a random number ν between 0 and 1 is selected. If $\nu \leq \exp(-\beta\Delta E)$, the spin is left reversed; otherwise it is returned to its previous value. Because the Boltzmann factor is used to decide the acceptance or rejectance of the new state, this procedure is called *Boltzmann sampling*.

9. Calculate f_1 for this configuration according to Eq. (6.18).

10. Repeat Steps 5–9 for the next site.

11. Add f_2 to f_1.

12. Repeat Steps 10 and 11 till all sites in the lattice have been visited. At this point f has been evaluated for N out of 2^N configurations. If desired, $\langle f \rangle$ can be evaluated by dividing by the number of configurations in the summation. This approaches Eq. (6.19) as the number of configurations approaches 2^N.

13. Repeat Steps 3–12 till $\langle f \rangle$ is obtained to the desired accuracy.

Step 8 is the heart of the Metropolis MC method. It is in this step that the Markovian chain with Boltzmann sampling is used to generate the next state. The Markovian chain is completely defined if the transition probability matrix **P** is given. Let us assume that the transition is from the kth state to the lth state. Then, we need to specify only the four elements P_{kk}, P_{ll}, P_{lk}, and P_{kl} of the matrix **P**, and Eq. (6.8) for the one-step transition becomes

$$p_k^{(n)} = P_{kk}p_k^{(n-1)} + P_{kl}p_l^{(n-1)} \tag{6.21}$$

for the kth state and

$$p_l^{(n)} = P_{lk}p_k^{(n-1)} + P_{ll}p_l^{(n-1)} \tag{6.22}$$

for the lth state. Also, for the transition from the kth state to the lth state, $p_l^{(n-1)} = 0$. This means that we need consider only P_{kk} and P_{lk} in Eqs. (6.21) and (6.22). According to the Metropolis MC method, these two one-step transition matrix elements are functions of the canonical distributions $\rho_k = \exp(-\beta E_k)/Z$ and $\rho_l = \exp(-\beta E_l)/Z$ and depend on whether $\rho_l \geq \rho_k$ or $\rho_l < \rho_k$. If $\rho_l \geq \rho_k$, $P_{lk} = 1$ and $P_{kk} = 0$. If $\rho_l < \rho_k$, the Boltzmann weighting applies, and $P_{lk} = \rho_l/\rho_k = \exp(-\beta\Delta E)$ and $P_{kk} = 1 - \rho_l/\rho_k = 1 - \exp(-\beta\Delta E)$. Of course, $\Delta E = E_l - E_k$. Briefly, the rule is always to accept a move resulting in the decrease of energy, and if the move results in an increase in energy, to accept it with the physically correct probability

$\exp(-\beta\Delta E)$. In other words, the moves are accepted with a probability that is the smaller of 1 and $\exp(-\beta\Delta E)$. Notice that it is never necessary to evaluate the partition function, since only ratios of statistical distribution functions are involved in the transition matrix. This observation is important, since numerical evaluation of partition functions is often not practical owing to the large number of states in the partition function.

A FORTRAN program ISING, which follows Steps 1–13 above for the calculation of the average energy of the two-dimensional Ising model, is given in Appendix D.5. The use of this program is illustrated in Figure 6.1 for a 20×20 lattice and $\beta J = 0.112$, which is for the paramagnetic phase. For convenience, the dimensionless quantity $\langle E \rangle / NJ$ is plotted. The initial configuration is one of completely parallel spins. The spins are varied in an ordered sequence beginning with site 1. An iteration consists of 400 sites. After about 100 iterations the dimensionless average energy per spin begins to approach its true value of –0.229 based on Onsager's solution (see Pathria, 1996, Section 12.3). The 500 iterations shown represent 200,000 configurations out of a possible $2^{400} \approx 2.58 \times 10^{120}$. Other values of βJ and N can be tried by the reader and compared with the results found in (Pathria, 1996, p. 384). The Boltzmann-sampling method discussed in this

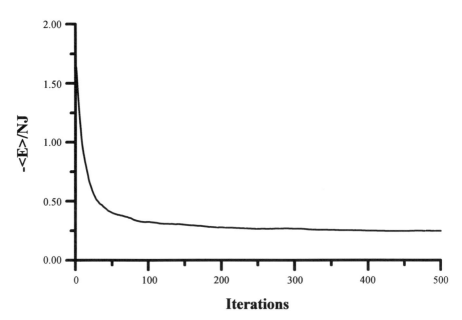

Figure 6.1   Ising model dimensionless energy per spin $\langle E \rangle / NJ$ for $\beta J = 0.112$.

section cannot be used to calculate the entropy or free energies, because that involves calculating the partition function. As explained in Chapter 5, special techniques have to be used for such quantities. One such method is discussed in Section 6.5.

6.5   NON-BOLTZMANN SAMPLING

6.5.1   Helmholtz Free Energy

The Helmholtz free energy is

$$A = -k_B T \ln Z \tag{6.23}$$

where Z is the canonical partition function, as usual. As noted above, Z cannot be evaluated numerically for most systems of interest. Hence, a direct calculation of A is not possible. However, if the free energy A^0 is known for some reference system, it is possible to evaluate A for a system that differs only slightly from the reference system, say by a perturbation. In this case,

$$A - A^0 = -k_B T \ln(Z/Z^0) \tag{6.24}$$

and the problem becomes one of evaluating a ratio of partition functions rather than the partition functions themselves. Accordingly, we focus attention on the evaluation of Z/Z^0.

If the perturbed system and the reference system differ only in the potential energy, as is usually the case, we need consider only configurational partition functions Q_N and Q^0 (see Chapter 4). In this case, the ratio Q_N/Q^0 is

$$\frac{Q_N}{Q^0} = \frac{\int \exp[-\beta\Phi(\mathbf{r})]d\mathbf{r}}{\int \exp[-\beta\Phi^0(\mathbf{r})]d\mathbf{r}} = \frac{\int \exp[-\beta\Phi(\mathbf{r}) + \beta\Phi^0(\mathbf{r})]\exp[-\beta\Phi^0(\mathbf{r})]d\mathbf{r}}{\int \exp[-\beta\Phi^0(\mathbf{r})]d\mathbf{r}}$$

$$= \int \rho^0(\mathbf{r})\exp[-\beta\Delta\Phi(\mathbf{r})]d\mathbf{r} = \langle\exp[-\beta\Delta\Phi(\mathbf{r})]\rangle_0 \tag{6.25}$$

where $\rho^0(\mathbf{r})$ is the normalized canonical distribution function for the reference system and $\Delta\Phi(\mathbf{r}) = \Phi(\mathbf{r}) - \Phi^0(\mathbf{r})$. The subscript 0 on the ensemble average emphasizes that the averages are taken with respect to the reference system. If the Metropolis MC method is used to evaluate $\langle\exp(-\beta\Phi)\rangle_0$, the result will be dominated by those parts of configuration space for which $\rho^0(\mathbf{r})$ is large. These are not necessarily the parts of configuration space that contribute most significantly to $\langle\exp(-\beta\Phi)\rangle_0$, because Eq. (6.25) is a product of $\rho^0(\mathbf{r})$ and $\exp[-\beta\Delta\Phi(\mathbf{r})]$, and, in general, these two functions will not peak in the same region of configuration space. The situation here is that parts of configuration space with small Boltzmann weighting may in fact contribute significantly to this average. Hence, there is a need for non-Boltzmann weighting when determining this average by a MC calculation.

6.5.2 Umbrella Sampling

One type of non-Boltzmann weighting is called *umbrella sampling* (Allen and Tildesley, 1989, Chapter 7). Umbrella sampling is more easily discussed if Q_N/Q^0 is integrated over $\Delta\Phi(\mathbf{r})$ rather than over $\mathbf{r}$.

$$\frac{Q_N}{Q^0} = \int \rho^0(\Delta\Phi)\exp(-\beta\Delta\Phi)d(\Delta\Phi) \tag{6.26}$$

Equation (6.26) is obtained from Eq. (6.25) by writing Eq. (6.25) as a sum over states and then converting to an integral over energy and multiplying by the density of states. The density of states has been incorporated into $\rho^0(\Delta\Phi)$ of Eq. (6.26). In order to get non-Boltzmann sampling, we weight the statistical distribution function $\rho^0(\Delta\Phi)$ in such a manner that it samples a much wider range of $\Delta\Phi$ than would $\rho^0(\Delta\Phi)$ alone. We are dealing here with a Markovian chain whose transition probability matrix elements for a transition from state k to state l are functions of the distribution functions ρ_k and ρ_l of Section 6.4.2 but with weighting factors w_k and w_l, respectively, such that $\rho_l = w_l\rho_l^0$ and $\rho_k = w_k\rho_k^0$. The distribution functions for the two systems are related by

$$\rho(\mathbf{r}) = \frac{w\rho^0(\mathbf{r})}{\int w\rho^0(\mathbf{r})\,d\mathbf{r}} = \frac{w\rho^0(\mathbf{r})}{\langle w\rangle_0} \tag{6.27}$$

where the denominators are required for normalization. In terms of the weighted distribution functions, Eq. (6.26) becomes

$$\frac{Q_N}{Q^0} = \frac{\langle\exp(-\beta\Delta\Phi)/w\rangle_w}{\langle 1/w\rangle_w} \tag{6.28}$$

where the subscript w indicates that the averages are taken with respect to the non-Boltzmann distribution functions, and the denominator is required for normalization. Comparing Eqs. (6.25) and (6.28),

$$\langle\exp(-\beta\Delta\Phi)\rangle_0 = \frac{\langle\exp(-\beta\Delta\Phi)/w\rangle_w}{\langle 1/w\rangle_w} \tag{6.29}$$

The right-hand side of Eq. (6.29) is obtained from the weighted (non-Boltzmann) MC sampling by using steps similar to those outlined in Section 6.4.2 for the Ising model. In the crucial Step 8, the move is accepted with a probability that is the smaller of 1 and $\rho_l/\rho_k = (w_l/w_k)(\rho_l^0/\rho_k^0)$. This procedure allows $A - A^0$ to be evaluated.

The weighting function $w(\Delta\Phi)$ must be chosen such that the $\Delta\Phi$ values are large and negative. There is no a priori method for doing this; the choice of $w(\Delta\Phi)$ depends on the system of interest, which requires a knowledge of the perturbed system, so that the method works best if the perturbed and reference systems are very similar. We would like the weighting function to be as wide ranging as possible, forming an "umbrella" over the perturbed and reference systems. For this reason, this type of non-Boltzmann sampling is called umbrella sampling.

The MC method is now so widespread that it is difficult even for specialized texts to do justice to it. A recent review is provided by Binder (Binder, 1995). One area of great importance is that of the Monte Carlo renormalization group (MCRG).

6.6   MONTE CARLO RENORMALIZATION GROUP

6.6.1   The Ising Model

It is very difficult to study numerically the Ising model near the critical point for two reasons. First, to obtain accurate results a very large lattice is required. Second, the spin–spin correlations have a very long range near the critical point, thus requiring a large number of coupling constants.

It was realized by Ma (Ma, 1976) that the MC and renormalization group (RG) methods can be combined to study critical phenomena. The advantages of such a combination accrue from the fact that only local properties and not global ones need to be calculated. We now discuss Ma's method as modified by Swendsen (Swendsen, 1979). It is based on the RG theory presented in Chapter 3, and it depends on the system's being close to criticality.

6.6.2   MCRG

In Chapter 3, we saw that a RG transformation R_b is one that changes the coupling constant $K = -\beta J$ in the Ising model Hamiltonian upon rescaling of the spins by a factor b to a new set of spins. Critical properties can be obtained by expanding the RG transformation near the fixed point of interest. In this section, we describe one way of obtaining the critical exponent for the correlation length of the Ising model by using MC techniques combined with the RG.

The operator R_b operates directly on the coupling constant such that $K_b = R_b K$. In general, the Ising Hamiltonian can have several coupling constants. Assume that we have a set of coupling constants K that are very close to the set K^* such that

$$K = K^* + \delta K \tag{6.30}$$

where δK represents a small variation in K. Recall that the superscript asterisk on K indicates the set of coupling constants associated with the fixed point. For each coupling constant K_n in the set, Eq. (6.30) gives

$$K_n = K_n^* + \delta K_n \tag{6.31}$$

If a RG transformation is applied to both sides of Eq. (6.31), we get

$$K_{nb} = K_n^* + \delta K_{nb} \tag{6.32}$$

We note that Eq. (6.32) is in the form of a Taylor's series for the expansion of K_{nb} in terms of K_n about $K_n = K_n^*$

$$K_{nb} = K_n^* + \sum_m \left(\frac{\partial K_{nb}}{\partial K_m} \right) (K_m - K_m^*) + \text{higher-order terms} \tag{6.33}$$

where the summation is over all the coupling constants in the set. Equation (6.33) illustrates the fact that each renormalized coupling constant is a function of all the original coupling constants. If we ignore the higher order terms, we have a linearized RG transformation, which is good only in the vicinity of the fixed point, but which allows us to treat the transformation by the methods of linear algebra.

The linearized transformation matrix is $\mathbf{R}$

$$\delta \mathbf{K}_b = \mathbf{R} \delta \mathbf{K} \tag{6.34}$$

or

$$\delta K_{nb} = \sum_m R_{nm} \delta K_{mb} \tag{6.35}$$

where we treat $\delta \mathbf{K}_b$ and $\delta \mathbf{K}$ as vectors and $\mathbf{R}$ as a square transformation matrix. From the theorems of linear algebra, we know that $\mathbf{R}$ has a set of eigenvalues and eigenvectors associated with it

$$\mathbf{R}^{(b)}\mathbf{e}_i = \lambda_i \mathbf{e}_i \tag{6.36}$$

or

$$\sum_m R_{nm}^{(b)} e_{im} = \lambda_i e_{in} \tag{6.37}$$

Also,

$$\mathbf{R}^{(b^2)}\mathbf{e}_i = \mathbf{R}^{(b)}\mathbf{R}^{(b)}\mathbf{e}_i = \lambda_i \mathbf{R}^{(b)}\mathbf{e}_i = \lambda_i^2 \mathbf{e}_i \tag{6.38}$$

The eigenvalues are functions of b such that $\lambda_i(b)\lambda_i(b') = \lambda_i(bb')$, which puts restrictions on the form of the function $\lambda_i(b)$, keeping in mind that $\lambda_i(b)$ must be a scalar. The only function that will do this is one of the form $\lambda_i(b) = b^{a_i}$ where both b and a_i are scalars.

The vector $\delta\mathbf{K}$ can be written

$$\delta\mathbf{K} = \sum_i c_i \mathbf{e}_i \tag{6.39}$$

where

$$c_j = \mathbf{e}_j \cdot \delta\mathbf{K} \tag{6.40}$$

from the orthonormality of the $\mathbf{e}_i$ basis. Thus,

$$\delta\mathbf{K}_b = \mathbf{R}\sum_i c_i \mathbf{e}_i = \sum_i c_i \lambda_i \mathbf{e}_i = \sum_i c_i' \mathbf{e}_i \tag{6.41}$$

The c_i' are the elements of $\delta\mathbf{K}_b$ in the basis $\mathbf{e}_i$ (the projection of $\delta\mathbf{K}_b$ along $\mathbf{e}_i$). From Eqs. (6.34) and (6.36), it is seen that the elements of $\delta\mathbf{K}_b$ are dependent on the eigenvalues λ_i. The absolute value of λ_i is $|\lambda_i| = (b^{a_i} b^{a_i^*})^{1/2}$, where a_i^* is the complex conjugate of a_i. If we require that a_i be real, $|\lambda_i| = (b^{2a_i})^{1/2}$. If a_i is positive, $|\lambda_i|$ increases as b increases so that $|\lambda_i| > 1$. If a_i is negative $|\lambda_i|$ decreases as b increases so that $|\lambda_i| < 1$. If $a_i = 0$, $|\lambda_i| = 1$.

If $|\lambda_i| > 1$, the corresponding components of δK_b will increase and eventually dominate the vector after many transformations. As $\delta\mathbf{K}_b$ increases, $\mathbf{K}_b$ will move away from the fixed point. Hence, such vectors are not on the critical hypersurface. Eigenvalues and eigenvectors associated with such vectors are said to be *relevant*. For initial vectors on the critical hypersurface, $|\lambda_i| < 1$ and the vectors will move to the fixed point. These eigenvalues and eigenvectors are said to be *irrelevant*. In case $|\lambda_i| = 1$, the eigenvalues and eigenvectors are said to be *marginal*.

For pedagogical reasons, we now consider only one coupling constant. Equation (6.34) becomes

$$\delta K_b = R\delta K = \lambda\delta K = b^a\delta K \tag{6.42}$$

from which we have

$$R = b^a = \left(\frac{\partial K_b}{\partial K}\right)_{K^*} \tag{6.43}$$

The linear transformation R, and therefore the exponent a, can be found using the following chain rule:

$$\frac{\partial\langle S_b\rangle}{\partial K} = \left(\frac{\partial\langle S_b\rangle}{\partial K_b}\right)\left(\frac{\partial K_b}{\partial K}\right) \tag{6.44}$$

where

$$S = \sum S_i S_j \tag{6.45}$$

and

$$S_b = \sum S_J S_K \tag{6.46}$$

The angle brackets denote an ensemble average. The derivatives in Eq. (6.44) can be found by straightforward differentiation

$$\frac{\partial\langle S_b\rangle}{\partial K} = \langle S_b S\rangle - \langle S_b\rangle\langle S\rangle \tag{6.47}$$

and

$$\frac{\partial\langle S_b\rangle}{\partial K_b} = \langle S_b S_b\rangle - \langle S_b\rangle\langle S_b\rangle \tag{6.48}$$

The averages in Eqs. (6.47) and (6.48) are evaluated at $K = K^*$ by Metropolis MC methods. The transformation matrix obtained from Eq. (6.37) gives the eigenvalue λ and the critical exponent a directly according to Eq. (6.42). If the eigenvalue is relevant, the critical exponent ν for the correlation length

ξ (see Chapter 3) can be obtained from the relation $\nu = 1/a$, which is proved as follows.

The correlation length goes to infinity as the critical temperature T_c is approached from above

$$\xi \approx A \left(\frac{T - T_c}{T_c} \right)^{-\nu} = A \left(\frac{K_c - K}{K} \right)^{-\nu} = A \left(\frac{\delta K}{K} \right)^{-\nu} \tag{6.49}$$

where A has dimensions of length. Similarly, the renormalized correlation length near the renormalized critical temperature is given by

$$\xi_b \approx A \left(\frac{K_{cb} - K_b}{K_b} \right)^{-\nu} = A \left(\frac{\delta K_b}{K_b} \right)^{-\nu} \tag{6.50}$$

At the fixed point, $K_{cb} = K_c = K^*$. Also, the original and the renormalized correlation lengths are related by $\xi_b = \xi/b$. Therefore, Eqs. (6.49) and (6.50) give

$$\frac{\delta K}{K} = \frac{1}{b^{1/\nu}} \frac{\delta K_b}{K_b} \tag{6.51}$$

Now, taking the limits $K, K_b \to K^*$ gives

$$\delta K_b = b^{1/\nu} \delta K \tag{6.52}$$

which, on using Eq. 6.42, yields

$$R = b^{1/\nu} \tag{6.53}$$

The relation $\nu = 1/a$ follows upon using Eq. (6.43).

This example has shown how the MC method can be combined with RG methods to provide a powerful tool in the study of Ising-type phase transitions. According to Swendsen (Swendsen, 1982, pp. 57–86), a major advantage of using the MC method with the RG method is the possibility of systematically improving the calculations with very little additional effort. Unlike the RG procedure, which must make approximations by eliminating coupling constants, the MC procedure can, in principle, keep all coupling constants. There are, of course, approximations that must be made in the MCRG approach, especially the finite-lattice approximation. However, the MCRG approximations can be checked easily by considering larger lattices. In this way, the combination of RG with MC methods has proven to be especially fruitful.

6.7   CONCLUDING REMARKS

In conclusion, we discuss some of the recent developments in computing, which are relevant to the simulation of liquids. We introduce the concepts of *vectorization*, *parallel processing*, and *massive parallel processing*. Our discussion is brief. For more details on vectorization and parallel processing, see the articles by Landau (Landau, 1995) and Heermann and Burkitt (Heermann and Burkitt, 1995), respectively. For massive parallel processing, see Robinson et al. (Robinson et al., 1996, Chapter 11).

In many computer programs, one finds that identical calculations are done on a set of data, piece by piece. It is as if doors are being attached to a set of cars. Henry Ford solved this problem by using an assembly line. A similar idea is used in vector computers. It is called *pipelining*. In this scheme, a whole set of data is moved on the pipeline, the data moving forward after each operation is performed on it. A good example is the operation of a typical DO loop. In the usual computer, called a *scalar machine*, a DO loop is performed on dimensioned arrays (in FORTRAN) on a line-by-line basis. In a vector machine, which uses pipelining, a new computer architecture is created, such that the DO loop is executed once and only once. Just as in the car assembly line, no results appear till the pipeline is full. Hence, this method is not too useful for DO loops involving a small number of steps. However, once the pipeline has become full, one result per clock cycle appears in spite of the fact that many more clock cycles may be needed to complete an operation.

Another common computational operation independently treats different parts of the physical system. For example, in the MC method, spins are flipped one by one to see whether a result is acceptable or not. In the case of the Ising model, only nearest neighbors interact, so that, upon flipping spins, most of the lattice is untouched. This procedure can be speeded up by having many computer processor elements rather than just one. With multiple processors, different parts of the physical system may be dealt with in parallel. Such parallel processing is possible, because all the parts of the physical system are not changing at the same time, especially for short-range interactions. Consequently, parallel processing can reduce computer time considerably and is able to deal with very large systems for very long times. Note that the developments described here are independent of any changes in the inherent speed of the individual processors. If one has 1024 or more processors, one obtains massive parallel computing.

With all these developments, one day, perhaps, Wilson's dream (Chapter 5) may become a reality.

6.8   EXERCISES AND PROBLEMS

1. Verify Eq. (6.18).

2. For Figure (6.2), which represents a two-dimensional Ising model, determine the exchange energy assuming cyclic boundary conditions.

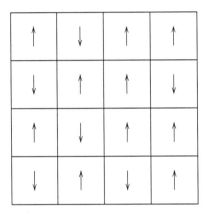

Figure 6.2   Two-dimensional Ising lattice.

3. Using the FORTRAN program MCINT in Appendix D.10 for the integral of the function $f(x) = 1/(1 + x^2)$, determine how many points must be sampled to get four-significant-figure accuracy between the limits of -4 and $+4$. How many points must be sampled to get two-significant-figure accuracy? By changing the program to read

```
x = -8.0+16.0*RAN2(IDUM)
AINT = AINT*16.0/N
```

determine the value of N for two-, three-, and four-significant-figure accuracy for integration between -8 and $+8$.

4. Use the FORTRAN program ISING in Appendix D.5 for $J/k_B T = 0.347$ and 0.602. How many iterations are necessary for the values of $\langle E \rangle/NJ$ to approach the calculated values of -0.868 and -1.912, respectively?

5. Assume that you have N items of data on which m operations are to be performed in a FORTRAN DO loop. Each operation on each item takes time t to be performed. Calculate how much time a serial and a parallel machine will take to finish the whole DO loop. For fixed m, calculate the

minimum number N for which the parallel machine will be faster. Now assume that the data are divided into r groups, which can be operated upon independently. Assuming p serial processors, calculate the time needed to finish the DO loop. Now replace these p serial processors by parallel ones and recalculate the time needed to finish the DO loop.

7

POLYMERS, PROTEINS, AND SPIN GLASS MODELS

7.1 INTRODUCTION

Even though crystalline solids are theoretically appealing, most solids in nature are disordered. The disorder may be in the positions of the molecules (*site disorder*) or in the interactions (*bond disorder*) or both. In the case of site disorder, the sites may not lie on a lattice, or they may be occupied by different molecules with varying probabilities. In the case of bond disorder, the interactions between molecules may be random variables. A great deal of progress has been made on the study of disordered systems in the last 20 years or so. The main focus here is on two of the most important disordered condensed systems: polymers and spin glasses.

Polymers are long-chain molecules that consist of a large number of smaller subunits linked together. The subunits are called monomers. If the monomers are identical, we have a homopolymer, otherwise we have a heteropolymer. An example of the former is *polystyrene*, which is made from styrene. Naturally occurring proteins are examples of the latter. Many polymers of interest have a highly specific and complex three-dimensional spatial structure. Since the beginning of polymer science, one of the fundamental problems has been the following. Given the nature of subunits and the interactions between them, can we predict the spatial structure of a polymer? In this chapter, the reader is introduced to some of the latest techniques and results in this polymer chain-folding problem.

Homopolymers are discussed first. The self-avoiding walk is used as a model of a polymer whose monomer elements repel one another through what is known as the *excluded-volume effect*. The end-to-end distance of the polymer shows critical behavior. This behavior is manifested in a continuous phase transition, called the coil-to-globule folding transition.

Section 7.3 discusses heteropolymers and, in particular, proteins and the phase transition between the natured and denatured forms of proteins. Several approaches are being used to understand the protein-folding problem. The ones we discuss are the HP lattice and the spin glass models.

The spin glass models were introduced originally to explain the puzzling behavior of dilute magnetic insulators. However, the techniques and insights arising from a study of spin glass Hamiltonians have had far reaching consequences in mathematics, computer science, and biology. [See the seven-part series of articles that appeared in *Physics Today* in 1988–1990, reproduced in Anderson (Anderson, 1994).] After reviewing the mean-field model of spin glasses, we end with an application of the spin glass model to the protein-folding problem.

7.2 HOMOPOLYMERS

7.2.1 Self-Avoiding Random Walk and Criticality

One of the simplest models for homopolymers is the "beads on a string" model to be discussed in Section 10.6. In its most general version, the string is flexible and there may be rotations of the string pieces around the beads. This gives what is known as a random coil polymer. A random coil is common for polymers at high temperatures or in good solvents, which implies that the short-range polymer–solvent interactions dominate at the expense of the long-range interactions within the polymer molecules. On the other hand, at low temperatures or in a poor solvent, the long-range attractive interactions within polymers dominate, leading to the folding of polymers into globular forms. We keep the discussion in this section simple and restrict ourselves to the case where equidistant beads are placed on a lattice, and the pieces of string are straight and rigid. In such a case, one may start from one end and walk around on the lattice points, thus generating a polymer. Where does one go from a given point of the lattice? If we allow the next step to land on any of the nearest neighbors, say, with equal probability, we get the random walk model of Chapter 1. This model was first considered by Kuhn (Kuhn, 1934). In this model, the shape of the polymer is what we get from a random walk. In order to compare theory with experiment, we need some measure of "foldedness." Perhaps the most important quantity is the root-mean-square end-to-end distance, R. This distance is related to the distribution of monomers around the center of mass (Richards, 1980, p. 75) and gives some idea of the size of the polymer. In the random walk model, from Eq. (1.20),

$$R = aN^{1/2} \tag{7.1}$$

where N is the number of links between beads or, alternatively, random steps, and a is the length of one step. The probability distribution of a given monomer at a distance $r(x, y, z)$ from the monomer at the origin is taken to be Gaussian [cf. Eq. (1.36)]:

$$p(r) = \left(\frac{3}{2\pi R^2}\right)^{3/2} \exp(-3r^2/2R^2) \tag{7.2}$$

where the independence of each of the three dimensions has been assumed. From this equation, we can calculate the entropy by using Eq. (1.59) if we remember that, in the microcanonical ensemble, Ω is the number of states accessible to the system consistent with its maintaining a fixed energy. Since the random walker has $2d$ directions available for each step, where d is the dimension of the lattice, there are a total of $(2d)^N$ possible configurations or states for N steps. For a distance r, the states are not all equally probable, but each state exists with a Gaussian probability given by Eq. (7.2). Accordingly, $\Omega(r) = (2d)^N p(r)$. This enables us to write the Helmholtz free energy from Eq. (1.84).

$$A(r) = A(0) + \left(\tfrac{3}{2}\right) k_B T r^2 / N a^2 \tag{7.3}$$

Equation (7.3) is found to be valid for many polymers under certain conditions (see below).

The random walk model is the simplest model, which is somewhat like an ideal gas model. It assumes that the folding arises owing to random, uncorrelated steps. It neglects many effects that arise in real polymers. One of the most important is the following. In real polymers, two monomers cannot be arbitrarily close to each other, thus giving rise to the *excluded-volume constraint*. In the lattice model, this constraint is ensured by making the walk self-avoiding. That is, a given lattice point is never allowed to be visited more than once. In contrast with the random walk problem, which can be solved exactly in any number of spatial dimensions, the self-avoiding random walk (SAW) cannot be solved exactly even in two dimensions, say on a square lattice. Of course, in one dimension, the SAW is not even random after the first step, because the self-avoiding constraint allows one to go in only one of the two available directions.

Usually, the first step toward including interactions in an ideal system is the mean-field theory. The van der Waals and the Landau theories, which were discussed in Chapter 3, are examples. The mean-field theory for the SAW was given by Flory (Flory, 1953). The following discussion of Flory's theory is based on de Gennes (de Gennes, 1979).

The expression for free energy given by Eq. (7.3) does not include the effects of interactions, the major effect being the excluded-volume constraint. To include these, Flory proceeded as follows. Since the excluded-volume effect has to do with the repulsion of monomers when they are very close, we introduce a local concentration $c(r)$ of monomers, which will be used as the order parameter. This order parameter has an average value, called the internal monomer concentration c_{int} given by

$$c_{int} = N/r^d \tag{7.4}$$

where, as above, d is the dimensionality ($d=3$ for real polymers). Assuming the repulsion to take place mainly between pairs, the free energy contribution (per unit volume) due to excluded volume is given by

$$A_{rep} = \left(\tfrac{1}{2}\right) k_B T v [c(r)]^2 \tag{7.5}$$

where v is called the *excluded-volume parameter*. The mean-field approximation consists in replacing the local concentration by its average c_{int} and neglecting all fluctuations. By doing this and integrating Eq. (7.5) over the whole volume (proportional to r^d), we get the total repulsive free energy

$$A_{rep,\,tot} = \tfrac{1}{2} k_B T v c_{int}^2 r^d = \tfrac{1}{2} k_B T v N^2 / r^d \tag{7.6}$$

The overall free energy is given by the sum of Eqs. (7.3) and (7.6).

$$A(r) = A(0) + \left(\frac{3}{2}\right) k_B T r^2 / N a^2 + \left(\frac{1}{2}\right) k_B T v N^2 / r^d \tag{7.7}$$

This free energy can be minimized as a function of r to find the optimal radius, R. We get

$$R \sim N^{3/(d+2)} \tag{7.8}$$

Equation (7.8) was deduced by Flory for $d = 3$. For general d, the exponent was obtained by Fisher (Fisher, 1969).

The behavior of R for large N, given by Eqs. (7.1) and (7.8), is reminiscent of the behavior of various quantities in critical theory. Since R is akin to a correlation length, we may compare Eqs. (7.1) and (7.8) to Eq. (3.83). On making the analogy that R is like a correlation length and $1/N$ is like $(T_c - T)/T_c$, we see that the exponent in Eqs. (7.1) and (7.8) is v, the correlation-length exponent. In fact, the analogy can be pushed even further, because it turns out that the SAW model arises as a limiting case when $n \to 0$, where n is the number of components of the order parameter in the Heisenberg model of ferromagnetism (de Gennes, 1979, pp. 272–275). Therefore, the SAW problem can be studied by the well-known techniques of critical phenomena as a special case of more general magnetic systems. The SAW problem satisfies the properties of scaling and universality and can be applied to a large number of polymer systems, which show remarkably similar behavior near their critical points.

Equation (7.8) gives a good estimate of the exponent of N. For $d = 1$, the exponent is 1, which is understandable because, for $d = 1$, the SAW is just a straight-line walk with r proportional to N. The values of the exponent, namely, $\frac{3}{4}$ and $\frac{3}{5}$, for $d = 2$ and 3, respectively, are within a few percent of the most accurate estimates obtained by other, more sophisticated, methods. These include the standard methods used in the study of critical phenomena, such as the renormalization group and Monte Carlo techniques. To top it all, the value 0.6 is quite close to the experimental value observed in many polymers.

Equation (7.8) helps us understand another general concept in critical phenomena. Let us calculate the repulsive free energy at the optimum R. From Eqs. (7.6) and (7.8), the free energy is given by

$$A(R)_{\text{rep}} \sim vN^{(4-d)/(d+2)} \tag{7.9}$$

which is obviously zero if the repulsive effects are neglected by taking the excluded-volume parameter v to be zero, as is done in the random-walk model. But there is another way the repulsive effects go to zero even for nonzero v. From Eq. (7.9), it is clear that for $d > 4$ these effects can be neglected in the limit of very large N. So the random walk model is an excellent approximation to the SAW for $d > 4$, even when excluded-volume effects are taken into account. In the language of critical phenomena, one says that 4 is the *upper critical dimensionality* for the SAW, which corresponds to the limit $n \to 0$ of the general Heisenberg model. It turns out that 4 is also the upper critical dimensionality for the liquid–gas and the Ising model critical behavior, both of which correspond to the case $n = 1$ of the general Heisenberg model.

7.2.2 The Coil-To-Globule Folding Transition

In Section 7.2.1, we saw that the root-mean-square end-to-end distance increases with the number of monomer subunits. In such cases the number of possible polymer configurations also increases with the number of monomer subunits. These polymer configurations are called coils. They have no fixed spatial structure and, in the case of biological polymers, no biochemical activity. In biochemical language, such states of polymers are called *denatured states*. Coils provide good experimental examples of SAW models.

At low temperatures or in a poor solvent, the coil states are unstable relative to close-packed globular states in which the polymer has fewer configurations available to it. The globular states are defined by the property that the range of monomer density correlations is finite even when the number of subunits goes to infinity. The transition between the coil and the globule states is a phase transition. Our discussion of this transition is based on Lifshitz et al. (Lifshitz et al., 1978).

The theory of the coil-globule phase transition proceeds in the usual three steps. First, one sets up an ideal polymer in which all the interactions are neglected. Then, one includes the interactions by using mean-field theory, in which the fluctuating order parameter is replaced by its average, and the free energy is calculated by locating its minimum. The final step is to bring in the heavy artillery of critical phenomena. Usually, this is a renormalization group or the Monte Carlo approach or a combination of the two.

As seen in Section 7.2.1, the ideal polymer chain, in which all interactions are neglected, can be described as a random walk. Since there are no interactions, the ideal polymer chain representing a globule is identical to the one describing a coil. Differences arise only when interactions are included. In the case of the globule, the ideal chain will be used as a reference chain.

Once again, it is convenient to use the Helmholtz free energy $A[c(r)]$, which is a function of the local concentration of monomers $c(r)$, the order parameter. The mean-field theory of the polymer chain is quite complex, involving many different regimes in many different parameters. We illustrate the theory by discussing the transition in large globules, where the total end-to-end distance is much greater than the length of a subunit and within what Lifshitz (Lifshitz, 1969) calls the *volume approximation*. In this approximation, one considers only the globular volume. For medium or small globules, one has to include the effects arising from surface structures. We assume a globule large enough that the surface is sharp, and the concentration $c(r)$ falls to zero abruptly when r reaches a certain cutoff value.

Within the mean-field approximation, we expand the various thermodynamic functions in terms of the average local order parameter, c. The free energy per unit volume, the chemical potential, and the pressure, respectively, are given by

$$A^*(c, T) = A - A_{id} \tag{7.10}$$

$$\mu^*(c, T) = \mu - \mu_{id} = \partial A^* / \partial c \tag{7.11}$$

$$p^*(c, T) = p - p_{id} = c\mu^* - A^* \tag{7.12}$$

where, for convenience, we have subtracted the quantities corresponding to the reference system of an ideal polymer. Expansion of Eq. (7.10) in powers of c gives, for the first two terms,

$$A^*(c, T) = k_B T B(T)c^2 + k_B T C(T)c^3 \tag{7.13}$$

Substitution of Eq. (7.13) into Eqs. (7.11) and (7.12) gives

$$\mu^*(c, T) = 2k_B T B(T)c + 3k_B T C(T)c^2 \tag{7.14}$$

$$p^*(c, T) = k_B T B(T)c^2 + 2k_B T C(T)c^3 \tag{7.15}$$

where $B(T)$ and $C(T)$ are virial coefficients representing the departure from ideality. The quantity to be minimized is the total free energy,

$$A^*(c, T)_{tot} = VA^*(c, T) = (N/c)A^*(c, T) \tag{7.16}$$

By using Eq. (7.13) in Eq. (7.16), we see that the derivative $\partial A^*(c, T)_{tot}/\partial c$ vanishes exactly where $p^*(c, T) = 0$. Therefore, the optimal c is given by

$$c = -B(T)/2C(T) \tag{7.17}$$

and the total free energy is given by

$$A^*(c, T)_{tot} = -Nk_B T [B(T)]^2/4C(T) \tag{7.18}$$

The vanishing of $B(T)$ gives the critical point, as usual. By assuming the simplest forms,

$$B(T) = B(T_c)(T - T_c) \tag{7.19}$$

$$C(T) = C(T_c) \tag{7.20}$$

we see that the critical point is at T_c.

The free energy, given by Eq. (7.18), is zero at the critical point, which implies from Eq. (7.10) that the system behaves as an ideal polymer, where interactions are neglected. In the theory of polymers, this point is called the *theta point* or the *Flory point*. Flory showed that, at this temperature, the random walk model gives excellent results when compared with experiments.

In summary, the polymer has the coil form above the critical temperature and undergoes a transition to the globule form below the critical temperature. As mentioned before, this simple theory neglects many important effects like surface effects and is not applicable to small globules. Small globules are discussed in the review by Lifshitz et al. (1978).

Our discussion in this section is not very suitable for protein folding. The theory is for homopolymers only and is a mean-field theory. To do justice to the folding problem, one must consider heteropolymers and go beyond mean-field theory, which is done in Section 7.3.

7.3   HETEROPOLYMERS: PROTEINS

7.3.1   Introduction

A polymer whose monomer units are not all identical and vary irregularly along the chain is a heteropolymer. The example we consider is that of a protein. The units of a protein consist of amino acid residues held together by a peptide linkage,

$$
\begin{array}{c}
\text{H} \\
| \\
-\text{C}-\text{N}- \\
\| \\
\text{O}
\end{array}
\tag{7.21}
$$

There are 20 amino acid residues available to naturally occurring proteins. Of the 20 residues, 7 are hydrophobic, 8 are polar, 4 are charged, and 1 (glycine) does not belong to any of these three categories. The sequence of amino acid residues along the chain, from the amine end to the carboxylic end, constitutes the primary structure of the protein. The twisting of the polypeptide chain into alpha (α) helices and beta (β) sheets constitutes the secondary structure. The formation of α helices and β sheets is associated with hydrogen bonding between local and nonlocal amino acid residues, respectively. Finally, the packing of the α helices and β sheets into domains of helices or sheets or combinations of these constitutes the tertiary structure. The tertiary structure involves nonlocal interactions between residues. These interactions include hydrophobic and polar interactions as well as the formation of disulfide bridges. It is the tertiary structure that determines the function of the protein.

Questions that arose among researchers early in the study of protein chemistry concerned protein specificity. In order to carry out its function in the cell, a protein needs a specific tertiary structure. Is this tertiary structure of a protein an artifact of the processes that produced it, or is there a more fundamental explanation of protein structure? This question was answered by Christian Anfinsen, who received the Nobel prize for his work. Anfinsen and co-workers denatured the enzyme ribonuclease in a urea, mercaptoethanol solution. Upon removal of the urea and mercaptoethanol, it was found that the ribonuclease renatured to its native state. The only explanation is that it is the amino acid sequence that dictates the tertiary structure of the protein. Furthermore, the native state of the protein represents the most thermodynamically stable state of the protein. However, this thermodynamic hypothesis has been questioned recently by Baker and Agard (Baker and Agard, 1994). If progress in protein engineering and design is to progress rapidly, it is essential to be able to predict a protein's tertiary structure from its one-dimensional amino acid sequence. Indeed, this is presently the major unsolved problem in structural molecular biology.

There are several approaches aimed at solving the protein-folding problem. Most of these are based upon secondary-structure prediction (Branden and Tooze, 1991, Chapter 16), which requires an intimate knowledge of amino acid structure and of the hydrogen bonding between the amino acid residues. We focus on those approaches that utilize statistical mechanical principles.

7.3.2   HP Lattice Model

The simplest approach that utilizes an amino acid sequence to determine the native structure of the protein is the HP lattice model. This model recognizes only hydrophobic (H) and polar (P) amino acid residues. The basic assumption in this approach is that it is the hydrophobic forces that determine the native structure of the protein. Other interactions, such as hydrogen bonding, then follow secondarily. There are very good arguments supporting this assumption (Kausmann, 1959), but these have little to do with the statistical mechanics of protein folding.

In the HP model the native configuration for a given HP sequence is the one with the most H–H contacts between distant (noncovalently bonded) neighbors of the folded chain. It is possible to use what Chan and Dill (Chan and Dill, 1991) call exhaustive computer enumeration to find all possible SAWs for short chains [up to 30 monomers according to Chan and Dill (Chan and Dill, 1993)]. Then, one counts the number of H–H contact pairs for a given HP sequence; the configuration (or configurations) with the most hydrophobic interactions is the native configuration (or configurations). In two dimensions, consider the FORTRAN program POLYMER (Appendix D.11). In this program, at each step, we choose only those sites that have not yet been visited. This algorithm does not generate self-avoiding walks, but something simpler called a "myopic" self-avoiding walk. The program is good enough to illustrate polymer folding. [For a mathematical discussion of the various MC methods, see Madras and Slade (Madras and Slade, 1993, Chapter 9).] Use the program POLYMER for the 14-monomer sequence (a 14-mer) H–H–P–P–H–P–H–H–P–H–P–H–H–H to generate several configurations in two dimensions. Then count the number of nonlocal, nearest-neighbor H–H contacts. The configuration with the most H–H contacts is the configuration of lowest energy. This configuration is probably not the global-energy minimum, since an exhaustive search would have to be done to find this state. The global minimum in the case of this 14-mer in two dimensions has seven H–H contacts. Exhaustive computer searches using statistically correct SAWs are useful for generating statistics for radii of gyration, amount of secondary structure, number of native states, and various correlations. However, such studies do not give insight into how the protein folds.

The folding process was addressed by Fiebig and Dill (Fiebig and Dill, 1993). The question they asked was how can a protein assemble a hydrophobic core without searching all possible hydrophobic pairings? This paradox was suggested by Levinthal (Levinthal, 1968), namely, how can a protein find a global optimal conformation without a globally exhaustive search? We can circumvent the necessity of an exhaustive search by considering what Fiebig and Dill call a "maximum entropy chain." To see how this works, consider the above 14-mer in two dimensions. We start at the left end and number the H monomers according to their positions in the chain. Then, we form H–H contacts in such a way as to preserve the maximum amount of entropy. The entropy is easily determined from $S = k_B \ln \Omega$ by counting the total number of states accessible to the chain. To maximize the entropy, we want to keep as many random-coil units as possible in the chain, which can be accomplished by making nonlocal contacts between the closest-lying residues. If we start at the left end, we see that the closest-lying residues for which a contact is possible are H_2 and H_5, which leaves a 10-mer random coil and 2034 possible states (Madras and Slade, 1993, Appendix C). We might guess that the next nonlocal contact is H_7–H_{10}, which leaves a 5-mer random coil and 13 possible states. Unfortunately, this procedure gives the correct native state only about 70% of the time. In our case, we would have ended up with six nonlocal contacts instead of seven. The correct choice in the second step should have been H_{10}–H_{13} (Chan and Dill, 1991). Fiebig and Dill worked out an alternative to the maximum entropy string that they call a "T-local string" or a "hydrophobic zipper." The details are to be found in their paper.

The HP model is not a realistic model in that the amino acid sequence is considered only to the extent that the residues are labeled hydrophobic or polar. More realistic models that include more interactions can be used, but the problem becomes one of finding the conformation with a free energy minimum. From studies of spin glass models, to be discussed in Section 7.4, it is known that conformations can get stuck in local minima. Molecular dynamics models have a difficult time escaping from these local minima. More success has been had with Monte Carlo methods. The simplest Monte Carlo method is one on a cubic or square lattice.

7.3.3 Monte Carlo Methods

Go and co-workers (Taketomi et al., 1975; Go and Taketomi, 1978; Go and Taketomi, 1979) set up a two-dimensional protein model on a square lattice as in Figure 7.1, where the first 23 monomers of Figure 7.1 are assigned labels A–D such that the interactive pairs of hydrophobic interactions are A–A, B–B, C–C, and D–D. If each interactive pair is assigned an energy $-\epsilon$, the energy of the conformation in Figure 7.1 is -11ϵ, which may not be the lowest energy conformation for a 23-monomer chain. The problem is to find the lowest energy conformation. The Monte Carlo method is well suited to this problem.

The standard Metropolis Monte Carlo procedure for proteins is very similar to that for the Ising model (Section 6.4). In the Ising model, a new configuration of spins is obtained by changing one of the lattice spins and accepting or rejecting the new configuration based on a Boltzmann distribution. For proteins, a new conformation is randomly chosen and the new conformation accepted or rejected in the same way. The rules for selecting a new conformation are somewhat arbitrary, but they must involve major portions of the chain so that the protein will not get stuck in a glassy state. The reader is referred to the papers by Go and co-workers for details of the Monte Carlo procedure. This simple model can be extended by including short-range interactions, which are dependent on the bond angles between peptide linkages. The conclusion to be drawn from this study is that long-range interactions are necessary for protein folding to the native state, but they are not sufficient. Apparently, the α helices and β sheets must be able to organize according to the short-range interactions (secondary structure) before the long-range interactions (tertiary structure) can be fully effective.

The most sophisticated approaches to protein folding consider the structure (bond lengths and angles) of the amino acids and the forces between adjacent amino acids. A full molecular dynamics simulation is not possible for two main reasons: (1) presently, computers are not able to handle large protein molecules, especially for folding studies, and (2) no one has figured out how to solve the multiminima problem. As mentioned above, Monte Carlo methods are better suited to such studies. However, the usual Metropolis Monte Carlo

Figure 7.1 The Go model.

methods are ineffective because they sample only a small part of configuration space near the initial conformation. To circumvent this problem, Paine and Scheraga (Paine and Scheraga, 1985) used a modified Monte Carlo algorithm that enabled them to sample all residues and concentrate on those that gave the lowest free energy. Even then, the study was restricted to a polypeptide composed of only five residues.

A model that requires no knowledge of the amino acid sequence or of the structural details of the polypeptide backbone is a random-energy model, known as the spin glass model. Since spin glass models are widely applied to systems other than polymers, we first discuss spin glass models in general and then discuss their application to protein folding.

7.4 SPIN GLASS MODELS

7.4.1 Introduction

Spin glass models were introduced to explain the occurrence of the magnetic spin glass phase in mixed systems such as CuMn or $Eu_xSr_{1-x}S$, where the magnetic atom is in the minority. In such systems a phase transition, as revealed by a peak in the susceptibility versus temperature curve, is observed from either a paramagnetic or a ferromagnetic phase to a spin glass. The spin glass models are characterized by having both ferromagnetic and antiferromagnetic interactions randomly distributed among the magnetic atoms. A spin glass phase occurs at low temperatures and at low concentrations of the magnetic atoms. The temperature at which the transition occurs is called the *freezing temperature*, T_f, because the random magnetic interactions are said to be "frozen in" below T_f.

Since spin glasses have both ferromagnetic and antiferromagnetic interactions, corresponding to the exchange constant J between a pair of spins being positive and negative, respectively, a unique ground state may not exist in these systems. For example, consider a simple square of four spins in which one of the nearest-neighbor bonds is antiferromagnetic, the other three being ferromagnetic. The three ferromagnetic bonds will force the four spins to be parallel to one another. The last antiferromagnetic bond cannot be satisfied without breaking one of the neighboring ferromagnetic bonds. But if we do that, then the ferromagnetic bond is unsatisfied. Technically, one says that the system is *frustrated*. Clearly, in this case a unique ground state that can satisfy all the bonds does not exist. In fact, finding the ground state of a frustrated random system is a very difficult optimization problem in the thermodynamic limit. Frustration is essential if a disordered system is to have spin glass properties.

7.4.2   EA and SK Models

One of the simplest spin glass models is the Edwards–Anderson (EA) model (Edwards and Anderson, 1975). It is of the form of the Heisenberg model, except that the exchange parameters governing the interactions between the spins are stochastic variables drawn from a Gaussian distribution. The Hamiltonian of the model is given by

$$\mathcal{H} = -\sum_{ij} J_{ij} \mathbf{S}_i \cdot \mathbf{S}_j - H \sum_i \mathbf{S}_i \tag{7.22}$$

where we take $\mathbf{S}_i$ to be a classical spin vector, the exchange parameters J_{ij} are randomly chosen according to a Gaussian distribution, and H is the applied magnetic field. The EA model is an example of a *quenched* random system as opposed to an *annealed* disordered system. In the former, the disorder is fixed and does not change appreciably with temperature or time, whereas in the latter, the entities causing disorder adjust themselves in time or with the change of temperature. The treatment of the two types of disorder is different. In the case of the annealed disorder, one typically calculates the average of the partition function Z over the randomness of the system, which is usually simpler than the case of the quenched disorder in which the free energy (essentially $\ln Z$) must be averaged.

Edwards and Anderson used the "replica trick" to deal with the averages required in quenched systems such as the spin glass. The trick is to write the log of the partition function as

$$\ln Z = \lim_{m \to 0} \frac{Z^m - 1}{m} \tag{7.23}$$

In other words, one must study m replicas of the given system, labeled by the replica index $\alpha = 1, 2, \ldots, m$, to calculate $Z^{(\alpha)}$. The partition functions of the m replicas are then multiplied to give Z^m, whose average is then calculated. The quantity $\langle \ln Z \rangle$ then gives the free energy and other properties. This trick is now a valuable instrument in the tool kit of all theorists.

Edwards and Anderson studied their model in the mean-field approximation and found that, just as in ordinary ferromagnets, there is a phase transition. This phase transition signals the appearance of a new phase with a cusp in the susceptibility mimicking the one seen in experiments. However, as contrasted with a ferromagnet, the spontaneous magnetization defined by

$$\mathbf{M} = \langle \langle \mathbf{S}_i \rangle_t \rangle_J \tag{7.24}$$

where the inner average is the usual thermal average and the outer one is over the distribution of exchange parameters, remains zero in the low-temperature phase in the case when the random exchange constants average to zero. So the spontaneous magnetization cannot be used as an order parameter for the spin glass transition. One has to introduce a new order parameter, now called the EA parameter. In the mean-field approximation, it is given by

$$q \equiv \langle \langle \mathbf{S}_i \rangle_t^2 \rangle_J \qquad (7.25)$$

and is nonzero in the spin glass phase.

The next important advance was by Sherrington and Kirkpatrick (SK) (Sherrington and Kirkpatrick, 1975), who introduced the mean-field version of the EA model but with Ising spins. The Hamiltonian of the SK model is given by Eq. (7.22), where the spins are just scalars equal to ± 1. The exchange parameters are distributed according to a Gaussian distribution such that the standard deviation is of the order of $1/\sqrt{N}$, where N is the number of spins. They solved this model exactly using the replica trick and found that it shows the spin glass phase transition with the EA order parameter. However, the low-temperature side of the solution was defective, as it showed a negative entropy near $T = 0$. The remedy for this defect was provided by Parisi and is described in the series of articles by Anderson in *Physics Today* (1988–1990) (see Anderson, 1994, pp. 526–538). For a review of spin glasses, see the review by Binder and Young (Binder and Young, 1986).

What Parisi found was that, for any temperature below the critical temperature, there is no single, stable thermodynamic state that is a solution to the mean-field theory. The SK assumption of such a state was not justified. Instead, there is an infinite number of such states that are similar to one another in different degrees. In other words, the phase space of the system is divided into regions, which are separated by barriers, which makes the system nonergodic in the thermodynamic limit. Also, it makes it very difficult to go from one state to another because of the necessity of passing through a large number of states in doing so. This difficulty occurs in experiments as well as in simulations, thus making the simulation of the spin glass phase a time-consuming exercise computationally. All these considerations make the spin glass phase different from the usual ordered phases and are responsible for its unique utility in so many diverse systems, as we now discuss.

7.4.3 Application of Spin Glasses in Other Fields

Spin glass Hamiltonians have been used in many problems in different areas of science. The following is based on the summary by Anderson (Anderson, 1994, pp. 495–499). The series of articles in *Physics Today* may also be consulted. In computer science, there is a famous optimization problem called the traveling salesperson problem. It is stated as follows. There are

N cities with given positions and distances r_{ij} between them. The problem is to minimize the total distance $L = \sum r_{ij}$ when optimizing a run of the salesperson through all the cities. New insights have been gained by applying the techniques learned in spin glass theory to the traveling salesperson and similar optimization problems.

In evolutionary biology, one studies populations of species distributed over a region. Whether new species arise suddenly or gradually is an important question that arises in such systems. Spin glass-like models have been used to support the view of sudden changes in species called "punctuated equilibrium."

In neuroscience, John Hopfield of Caltech has used spin glass ideas to study neural networks and brain function. The neurons are represented by spins and the coupling synapses by random exchange parameters. We mention just one result. The fact that humans are able to recall many details of an event starting from fragmented information is due to what is called "content addressable" memory. The spin glass model can be mapped into a neural network model that incorporates this kind of memory by a suitable choice of exchange constants.

More examples and references to original work can be found in Anderson (Anderson, 1994). We now apply spin glass ideas to polymers.

7.4.4 Spin Glasses and Polymers

Experimental protein-folding studies have shown that globular proteins in vivo fold into their native states within microseconds to minutes. This time is very rapid compared to the time it would take an equivalent homopolymer to fold to the same conformation. Studies have shown also that the folding process is such that these proteins fold relatively rapidly into a collapsed globule, which represents most of the secondary-structure formation. Domains of secondary structures then slowly form till enough of the tertiary structure exists that the protein rapidly condenses to its native state. Occasionally, the protein will end up in a higher energy nonnative state, which can happen, for example, with myoglobin upon flash photolysis. The protein in this case is said to be in a *glassy state*.

We first point out the similarities between proteins and spin glasses. First, spin glasses undergo a freezing transition from a disordered state to a spin glass state at T_f, while globular proteins undergo a transition from a denatured state to a glassy state or to a native state. Second, many local free energy minima exist in spin glasses giving rise to frustration. The kinetics of protein folding indicates that many free energy minima exist in protein conformations, which are also frustrated. Third, just as the magnetic states are randomly distributed in spin glasses, it is assumed that the residues in native and nonnative conformations are randomly distributed along the polypeptide chain. Finally, the order parameter in a spin glass is the square of the local average magnetic moment, and the order parameter in a protein is the fraction of residues in their native states.

A simple spin glass model (Bryngelson and Wolynes, 1987) applicable to polymers takes the energy of the protein as

$$E = -\sum_i \epsilon_i - \sum_i J_{i,i+1} - \sum_{ij} K_{ij} \qquad (7.26)$$

In Eq. (7.26), ϵ_i is the energy of the ith residue and is a function of the states of the peptide backbone; it is thus sensitive to the protein's primary structure. The $J_{i,i+1}$ term represents the nearest-neighbor interaction energy, which is responsible for hydrogen-bond formation (or the protein's secondary structure). Finally, the K_{ij} term represents the energy of short-range interactions between residues widely spaced along the chain. These are primarily hydrophobic interactions and are responsible for the tertiary structure. The simple model takes the terms ϵ_i, $J_{i,i+1}$, and K_{ij} equal to zero when the respective residues are not in their native states. The model is further simplified by using mean-field energies with a random distribution of native and nonnative residues:

$$\langle E \rangle = N - \epsilon\rho - J\rho^2 - \eta[\langle K \rangle + (K - \langle K \rangle)\rho^2] \qquad (7.27)$$

where N is the number of residues, ϵ, J, and K are fixed primary, secondary, and tertiary energies for native states, ρ is the fraction of residues in their native states, and η is proportional to the ratio of the total monomer volume to the volume of the protein (the excluded-volume effect).

The entropy is found as $S = k \ln \Omega(E, \rho)$, where $\Omega(E, \rho)$ is the average number of protein states with given energy E and residue fraction ρ. The Helmholtz free energy is found (Bryngelson and Wolynes, 1990) from Eq. (1.84). A plot of free energy versus ρ shows a double well with a high-energy unfolded protein ($\rho \approx 0.2$) and a low-energy folded protein ($\rho \approx 0.88$). This calculation depends on the *principle of minimal frustration* in order to get a native structure instead of being trapped in the glassy state. This principle states that the values of the mean interaction parameters in Eq. (7.27) must be such as to reinforce one another in the native structure, so that the energy gap between the native structure and misfolded structures is as large as possible. As Frauenfelder and Wolynes (Frauenfelder and Wolynes, 1994, p. 58) note, the principle of minimal frustration asserts that proteins exist on the border between the simple and the complex. In complexity theory, this is known as "living on the edge of chaos," which is common for systems far from equilibrium (see Chapters 12 and 13) such as biological systems.

Work along these lines by Onuchic et al. (Onuchic et al., 1995) has given rise to the picture of an energy funnel for folding in a multidimensional hyperspace. In this picture, the denatured protein undergoes a continuous transition at the Flory temperature to a molten globule, which contains perhaps 65% helical secondary structure. As the temperature (energy) is lowered, the percent of tertiary structure increases. At a low temperature T_f, where the protein is about 90% native structure, a glass transition is possible. Otherwise, the protein continues on down the funnel to the native state at the bottom.

7.5 EXERCISES AND PROBLEMS

1. Show that Eq. (7.8) follows from Eq. (7.7) by minimizing the free energy as a function of r.

2. Define and evaluate the critical exponents β, γ, and δ for the mean-field model in Section 7.2.2.

3. Use the FORTRAN program POLYMER for the 14-mer example in Section 7.3.2 to generate several configurations using a different random-number-generator seed for each configuration. Count the number of H–H contacts in each case. Do you agree that an exhaustive search is necessary to find the global-energy minimum?

4. Continue the maximum entropy search started in Section 7.3.2 and verify that it gives six nonlocal contacts.

5. Use the POLYMER program to generate several 23-mer examples. Use the assignments in Figure (7.1) to determine the energies of the examples if each A–A, B–B, C–C, and D–D pair has an energy $-\epsilon$.

6. Consider a number of noninteracting spins in a nonuniform field with the Hamiltonian

$$\widehat{\mathcal{H}} = -\sum_i H_i \widehat{S}_i \qquad (7.28)$$

where $s_i = \pm 1$. Calculate the partition function of this system. Assume that the field is random with the distribution

$$p(H_i) = \tfrac{1}{2}\delta(H_i + H) + \tfrac{1}{2}\delta(H_i - H) \qquad (7.29)$$

so that the field is $\pm H$ with equal probability. By considering the system to have quenched disorder, calculate the order parameters M and q defined in Eqs. (7.24) and (7.25), respectively.

7. In the model of Problem 6, consider the disorder to be annealed. Calculate the average free energy and the average magnetization.

8. Take a square lattice of four spins and four bonds among nearest neighbors. If each bond can take the value $\pm J$, there are 16 possible ground states. Find out how many are ferromagnetic, antiferromagnetic, and frustrated, respectively.

PART III

NONEQUILIBRIUM STATISTICAL MECHANICS

8

THE BOLTZMANN EQUATION

8.1 INTRODUCTION

In our quest for universals in many-body systems, we now turn our attention to nonequilibrium phenomena. First, we point out that the concept of equilibrium is as much an idealization as the concept of a frictionless pendulum or a square well. The reason is that experimental systems are never really in equilibrium. Consider the following discussion based on Münster (Münster, 1969, pp. 60–62). To describe a physical system completely, one needs the thermodynamic variables and a complete set of additional internal variables, say ξ_i. Assume that these internal variables relax to their equilibrium values ξ_{0i} after a long time, the relaxation times being given by τ_i. Let the time t_e be the duration of the experimental observation of the system. The variables ξ_i can be divided into three classes depending on whether $\tau_i \gg t_e, \tau_i \approx t_e, \tau_i \ll t_e$. The idealization involved in the definition of equilibrium involves the assumptions that (1) the variables of Class 1 have an infinite relaxation time, (2) the variables of Class 2 do not exist and, (3) the variables of Class 3 have a zero relaxation time. With these assumptions, the variables of both Classes 1 and 3 are constants, and it becomes possible to consider the system to be in equilibrium for the duration of the experiment.

By definition, nonequilibrium phenomena are those for which the physical observables change in space or time or both. The study of nonequilibrium systems is considered to be the more fundamental, since the equilibrium systems form a limiting case, which the nonequilibrium systems may approach after a long time. Still, equilibrium systems are usually discussed first because they are easier to grasp, much more is known about them, and they do show such wonderful and basic phenomena as phase transitions and polymer folding.

Equilibrium systems are very similar to one another in that they can be described by the universal laws of thermodynamics. By contrast, no such general laws exist to describe nonequilibrium systems. However, a broad classification can be made in terms of how far the nonequilibrium systems are from equilibrium. Functions or equations for systems not too far from equilibrium can usually be expanded about the equilibrium state and studied using linear mathematics, which gives exact solutions. The two best-known results in the linear regime are the Onsager reciprocity relations and the minimum entropy production theorem (see Prigogine, 1980, pp. 86–88).

In Chapters 8–10, we discuss systems near equilibrium. We start with a discussion of the Boltzmann equation, which describes the average behavior of a dilute system as it approaches equilibrium. The role of fluctuations in near-equilibrium systems is examined in Chapter 9. The concept of a stochastic, Markovian process is introduced and the various diffusion equations, following the studies of Einstein, Fokker–Planck, and Langevin, are then discussed. The central idea in all these approaches is to focus attention on a few variables of interest and eliminate all other variables from discussion, an approach that has been called a contracted description. The general approach of Zwanzig and Mori is the subject of Chapter 10. Here we show how the contracted approach can be used in a formal setting to provide the foundations of generalized Langevin-type equations involving processes with memory, or non-Markovian processes.

If the system is not too far from equilibrium, a state may arise in which the various observables do not depend on time but do depend on space. Such a state is called a *nonequilibrium steady state*. As an example, consider a copper rod heated at one end. After a while, the temperature of the rod will become independent of time, but will decrease with distance away from the heat source. An example of steady-state theory is the Kramers model of chemical reactions, which is discussed in Chapter 11. We include a discussion of the transition state, the Kramers, and the Grote–Hynes theories. We shall see how, in the long relaxation time limit, the properties of scaling and universality arise naturally in the Grote–Hynes theory.

In Chapters 12 and 13, we study systems far from equilibrium where nonlinearity enters the picture. Nonequilibrium statistical mechanics, in particular, and nonlinear science, in general, are now parts of a new subject called *complexity theory*. According to Michael C. Taylor, author of the FAQ (frequently asked questions) of the USENET newsgroup *sci.fractals* on the internet, "Emerging paradigms of thought encompassing fractals, chaos, nonlinear science, dynamic systems, self-organization, artificial life, neural networks, and similar systems comprise the science of complexity." (The information on how this FAQ may be obtained and other information on *World Wide Web* and *ftp* sites of interest to students of Statistical Mechanics may be found in Appendix A.)

In our study of complex many-body systems in Part II, we saw how relatively simple properties like scaling and universality emerge in certain limits for systems at equilibrium. In contrast, in the study of nonlinear systems far from equilibrium, we shall see that the ultimate fate of a system is not always thermodynamic equilibrium. We shall see also how very simple mathematical equations, describing real physical and chemical systems, give rise to complicated behavior like nonlinear transients, oscillations, fractals, and chaos.

In Chapter 12, we discuss oscillating chemical reactions and provide examples of phase trajectories, limit cycles, bifurcation, strange attractors and chaos. In Chapter 13, we focus on cellular automata and show how simple mathematical rules can lead to self-organization and chaos. Finally, the fractal nature of cellular automata, Brownian motion, and polymer folding is discussed.

8.2 BOLTZMANN THEORY

The work of Ludwig Boltzmann laid the foundations of statistical mechanics. Historically, the Boltzmann equation was instrumental in proving the *H*-theorem (see Section 1.7.3) and in explaining irreversible behavior. It is important that students of statistical mechanics have some exposure to the Boltzmann equation and to Boltzmann's *stosszhalansatz*. It is the purpose of this chapter to provide that exposure. The general solution of the Boltzmann equation is not easy; this chapter discusses only the equilibrium solution. The chapter ends with a derivation and discussion of transport properties in dilute gases.

Maxwell and Boltzmann are the architects of the kinetic molecular theory of gases. This amazing theory allows many observations to be explained from a few postulates. The simple theory deals with gases at equilibrium. From this theory has emerged, on the molecular level, concepts of velocity distributions, effusion of gases, mean free path, and the equipartition of energy principle. Discussions of the simple kinetic molecular theory can be found in all physical chemistry textbooks (e.g., Levine, 1995, Chapter 15). Hence, we do not dwell on it here. Rather, in this chapter we pursue the nonequilibrium behavior of gases based on the work of Boltzmann. This work is an extension of the simple kinetic molecular theory and takes into account explicitly the collisions between molecules. Historically, this was the first successful attempt to account for irreversibility. It is not surprising that this was done first for gases, since the many-body problem in condensed phases, which was considered in Part II and will be considered again in subsequent chapters, is very difficult.

Boltzmann's equation represents an attempt to get an expression for the time dependence of the molecular statistical distribution function. The phase space of interest for Boltzmann's equation is μ space (see Section 1.6.3). Boltzmann neglected internal molecular degrees of freedom and assumed that the gas was so dilute that only two-body, or binary collisions, were important. Also, only short-range forces were considered; this eliminated considerations of gaseous ions.

There are two approaches to Boltzmann's equation. The approach Boltzmann took is based on kinetic molecular theory and considers collisions of hard spheres based on classical mechanics. The other, which was developed much later and is based on the work of Kirkwood, derives the Boltzmann equation from the Liouville equation. This latter derivation, which is beyond the scope of this book, leads to a hierarchy of distribution functions called the BBGKY hierarchy for Bogoliubov, Born, Green, Kirkwood, and Yvon (BBGKY). For details of the derivation using the BBGKY hierarchy, see Huang (Huang, 1987, pp. 65–71).

Before we turn to Boltzmann's derivation, which we consider to be simpler and more informative, pedagogically speaking, it is first necessary to understand the mechanics of binary collisions.

8.3   BINARY COLLISIONS

8.3.1   Elastic Collisions

Because three-body collisions in the gas phase are rare and higher body collisions are essentially nonexistent, we restrict our discussion to two-body interactions with a central force potential. We assume two molecules of masses m_1 and m_2, position vectors $\mathbf{r}_1$ and $\mathbf{r}_2$, initial velocities $\mathbf{v}_1$ and $\mathbf{v}_2$, and final velocities $\mathbf{v}_1'$ and $\mathbf{v}_2'$. The total momentum and total energy are always conserved. An elastic collision is defined as one in which the potential and kinetic energies are conserved separately. For inelastic collisions, some of the kinetic energy is converted to internal motion and is not conserved. We consider only elastic collisions.

The conservation of momentum requires that the center of mass of the binary system move with constant velocity. The center of mass velocity is

$$\mathbf{V} = \frac{m_1 \mathbf{v}_1 + m_2 \mathbf{v}_2}{m_1 + m_2} = \frac{m_1 \mathbf{v}_1' + m_2 \mathbf{v}_2'}{m_1 + m_2} \tag{8.1}$$

where $\mathbf{V} = d\mathbf{R}/dt$ is the velocity of the center of mass and $\mathbf{R}$ is the position of the center of mass. The relative position vector $\mathbf{r} = \mathbf{r_1} - \mathbf{r_2}$ gives a relative velocity $\mathbf{v} = \mathbf{v_1} - \mathbf{v_2}$. The dynamics of the collision process are best expressed in terms of the relative velocity. We first rewrite Eq. (8.1) to get

$$(m_1 + m_2)\mathbf{V} = m_1\mathbf{v_1} + m_2\mathbf{v_2} \tag{8.2}$$

Expressing $\mathbf{v_2}$ in terms of $\mathbf{v}$ and solving for $\mathbf{v_1}$,

$$\mathbf{v_1} = \mathbf{V} - \mu\mathbf{v}/m_1 \tag{8.3}$$

where $\mu = m_1 m_2/(m_1 + m_2)$ is the reduced mass of the binary system. Likewise,

$$\mathbf{v_2} = \mathbf{V} + \mu\mathbf{v}/m_2 \tag{8.4}$$

The conservation of kinetic energy requires that

$$\tfrac{1}{2}m_1 v_1^2 + \tfrac{1}{2}m_2 v_2^2 = \tfrac{1}{2}m_1 v_1'^2 + \tfrac{1}{2}m_2 v_2'^2 \tag{8.5}$$

or

$$\tfrac{1}{2}(m_1 + m_2)\mathbf{V}^2 + \tfrac{1}{2}\mu\mathbf{v}^2 = \tfrac{1}{2}(m_1 + m_2)\mathbf{V}^2 + \tfrac{1}{2}\mu\mathbf{v}'^2 \tag{8.6}$$

It immediately follows from Eq. (8.6) that

$$|\mathbf{v}| = |\mathbf{v'}| \tag{8.7}$$

Equation (8.7) states that the magnitude of the relative velocity vector cannot change. However, there is no restriction on the direction of this vector.

Expressed in terms of the relative velocity, the binary collision can be reduced to a one-body, central-force problem with scattering into a solid angle $d\Omega = \sin\Theta d\Theta d\phi$. The cylindrical symmetry of the problem (along the z axis, say) enables the solid angle to be written as $d\Omega = 2\pi \sin\Theta d\Theta$. The angle Θ is usually referred to as the scattering angle. It is the angle through which a particle is scattered by a central force as shown in Figure 8.1. The angle Θ is not to be confused with the spherical polar angle θ, which is used to follow the motion of the particle. In the case of two-particle scattering, the relative velocity can be expressed in either the space-fixed (laboratory) coordinate system or the center-of-mass coordinate system; it is immaterial which system is used. One should be aware that scattering experiments are done in the laboratory coordinate system, where the one-particle experimental scattering angles χ_1 and χ_2 are, in general, unequal to Θ. Two-particle

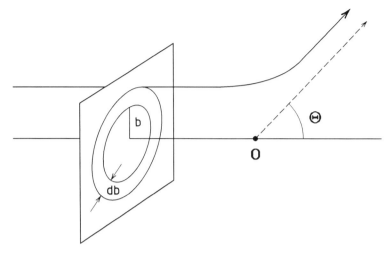

Figure 8.1   One-body, central-force scattering by a center at O.

scattering is shown in Figure 8.2, where particle 2 is originally at rest. In Figure 8.2, Θ is now the angle between the initial and final relative position vectors, $\mathbf{r}$ and $\mathbf{r}'$. If $m_1 \ll m_2$, we have essentially a one-particle scattering problem, and $\chi(\equiv \chi_1) = \Theta$. If $m_1 = m_2$, then $\chi_1 = \chi_2$ and $\Theta = 2\chi$. For other situations, $\chi_1 \neq \chi_2$ and the relation between Θ and χ can be worked out in the center-of-mass coordinate system (see Goldstein, 1980, pp. 105–119).

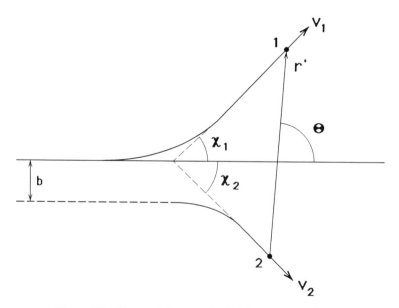

Figure 8.2   Two-particle scattering in laboratory coordinates.

8.3.2   Scattering Cross Sections

The dynamical aspects of a binary collision are contained in the *scattering cross section*. The number of molecules per unit time dN/dt scattered from a scattering center is directly proportional to the incident flux of molecules J (molecules/unit area/unit time) and the element of solid angle $d\Omega$ into which they are scattered. The constant of proportionality is the *differential cross section* $\sigma(\Omega)$ (sometimes written as $d\sigma/d\Omega$ and having dimensions of area):

$$dN/dt = J\sigma(\Omega)\,d\Omega = 2\pi J\sigma(\Theta)\sin\Theta\,d\Theta \tag{8.8}$$

for molecules scattered through an angle Θ as shown in Figure 8.1. The total number of molecules scattered in time t is

$$N = \int J\sigma(\Omega)\,d\Omega\,dt = J\sigma_t t \tag{8.9}$$

where σ_t is the *total* (or *integral*) *cross section*.

At a large distance from the scattering center, the energy in center-of-mass coordinates is $\frac{1}{2}\mu v^2$ (McQuarrie, 1976, p. 368). Near the scattering center, the intermolecular potential energy $\phi(\mathbf{r})$ becomes effective, and the scattering event must be expressed in terms of r and θ. The kinetic energy near the scattering center is $\frac{1}{2}\mu(\dot{r}^2 + r^2\dot{\theta}^2)$. The conservation of energy requires that

$$\tfrac{1}{2}\mu v^2 = \tfrac{1}{2}\mu(\dot{r}^2 + r^2\dot{\theta}^2) + \phi(\mathbf{r}) \tag{8.10}$$

Defining an impact parameter b (see Figure 8.1) as the perpendicular distance between the center of force and the incident velocity, the angular momentum outside the effective range of the potential is $\mu b v$. The magnitude of the angular momentum during collision is $\mu r^2\dot{\theta}$. Angular momentum must be conserved, therefore,

$$\mu b|\mathbf{v}| = \mu r^2\dot{\theta} \tag{8.11}$$

From Figure 8.1, by equating the number of particles lying between angles Θ and $\Theta + d\Theta$ and the ones with impact parameter lying between b and $b + db$, we get

$$2\pi J b|db| = 2\pi J\sigma(\Theta)\sin\Theta|d\Theta| \tag{8.12}$$

Since the impact parameters and angles may vary in opposite directions, absolute signs have been used. The differential scattering cross section in center-of-mass coordinates is

$$\sigma(\Theta) = \frac{b}{\sin \Theta} \left| \frac{db}{d\Theta} \right| \tag{8.13}$$

The procedure for calculating the differential cross-section is first to calculate the impact parameter b as a function of the scattering angle and total energy and then use Eq. (8.13) to obtain $\sigma(\Theta)$.

As shown by Goldstein (Goldstein, 1980, p. 117) the relation between the cross-section for scattering in laboratory coordinates $\sigma(\chi)$ and in the center-of-mass system $\sigma(\Theta)$ is

$$\sigma(\chi) = \sigma(\Theta) \frac{\sin \Theta}{\sin \chi} \left| \frac{d\Theta}{d\chi} \right| \tag{8.14}$$

8.4   BOLTZMANN EQUATION

8.4.1   Derivation of Boltzmann Equation

For the purpose of deriving the Boltzmann equation, it is necessary to define a one-molecule distribution function $f(\mathbf{r}, \mathbf{v}, t)$ in coordinate and velocity space, or μ space. This function is defined such that the number of molecules between $\mathbf{r}$ and $\mathbf{r} + d\mathbf{r}$, with velocities between $\mathbf{v}$ and $\mathbf{v} + d\mathbf{v}$ at time t, is

$$dN = f(\mathbf{r}, \mathbf{v}, t) \, dr \, dv \tag{8.15}$$

The distribution function $f(\mathbf{r}, \mathbf{v}, t)$ refers to μ space. Recall that, in μ space,

$$dr \, dv = dx \, dy \, dz \, dv_x \, dv_y \, dv_z$$

Our task is to find an equation for $f(\mathbf{r}, \mathbf{v}, t)$ as it changes in time and space. The following discussion is based on Huang (Huang, 1987, Chapter 3).

We focus on a given volume element $dr \, dv$. This volume element cannot be microscopic, since the distribution function then would fluctuate too wildly. On the other hand, the volume element cannot be too large, because we want to be able to write the derivatives of the distribution function. Consequently, the size of the volume element must be much greater than atomic dimensions yet much smaller than the size of the macroscopic box containing the system. This contracted description is typically applied to macroscopic systems that are treated by hydrodynamics and classical electrodynamics.

The distribution function changes because molecules are constantly leaving and entering the given volume element. The motion is of two kinds, a

streaming motion due to external forces and collisions due to mutual interactions. We include these two motions one at a time. If an external force $\mathbf{F}$
acts on a system of N molecules, each of mass m, a molecule occupying the
position $(\mathbf{r}, \mathbf{v})$ in μ space at time t will move to $(\mathbf{r} + \mathbf{v}\,dt, \mathbf{v} + \mathbf{F}\,dt/m)$ at time
$t + dt$. The new volume element in μ space is denoted by $dr'\,dv'$. It follows
that

$$f(\mathbf{r}, \mathbf{v}, t)\,dr\,dv = f(\mathbf{r} + \mathbf{v}\,dt, \mathbf{v} + \mathbf{F}\,dt/m, t + dt)\,dr'\,dv' \qquad (8.16)$$

It can be shown by using Liouville's theorem and the incompressibility of
μ space that $dr\,dv = dr'\,dv'$ so that

$$f(\mathbf{r}, \mathbf{v}, t) = f(\mathbf{r} + \mathbf{v}\,dt, \mathbf{v} + \mathbf{F}\,dt/m, t + dt) \qquad (8.17)$$

If a molecule, say Molecule 1, undergoes a collision with another molecule,
say Molecule 2, then Molecule 1 may not end up at $(\mathbf{r} + \mathbf{v}\,dt, \mathbf{v} + \mathbf{F}\,dt/m)$ in
a time $t + dt$. When collisions are included, Eq. (8.17) changes to

$$f(\mathbf{r} + \mathbf{v}\,dt, \mathbf{v} + \mathbf{F}\,dt/m, t + dt) = f(\mathbf{r}, \mathbf{v}, t) + dt\left(\frac{\partial f_{\text{coll}}}{\partial t}\right) \qquad (8.18)$$

where this equation defines the collision term.

The assumption that the volume element $dr\,dv$ is small allows us to differentiate the distribution function. We expand the left-hand side of Eq. (8.18)
in multiple Taylor's series to get

$$f(\mathbf{r} + \mathbf{v}\,dt, \mathbf{v} + \mathbf{F}\,dt/m, t + dt)$$
$$= f(\mathbf{r}, \mathbf{v}, t) + dt\left(\mathbf{v} \cdot \nabla_r + \frac{\mathbf{F}}{m} \cdot \nabla_v + \frac{\partial}{\partial t}\right) f(\mathbf{r}, \mathbf{v}, t) + \cdots \qquad (8.19)$$

Combining Eqs. (8.18) and (8.19), we get, in the limit as dt approaches zero,

$$\left(\mathbf{v} \cdot \nabla_r + \frac{\mathbf{F}}{m} \cdot \nabla_v + \frac{\partial}{\partial t}\right) f(\mathbf{r}, \mathbf{v}, t) = \left(\frac{\partial f_{\text{coll}}}{\partial t}\right). \qquad (8.20)$$

This equation is almost the Boltzmann equation. What remains to be done is to
express the collision term explicitly in terms of the differential cross-section.

8.4.2 Collision Term

The general treatment of the collision term is difficult. The following assumptions are usually made in the treatment of collisions:

1. The region $dr\,dv$ in μ space contains many molecules, even in the case of a dilute gas.
2. The time during which molecules are in collision is short relative to the time between collisions.
3. The intermolecular forces are all short range and are restricted to one region dr.
4. Because there are a large number of molecules per phase cell, a molecule does not move far spatially before experiencing a collision, which prevents motion along the $\mathbf{r}$ axis and gives the situation shown in Figure 8.3.
5. There can be large velocity changes upon collision, which in conjunction with item 4, restricts motion to the vertical, $\mathbf{v}$, direction in Figure 8.3.

Consider a Molecule 1 in the volume $dr\,dv_1$ (see Figure 8.3). The other Molecules 2 that are in the volume $dr\,dv_2$ and collide with Molecule 1 form an incident flux given by

$$J = f(\mathbf{r}, \mathbf{v}_2, t)|\mathbf{v}|\,dv_2 \tag{8.21}$$

where $f(\mathbf{r}, \mathbf{v}_2, t)\,dv_2$ is the density dN_2/dr and $\mathbf{v} = \mathbf{v}_2 - \mathbf{v}_1$. The number of collisions between Molecules 1 and 2, or looked upon another way, the number of 2 molecules scattered by Molecule 1 is, from Eq. (8.8),

$$dN_2 = J\sigma(\Omega)\,d\Omega\,dt \tag{8.22}$$

in the time interval dt. Substitution of Eq. (8.21) into Eq. (8.22) gives

$$dN_2 = f(\mathbf{r}, \mathbf{v}_2, t)|\mathbf{v}|\,dv_2\sigma(\Omega)\,d\Omega\,dt \tag{8.23}$$

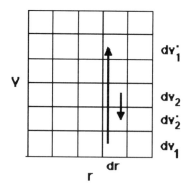

Figure 8.3 Phase cell in $\mathbf{r}, \mathbf{v}$ space. The arrows show the movement of Molecules 1 and 2 upon collision.

The rate for all Molecules 2 scattering off one molecule is obtained by dividing Eq. (8.23) by dt and summing over all Molecules 2.

$$\frac{dN}{dt} = \sum_2 \frac{dN_2}{dt} = \int_{v_2} f(\mathbf{r}, \mathbf{v}_2, t)|\mathbf{v}|\sigma(\Omega)\, d\Omega\, dv_2 \tag{8.24}$$

The total rate R^- for molecules scattered into $d\Omega$ is obtained by multiplying by the number of scattering Molecules 1 in $dr\, dv_1$.

$$R^- = \frac{dN_1}{dr\, dv_1}\frac{dN}{dt} = f(\mathbf{r}, \mathbf{v}_1, t)\int_{v_2} f(\mathbf{r}, \mathbf{v}_2, t)|\mathbf{v}|\sigma(\Omega)\, d\Omega\, dv_2 \tag{8.25}$$

where Eq. (8.15) has been used.

The writing of Eq. (8.25) has introduced a very important assumption into the derivation. The product $f(\mathbf{r}, \mathbf{v}_1, t)f(\mathbf{r}, \mathbf{v}_2, t)$ assumes independence of these distribution functions. This implies no correlation between velocities, and is Boltzmann's collision hypothesis (*stosszahlansatz*), which is strictly true only if the gas is in a state of molecular chaos. It is this assumption that introduces irreversibility into the derivation (see Cole, 1967, pp. 169 and 190; Rice and Gray, 1965, p. 270).

The rate at which particles enter $dr\, dv$ from $dr\, dv'$ is equal to the number of collisions of all particles with velocities $\mathbf{v}_2'$ with all particles per unit volume with velocity $\mathbf{v}_1'$, the scattering being into $d\Omega'$. As above, the incident flux is

$$J' = f(\mathbf{r}, \mathbf{v}_2', t)|\mathbf{v}'|\, dv_2' \tag{8.26}$$

Following the same line of reasoning as above, the rate at which molecules are scattered into $dr\, dv$ is

$$R^+ = \frac{dN_1'}{dr\, dv_1'}\frac{dN'}{dt} = f(\mathbf{r}, \mathbf{v}_1', t)\int_{v_2} f(\mathbf{r}, \mathbf{v}_2', t)|\mathbf{v}'|\sigma(\Omega')\, d\Omega'\, dv_2 \tag{8.27}$$

From Liouville's theorem, $dr\, dv = dr\, dv'$ and $\sigma(\Omega) = \sigma(\Omega')$. The sum of the rates at which molecules enter and leave $dr\, dv$ is $\partial f_{\text{coll}}/\partial t$

$$\frac{\partial f_{\text{coll}}}{\partial t} = R^+ - R^-$$

$$= \int_{v_2} [f(\mathbf{r}, \mathbf{v}_2', t)f(\mathbf{r}, \mathbf{v}_1', t) - f(\mathbf{r}, \mathbf{v}_1, t)f(\mathbf{r}, \mathbf{v}_2, t)]|\mathbf{v}|\sigma(\Omega)\, d\Omega\, dv_2 \tag{8.28}$$

for scattering into $d\Omega$, where R^- represents a net loss and so must enter this equation with a minus sign. If we now sum over all Ω and use Eq. (8.20), the final equation is

$$\left(\mathbf{v}_1 \cdot \nabla_r + \frac{\mathbf{F}}{\mathbf{m}} \cdot \nabla_v + \frac{\partial}{\partial t}\right) f_1 = \int_{v_2} \int_\Omega (f_2' f_1' - f_2 f_1) \sigma(\Omega) |\mathbf{v}| \, d\Omega \, dv_2 \quad (8.29)$$

where $f_i' = f(\mathbf{r}, \mathbf{v}_i', t)$ and $f_i = f_i(\mathbf{r}, \mathbf{v}_i, t)$. This equation is the celebrated *Boltzmann's equation*.

8.4.3   Equilibrium

At equilibrium, the rates R^+ and R^- are equal, which is an example of the *principle of detailed balance*. This principle is based upon the *principle of microscopic reversibility*, which in this case states that at equilibrium the rate of the process that sends a molecule from $dr\, dv$ to $dr'\, dv'$ equals the rate for the reverse process $dr'\, dv'$ to $dr\, dv$. At equilibrium the collision term vanishes. From Eq. (8.29), we see that this can happen if

$$f_2 f_1 = f_2' f_1' \quad (8.30)$$

This condition is then sufficient for equilibrium. Interestingly, Eq. (8.30) is independent of the differential cross-section and, therefore, of the nature of the system. We solve Eq. (8.30) for an ideal gas. The solution will be applicable to all classical systems.

The following is based on Huang (Huang, 1987, pp. 75–78). We solve Eq. (8.30) for a one-dimensional ideal gas. The three-dimensional case is very similar. By taking the log of Eq. (8.30), we get

$$\ln f_1 + \ln f_2 = \ln f_1' + \ln f_2' \quad (8.31)$$

which indicates that the quantity $\ln f$ is conserved in any collisions between the particles. The only quantities conserved in the elastic collisions of an ideal gas are the kinetic energy, momentum, and mass. Therefore, we can take $\ln f$ to be a linear combination of these conserved quantities:

$$\ln f = \alpha p^2 / 2m + \beta p + \gamma m \quad (8.32)$$

where α, β, and γ are constants. The constant α must be negative in order that the distribution function can be normalized. The normalization integral is

$$\int_V \int_{-\infty}^{\infty} f \, dx \, dp = N \quad (8.33)$$

Since the mass of each ideal gas molecule is fixed, the mass is constant and can be combined with the constants α and γ to give

$$\ln f = -ap^2 + bp + c \tag{8.34}$$

By writing Eq. (8.34) as

$$f = C \exp(-ap^2 + bp) \tag{8.35}$$

where $C = \exp c$, and substituting it in Eq. (8.33), we get

$$\int_V \int_{-\infty}^{\infty} C \exp(-ap^2 + bp)\, dx\, dp = N \tag{8.36}$$

Since f does not depend on x, the volume integral gives the total volume V, which is just the length L of the one-dimensional system, and the momentum integral is a Gaussian function. The integration can be performed by completing the square in the argument of the exponential. We get

$$C = \rho \sqrt{\frac{a}{\pi}} \exp\left(\frac{-b^2}{4a}\right) \tag{8.37}$$

where $\rho = N/V$ is the particle density.

The average momentum of the system is given by

$$\langle p \rangle = \frac{1}{N} \int_V \int_{-\infty}^{\infty} p f\, dx\, dp \tag{8.38}$$

and can be shown to be $\langle p \rangle = b/2a$. Assuming that the center of mass of the ideal gas is at rest implies that its average total momentum vanishes, which makes $b = 0$. Hence, the distribution function is now given by

$$f = \rho \sqrt{\frac{a}{\pi}} \exp(-ap^2) \tag{8.39}$$

which is beginning to look like the Maxwell–Boltzmann distribution. We still need to evaluate a, which is done by evaluating the pressure and comparing the result to the ideal gas equation of state, $P = \rho k_B T$.

Imagine a "line" segment of length v along the one-dimensional system. Let one end of the segment be a "wall" representing a perfectly reflecting point. The total number of molecules in the line segment is $vf\, dp$, by definition of f, which is also the number of molecules hitting the wall per second, assuming v to be positive. Since each molecule experiences a momentum change $2p$ on hitting the wall, the momentum change per second is $2vpf\,dp$. The total "pressure," which is just the force at one end in this case, is obtained by integration for all $p > 0$. The pressure is

$$P = \int_0^\infty 2vpf\,dp \tag{8.40}$$

which can be evaluated by substituting for f from Eq. (8.39) and integrating the Gaussian function. The result is

$$P = \frac{\rho}{2ma} \tag{8.41}$$

where m is the mass of each molecule. Comparing Eq. (8.41) with the one-dimensional ideal gas equation, $P = \rho k_B T$, we see that $a = 1/2mk_B T$, whence

$$f = \frac{\rho}{\sqrt{2\pi mk_B T}} \exp(-mv^2/2k_B T) \tag{8.42}$$

which is the one-dimensional Maxwell–Boltzmann distribution.

The solution of the Boltzmann equation for interacting systems not in equilibrium is a very difficult problem. A systematic approach for doing this was pioneered by Chapman and Enskog (see Hirschfelder et al., 1966, Section 7.3).

8.5   TRANSPORT PROPERTIES

8.5.1   Hydrodynamic Equations of Change

Whenever nonequilibrium conditions exist in a system, the system will react in such a manner as to establish equilibrium. The movement of a system toward equilibrium gives rise to the transport of those properties that contribute to the nonequilibrium. Empirically, transport has been associated with a gradient in one of the macroscopic physical properties of the system. Thus, transport of mass, momentum, and kinetic energy are associated with gradients in density, velocity, and temperature, respectively. The following discussion is based on Huang (Huang, 1987, Chapter 5).

We let $\chi(\mathbf{r}, \mathbf{v}, t)$ represent one of the conserved transport properties, and we seek an equation for the change of $\langle \chi(\mathbf{r}, t) \rangle$ as a function of $\mathbf{r}$ and t, where

$$\langle \chi(\mathbf{r}, t) \rangle = \frac{1}{\rho(\mathbf{r}, t)} \int f(\mathbf{r}, \mathbf{v}, t) \chi(\mathbf{r}, \mathbf{v}, t)\,dv \tag{8.43}$$

$\rho(\mathbf{r}, t)$ being the particle density. The equation of change can be found by multiplying the Boltzmann equation by $\chi(\mathbf{r}, \mathbf{v}, t)$ and integrating over dv. The integration is straightforward (Huang, 1987, pp. 96–97). The result is

$$\frac{\partial \langle \rho \chi \rangle}{\partial t} + \rho \left\langle \frac{\partial \chi}{\partial t} \right\rangle + \nabla_r \cdot \langle \rho \chi \mathbf{v} \rangle - \rho \langle \mathbf{v} \cdot \nabla_r \chi \rangle - \frac{\rho}{m} \langle \mathbf{F} \cdot \nabla_v \chi \rangle - \frac{\rho}{m} \langle \nabla_v \cdot \mathbf{F} \chi \rangle = 0$$

(8.44)

where ∇_r and ∇_v represent derivatives with respect to the components of the position coordinate and the velocity, repectively. Equation (8.44) is *Enskog's conservation theorem*.

8.5.2 Transport Equations

Now we illustrate the use of Eq. (8.44) by considering the equations for transport of mass and momentum. If χ is equal to the mass m, then many terms in Eq. (8.44) vanish, and it reduces to

$$m \frac{\partial \rho}{\partial t} + \nabla_r \cdot \langle \rho m \mathbf{v} \rangle = 0 \qquad (8.45)$$

or

$$\frac{\partial \rho_m}{\partial t} + \nabla_r \cdot (\rho_m \mathbf{u}) = 0 \qquad (8.46)$$

which is the equation of continuity for mass density, $\rho_m = m\rho$, where we have introduced the average velocity

$$\mathbf{u} = \langle \mathbf{v} \rangle \qquad (8.47)$$

It states that mass is conserved for fluid flow. Note that $\mathbf{u} = \mathbf{u}(\mathbf{r}, t)$ but that $\mathbf{r}$, $\mathbf{v}$, t are independent variables.

If we let $\chi = m\mathbf{v}$ in Eq. (8.44), we get

$$\frac{\partial \langle \rho_m \mathbf{v} \rangle}{\partial t} + \nabla_r \cdot \langle \rho_m \mathbf{v} \mathbf{v} \rangle - \frac{\rho_m}{m} \mathbf{F} = 0 \qquad (8.48)$$

where velocity-dependent forces have not been considered. This equation can be reduced further by introducing the average velocity $\mathbf{u}$, in which case the tensor $\mathbf{v}\mathbf{v}$ can be written in terms of the *pressure tensor* $\mathcal{P}$, whose components are given by

$$P_{ij} = \rho_m \langle (v_i - u_i)(v_j - u_j) \rangle \qquad (8.49)$$

The pressure tensor is a generalization of the usual hydrostatic pressure P. In general, this tensor is needed for the following reason. Pressure is defined as the force per unit area, both of which are vectors. So, in general, their "ratio" is a tensor. This tensor is analogous to the emergence, in mechanics, of the moment of inertia tensor, which is a ratio of the angular momentum and the angular velocity vectors. The equation one gets after reduction is

$$\rho_m \frac{\partial \mathbf{u}}{\partial t} + \rho_m \mathbf{u} \cdot \nabla_r \mathbf{u} = -\nabla_r \cdot \mathcal{P} + \frac{\rho_m}{m} \mathbf{F} \qquad (8.50)$$

Equation (8.50) is called *Euler's equation*. It is a fundamental equation in fluid dynamics. If one makes the local equilibrium assumption, in which case the distribution of velocities is given by the Maxwell–Boltzmann distribution [Eq. (1.103)], but the temperature and the average velocity are slowly varying functions of $\mathbf{r}$ and t, the pressure tensor becomes a diagonal tensor, all of whose diagonal terms become equal to the usual pressure P. In that case, Eq. (8.50) simplifies to

$$\rho_m \frac{\partial \mathbf{u}}{\partial t} + \rho_m \mathbf{u} \cdot \nabla_r \mathbf{u} = -\nabla_r P + \frac{\rho_m}{m} \mathbf{F} \qquad (8.51)$$

which is sometimes called the Euler equation.

The inclusion of shear stresses would have given us the Navier–Stokes equation for viscous flow (see Section 10.6.2). For a derivation of the Navier–Stokes equation and applications of hydrodynamics, see Huang (Huang, 1987, Chapter 5).

8.6   CONCLUDING REMARKS

The Boltzmann equation is important because it represents the first attempt to understand irreversibility. It is to kinetic molecular theory what the Liouville equation is to statistical mechanics. Many of our ideas concerning irreversible processes have originated with the Boltzmann equation. The irreversible nature of the Boltzmann equation comes from Boltzmann's *stosszahlansatz*. By using this approach, Boltzmann was able to prove his famous H theorem (see Huang, 1987, pp. 73–75). Even so, Boltzmann had many opponents. Foremost among these was Lochschmidt, who argued that Boltzmann's equations, being microscopic classical mechanical equations, should be reversible and Zermelo, who argued that the Poincaré recurrence theorem demands that a system ultimately return to its original state. It was left to the Ehrenfests and to Smoluchowski, after Boltzmann's death, to correctly explain the situation. They were able to show that, on average, the H function decreases in time, although fluctuations allow both increases and decreases in H. It was also shown that, if the Poincaré recurrence time is long enough, the process will be irreversible. As a practical matter, Poincaré recurrence times are longer than the age of the universe!

As discussed above, the Boltzmann equation leads directly to the transport equations for gaseous diffusion, viscosity, and thermal conductivity. Attempts have been made to apply the Boltzmann equation to dense gases and liquids. However, to go beyond the first-order solution is very difficult and requires approximations of the intermolecular potential. Several approaches to irreversible behavior have been developed since the time of Boltzmann. In Chapters 9–13, we discuss some of these approaches.

8.7   EXERCISES AND PROBLEMS

1. For a hard-sphere potential, what is the integral scattering cross-section?

2. Boltzmann defined the function H as $H = \int f \ln f \, d\mathbf{v}$. Differentiate H with respect to time. Noting that the left side of Boltzmann's equation (8.29) can be written as $\partial f_{\mathrm{coll}}/\partial t$, make appropriate substitutions in the dH/dt equation for f_1. Do the same for f_2, f_1', and f_2'. Then show that $dH/dt \leq 0$. This procedure is Boltzmann's proof of the H theorem.

3. Derive Eq. (8.37).

4. By performing the integral in Eq. (8.38), calculate the average momentum.

5. Derive Eq. (8.41).

6. Derive Eq. (8.46).

7. For an incompressible fluid, the density ρ_m is constant. From Eqs. (8.46) and (8.51), derive an equation of motion for an ideal, incompressible fluid. Now assume that the only external force is due to gravity

$$\mathbf{F} = -mg\hat{\mathbf{z}} \tag{8.52}$$

where $\hat{\mathbf{z}}$ is a unit vector in the upward direction. Show that the equation of motion in the z direction becomes

$$\frac{dP}{dz} = -g\rho_m \tag{8.53}$$

Use the ideal gas equation $P = \rho_m k_B T/m$ in Eq. (8.53) to obtain the equation

$$k_B T \frac{d\rho_m}{dz} + mg\rho_m = 0 \tag{8.54}$$

and solve the resulting equation for pressure to obtain the barometric equation

$$P(z) = P(0)\exp[-mgz/(k_B T)] \tag{8.55}$$

9

APPROACHES TO
BROWNIAN MOTION

9.1 INTRODUCTION

In 1827 the botanist Robert Brown, while studying pollen grains and other particles suspended in water and other fluids under a microscope, observed that their motion was irregular. The motion of pollen grains and the random walk (see Section 1.2.1) represent what are known as stochastic processes. That is, in time some variable such as the position x traces out an irregular path that can be treated by statistical methods. This motion, called *Brownian motion*, has served as a model for treating a host of irreversible processes that occur in nature.

In this chapter, Brownian motion is studied as an example of a nonequilibrium phenomenon not too far from equlibrium. It is examined from various points of view. The concept of a stochastic, Markovian process is introduced, and the Chapman–Kolmogorov and Fokker–Planck equations are derived. Brownian motion is then treated as diffusion, and its statistical nature is explored further by means of the Fokker–Planck and Langevin equations. Then Brownian motion is studied from the viewpoint of time series analysis, which is popular in such diverse fields as economics and engineering. Finally, a microscopic derivation of the Langevin equation is given by modeling the fluid as a collection of independent harmonic oscillators, following Ford, Kac, Mazur, and Zwanzig.

9.2   NATURE OF BROWNIAN MOTION

In the case of pollen grains suspended in water, the water molecules are constantly in motion, bombarding the pollen grains from all sides with very high frequency. The grains respond to the net effect of the bombardment by behaving in a seemingly erratic manner. However, the very fact that there are a very large number of water molecules colliding with the pollen grains per unit time means that the problem of understanding the random motion is susceptible to a statistical treatment. In this way, it should be possible to relate macroscopic properties such as viscosity to random displacements of a Brownian particle. Einstein was one of the first scientists to derive such relations, which Jean Perrin used to obtain an experimental value for Avogadro's number. Perrin was awarded the Nobel prize for his work.

In terms of stochastic processes (described in Section 1.3), we shall see later that Brownian motion is a stationary Markovian process. In addition, the various probability distributions that are involved in the description of Brownian motion are Gaussian probability distributions. This makes Brownian motion a Gaussian process.

Brownian motion is also called a Wiener process, since the first concise mathematical discussion of Brownian motion was given by Norbert Wiener as a special case of a Markovian stochastic process. Accordingly, we now study the mathematics of Markovian processes.

9.3   MATHEMATICAL PRELIMINARIES

9.3.1   Markovian Processes

As mentioned in Section 1.3, a stochastic process is the development of a random variable $X(t)$. The probability $P[x(t_i)] \, dx(t_i)$ that the value of the stochastic variable at time t_i lies in the range between $x(t_i)$ and $x(t_i) + dx(t_i)$ is known for all x and t_i. For the purpose of mathematical discussion, the exact physical nature of the random variable $X(t)$ is immaterial. For the sake of concreteness, however, one may think of it as one component of the position or velocity of a Brownian particle, in which case the range of allowed values is $(-\infty, \infty)$. Let us assume that the stochastic variable has attained certain values at earlier times $t_{i-1} > t_{i-2} > \cdots > t_0$. We may say that the system, presumably described by the random variable, has been in this sequence of states. We then ask for the probability of a value for the stochastic variable at the next time step, given that the earlier values have been attained. This probability naturally involves the conditional probability density denoted by $P[x(t_i)|x(t_{i-1}), x(t_{i-2}), \ldots, x(t_0)]$. A Markovian process is one whose probability densities satisfy the equation

$$P[x(t_i)|x(t_{i-1}), x(t_{i-2}), \ldots, x(t_0)] = P[x(t_i)|x(t_{i-1})] \tag{9.1}$$

In other words, in a Markovian process the conditional probability density for obtaining the value $x(t_i)$, at a future time t_i, depends only on the conditions existing at the present time t_{i-1}, but not on any of the previous times. Such a conditional probability density is called a transition probability density. By using definition (9.1) and the laws of probability theory discussed in Chapter 1, we shall obtain an equation satisfied by the Markovian process. In general, one may consider Markovian processes involving many random variables. For simplicity, we restrict our attention to the case of a single stochastic variable.

The law of conditional probability can be derived by using the definitions of the conditional and the joint probability densities (Chapter 1). For the special case $P[x(t_i)|x(t_{i-1}), x(t_{i-2})]$, involving three successive times, this law implies that

$$P[x(t_i), x(t_{i-1}), x(t_{i-2})] = P[x(t_i)|x(t_{i-1}), x(t_{i-2})]P[x(t_{i-1}), x(t_{i-2})] \tag{9.2}$$

where the left-hand side denotes the joint probability of the three indicated values of the random variable. Now, on the right-hand side, we make the Markovian assumption on the first factor and use the conditional probability law again for the second factor to get

$$P[x(t_i), x(t_{i-1}), x(t_{i-2})] = P[x(t_i)|x(t_{i-1})]P[x(t_{i-1})|x(t_{i-2})]P[x(t_{i-2})] \tag{9.3}$$

This equation clearly indicates that the evolution of a stochastic variable from an initial state t_{i-2} to t_i, as determined by the joint probability on the left-hand side, is just a product of the initial probability and of the conditional transition probabilities from one state to the next. This "chain rule," which can be generalized to n time steps, is also used to define a Markovian process.

Integration of Eq. (9.3) over all possible values of $x(t_{i-1})$, gives the joint probability for two values of the random variable at times separated by two steps:

$$P[x(t_i), x(t_{i-2})] = \int P[x(t_i)|x(t_{i-1})]P[x(t_{i-1})|x(t_{i-2})]P[x(t_{i-2})] \, dx(t_{i-1}) \tag{9.4}$$

Of course, if the stochastic variable is discrete valued, the integral in Eq. (9.4) should be replaced by a sum. Also, in that case, the probabilities rather than the probability densities are involved. Applying the conditional probability law once more to the left-hand side of this equation and cancelling $P[x(t_{i-2})]$ from both sides, we obtain

$$P[x(t_i)|x(t_{i-2})] = \int P[x(t_i)|x(t_{i-1})]P[x(t_{i-1})|x(t_{i-2})]\,dx(t_{i-1}) \qquad (9.5)$$

This basic equation of Markovian theory is called the *Chapman–Kolmogorov* or the *Smoluchowski* equation. For the discussion to follow, we change notation slightly and write Eq. (9.5) as

$$P[x_2(t_2)|x_0(t_0)] = \int P[x_2(t_2)|x_1(t_1)]P[x_1(t_1)|x_0(t_0)]\,dx_1(t_1) \qquad (9.6)$$

The Chapman–Kolmogorov equation implies that the conditional probability density for a Markovian system to be in the state x_2 at time t_2 when it began in the state x_0 at t_0 is equal to the "sum" of the conditional probability densities of all the ways the system can get from x_0 to x_2. The Markovian nature of this equation is embodied in the occurrence of independent conditional probabilities.

9.3.2 Fokker–Planck Equation

In general, the transition probability density, such as $P[x_1(t_1)|x_0(t_0)]$ in Eq. (9.6), may depend on all four quantities entering into its definition, x_1, x_0, t_1, and t_0. At the other extreme, it may be just a constant, independent of all four of these. This latter case was true for the symmetric random walk in Chapter 1, which makes the random walk a Markovian process. Here, we consider an intermediate case, where the time dependence is such that the transition probability densities are functions only of the difference $t_1 - t_0$. To remove any further dependence on absolute time, we shall assume that the initial probability density for the starting state $X(t_0)$ is independent of time. According to our earlier discussion, Markovian processes that satisfy these two conditions are stationary Markovian processes for which Eq. (9.6) becomes

$$P[x_2(t + \tau)|x_0] = \int P[x_2(\tau)|x_1]P[x_1(t)|x_0]\,dx_1 \qquad (9.7)$$

where $t = t_1 - t_0$, and the time interval $\tau = t_2 - t_1$ is small. One process to which Eq. (9.7) will be applied is Brownian motion, whereby the Brownian particle experiences rapid, random fluctuations of its position variable $x(t)$. To study these fluctuations, let us write $x_2 - x_1 = \xi$, where ξ is considered to be small. By eliminating x_1 in favor of ξ in Eq. (9.7),

$$P[x_2(t + \tau)|x_0] = \int P[x_2(\tau)|x_2 - \xi]P[x_2(t) - \xi|x_0]\,d\xi \qquad (9.8)$$

Now, we expand Eq. (9.8) for small τ and small ξ in order to obtain a differential equation for the probability density. Expanding the second factor in the integrand of Eq. (9.8) in a Taylor's series in $x_2 - \xi$ about $\xi = 0$ up to second order reduces the right-hand side to

$$\int P[x_2(\tau)|x_2 - \xi]\left[P[x_2(t)|x_0] - \xi\left(\frac{\partial P[x_2 - \xi|x_0]}{\partial(x_2 - \xi)}\right)_0 \right.$$
$$\left. + \frac{1}{2}\xi^2\left(\frac{\partial^2 P[x_2 - \xi|x_0]}{\partial(x_2 - \xi)^2}\right)_0 + \cdots\right]d\xi \qquad (9.9)$$

where we have omitted the argument t of x_2 in the derivatives for simplicity. We further simplify the notation by introducing the quantities

$$\langle \xi^n \rangle_\tau = \int \xi^n P[x_2(\tau)|x_2 - \xi]\,d\xi \qquad (9.10)$$

The special case $n = 0$ of Eq. (9.10) assures that the probabilities are normalized, and the subscript τ serves to remind us that the averages of powers of ξ are taken at fixed τ. Another subtle point about Eq. (9.10) must be noted. The averages given by this equation depend on x_2, in principle, but including this dependence will take us into considerations such as a space-dependent diffusion constant and other complications. Therefore, we assume that the averages defined by Eq. (9.10) are independent of x_2 and introduce these averages into the expression (9.9), which is the right-hand side of Eq. (9.8). Equation (9.8) then becomes

$$P[x_2(t + \tau)|x_0] = P[x_2(t)|x_0] - \langle\xi\rangle_\tau\left(\frac{\partial P[x_2 - \xi|x_0]}{\partial(x_2 - \xi)}\right)_0$$
$$+ \frac{1}{2}\langle\xi^2\rangle_\tau\left(\frac{\partial^2 P[x_2 - \xi|x_0]}{\partial(x_2 - \xi)^2}\right)_0 + \cdots \qquad (9.11)$$

where the subscripts on the derivatives imply $\xi = 0$. Expanding the left-hand side of Eq. (9.11) about $\tau = 0$ gives

$$P[x_2(t + \tau)|x_0] = P[x_2(t)|x_0] + \tau\left(\frac{\partial P[x_2 + \tau|x_0]}{\partial(t + \tau)}\right)_0 + \cdots, \qquad (9.12)$$

where the subscript 0 indicates $\tau = 0$. By subtracting Eq. (9.11) from Eq. (9.12), dividing by τ, and taking the limit $\tau \to 0$ in the resulting expression, we obtain

$$\left(\frac{\partial P[x_2 + \tau|x_0]}{\partial(t + \tau)}\right)_{\tau=0} = -A\left(\frac{\partial P[x_2 - \xi|x_0]}{\partial(x_2 - \xi)}\right)_{\xi=0}$$

$$+ B\left(\frac{\partial^2 P[x_2 - \xi|x_0]}{\partial(x_2 - \xi)^2}\right)_{\xi=0} + \cdots \quad (9.13)$$

where the constants A and B are defined by the following limits, which are assumed to exist.

$$A = \lim_{\tau \to 0} \frac{\langle \xi \rangle_\tau}{\tau} \quad (9.14)$$

$$B = \lim_{\tau \to 0} \frac{1}{2} \frac{\langle \xi^2 \rangle_\tau}{\tau} \quad (9.15)$$

Usually, terms higher than the second derivative on the right-hand side of Eq. (9.13) are ignored. The equation obtained by neglecting these terms is called the *Fokker–Planck equation*. Recognizing that the probability density depends on t and x_2, and writing x for the latter, the Fokker–Planck equation may be written in the more familiar form

$$\frac{\partial P(x,t)}{\partial t} = -A\frac{\partial P(x,t)}{\partial x} + B\frac{\partial^2 P(x,t)}{\partial x^2} \quad (9.16)$$

where A and B are the constants defined above. It should be remembered that the probability density in Eq. (9.16) is a conditional one, even though the notation has been simplified. In practice, the fact that the probability density $P(x, 0)$ is given translates into an initial or a boundary condition for the partial differential equation (9.16).

To summarize, we started with a general random process. The Markovian assumption led to the Chapman–Kolmogorov equation (9.6). The assumption of stationarity reduced this equation to Eq. (9.7). Equation (9.7) is difficult to solve, since it is a nonlinear integral equation containing no specific physical information. Physical content was supplied above by assuming a heavy Brownian particle bombarded by very light particles. The initial position was fixed at x_0, and the heavy particle underwent a large number of collisions of very brief duration compared to the macroscopic time t. At one of the arbitrary positions x_1, the Brownian particle moved a very short distance ξ in a very brief time τ to arrive at x_2, so that we considered only those positions

x_1 that were within ξ of x_2. Smallness of ξ and τ made it possible for us to derive the partial differential equation (9.13). However, the magnitudes of ξ and τ had to be large enough for Markovian behavior to apply. It is this requirement on the motion of the Brownian particle that renders the process irreversible. The assumption that there exist space and time intervals ξ and τ small enough for the derivatives to exist and large enough for the validity of coarse graining is the basis of the hydrodynamic description of Brownian motion. We have already encountered this assumption in the discussion of the Boltzmann equation in Chapter 8. Finally, the neglect of spatial derivatives higher than second gave rise to the Fokker–Planck equation (9.16). This equation describes the continuous development of the transition probability density of a stochastic Markovian process from an initial value of this density. In Section 9.4, we consider Einstein's equation for diffusion and show that it satisfies Eq. (9.16).

9.4 DIFFUSION OF A BROWNIAN PARTICLE

9.4.1 Fick's Laws

If a system is prepared with a nonuniform density, it is an experimental fact that an irreversible flow of mass takes place from regions of high density to ones of lower density till a uniform density is obtained. This process is called diffusion, and the diffusion flux, or the rate of flow of mass per unit area per unit time, is found to be proportional to the gradient of the density,

$$\rho_m \mathbf{v} = -D\nabla \rho_m \tag{9.17}$$

In this equation, which is called *Fick's first law*, $\mathbf{v}$ is the velocity of flow and the constant D is the diffusion coefficient. This equation is one of many phenomenological transport equations (see Chapter 10). If this equation is substituted into the right-hand side of the equation of continuity,

$$\frac{\partial \rho_m}{\partial t} = -\nabla \cdot (\rho_m \mathbf{v}) \tag{9.18}$$

which expresses the conservation of mass [Eq. (8.46)], we get

$$\frac{\partial \rho_m}{\partial t} = D\nabla^2 \rho_m \tag{9.19}$$

This equation is *Fick's second law*, which is also called *the diffusion equation*.

9.4.2   Einstein's 1905 Equation

If particle density is substituted for mass density in the diffusion equation and if it is assumed that the mass density is proportional to the probability density P, the diffusion equation becomes

$$\frac{\partial P}{\partial t} = D\nabla^2 P \tag{9.20}$$

Of course, the probability density is assumed to be suitably normalized. If one-dimensional Brownian motion is characterized by its position coordinate z, Eq. (9.20) becomes

$$\frac{\partial P}{\partial t} = D\frac{\partial^2 P}{\partial z^2} \tag{9.21}$$

This equation is identical to the Fokker–Planck equation (9.16), with $A = 0$, and $B = D$. In analogy to Eq. (9.15), D is defined as

$$D = \lim_{\tau\to 0}\frac{1}{2}\frac{\langle \Delta z^2\rangle_\tau}{\tau} \tag{9.22}$$

where Δz is the displacement caused by the fluctuations in z, with $\langle \Delta z\rangle = 0$ in agreement with $A = 0$. Equation (9.22) was first derived by Einstein in 1905 (Einstein, 1956, p. 17). This paper, along with other works of Einstein on Brownian motion, have been translated by A. D. Cowper and edited by R. Fürth and is now in a Dover edition (Einstein, 1956). It shows that the second moment of the displacement is directly proportional to the displacement time, the constant of proportionality being twice the phenomenological diffusion constant. For a historical introduction and discussion of Einstein's work on Brownian motion and other subjects, in the context of contemporary physics, see the book by Pais (Pais, 1982).

The solution of Eq. (9.21), subject to the initial condition that the Brownian particle is present at the position z_0 at $t = 0$ or $P(z, 0) = \delta(z - z_0)$, is given by the Gaussian function

$$P(z, t) = \frac{1}{(4\pi Dt)^{1/2}} \exp\left[\frac{-(\Delta z)^2}{4Dt}\right] \tag{9.23}$$

with $\Delta z = z - z_0$. Equation (9.23) satisfies Eq. (9.21), as can be verified by direct substitution. The quantity $P(z, t)$ is the conditional probability density that, at a given time t, the stochastic position variable $Z(t)$ has a value z. Since the probability density is Gaussian, one can calculate the various moments and correlations by using the standard properties of Gaussian integrals. Thus,

$$\langle z(t) \rangle = z_0 \tag{9.24}$$

and

$$\langle z(t')z(t'') \rangle = \langle z(t') \rangle \langle z(t'') \rangle = z_0^2 \qquad t' \neq t'' \tag{9.25}$$

$$\langle [z(t)]^2 \rangle = z_0^2 + 2Dt \tag{9.26}$$

which yield

$$\langle \Delta z(t) \rangle = 0 \tag{9.27}$$

and

$$\langle \Delta z(t') \Delta z(t'') \rangle = 0 \qquad t' \neq t'' \tag{9.28}$$

$$\langle [\Delta z(t)]^2 \rangle = \langle [z(t)]^2 \rangle - z_0^2 = 2Dt \tag{9.29}$$

Equation (9.23) describes what happens to a Brownian particle at $t > 0$ if it is introduced into a fluid at $t = 0$. It spreads out into the fluid with the position coordinate satisfying the following. The average displacement from the initial position is zero. The displacements at different times are entirely uncorrelated as seen in Eq. (9.28), which is a consequence of the Brownian motion being a Markovian process. The squared displacement at a given time, which is the variance, is proportional to t. The spreading also depends on the diffusion constant D, which can be determined experimentally by examining the displacement of Brownian particles in the fluid. What properties of the system determine this constant? Can we relate it to certain microscopic properties of the fluid? To answer these questions, we turn to diffusion or Brownian motion under gravity, just as Einstein did in his 1906 paper (Einstein, 1956, p. 19).

9.4.3 Brownian Particle in a Gravitational Field

Consider a particle of mass m suspended in a viscous fluid. Let there be a gravitational field in the $-z$ direction with an acceleration g due to gravity. Because of the gravitational force, there is an overall drifting of the Brownian particle in the downward direction superimposed on its random motion. Because of the downward drift, there is an upward frictional force caused by the fluid viscosity. According to Stokes' law, this frictional force is directly proportional to the velocity of the particle, the constant of proportionality

γ being called the coefficient of friction. By using hydrodynamics, one can show that $\gamma = 6\pi\eta a$, where a is the radius of a spherical Brownian particle and η is the coefficient of viscosity of the fluid. [For a derivation, see (Huang, 1987, pp. 119–122).] If we neglect the force of buoyancy due to the difference in the density of the fluid and the particle, the total force on the particle is the sum of the gravitational and the frictional forces.

Initially, the downward velocity of the particle will increase because of the constant force of gravity and decrease owing to the frictional force, which increases with velocity. Therefore, after a certain time, the two forces will balance, and the particle will move with a constant terminal or drift velocity $\mathbf{v}_d$. When this velocity is attained, the downward gravitational force must be equal to the upward frictional force:

$$- mg\widehat{\mathbf{z}} = \gamma\mathbf{v}_d \tag{9.30}$$

where $\widehat{\mathbf{z}}$ is a unit vector in the upward direction. Because of the drift velocity, there is an extra mass flow per unit area per unit time given by $\rho_m\mathbf{v}_d$, which must be added to the right-hand side of Eq. (9.19). The extra term is given by

$$\frac{\partial\rho_m}{\partial t} = \frac{\partial\rho_m}{\partial z}\frac{\partial z}{\partial t} = v_d\frac{\partial\rho_m}{\partial z} \tag{9.31}$$

so that Eq. (9.19) is replaced by

$$\frac{\partial\rho_m}{\partial t} = D\frac{\partial^2\rho_m}{\partial z^2} + \frac{mg}{\gamma}\frac{\partial\rho_m}{\partial z} \tag{9.32}$$

where we recognize that the space variations are in the z direction only. Once again, replacing the mass density with the probability density, we obtain

$$\frac{\partial P}{\partial t} = D\frac{\partial^2 P}{\partial z^2} + \frac{mg}{\gamma}\frac{\partial P}{\partial z} \tag{9.33}$$

This equation generalizes the diffusion equation (9.21) to the case of a uniform field and is exactly the Fokker–Planck equation (9.16), with $B = D$, as before, and $A = -mg/\gamma$. The drift velocity has the interpretation that it is the average displacement in the particle's position $\langle\Delta z\rangle$ during a short time interval τ:

$$v_d = \lim_{\tau\to 0}\frac{\langle\Delta z\rangle_\tau}{\tau} \tag{9.34}$$

The diffusion constant is still given by Eq. (9.22). The solution of Eq. (9.33), which can be found in the review by Chandrasekhar (Chandrasekhar, 1943, pp. 57–59), tells us how the probability density changes with height and time for the Brownian particle in a gravitational field. After a very long time, when equilibrium has been reached, P no longer depends on time, and Eq. (9.33) reduces to

$$0 = D\frac{d^2P}{dz^2} + \frac{mg}{\gamma}\frac{dP}{dz} \qquad (9.35)$$

By integrating with respect to z and using the condition that, for $z \to \infty$, both P and $dP/dz \to 0$, we get

$$0 = D\frac{dP}{dz} + \frac{mg}{\gamma}P \qquad (9.36)$$

Going back to mass density,

$$0 = D\frac{d\rho_m}{dz} + \frac{mg}{\gamma}\rho_m \qquad (9.37)$$

We can obtain an equation for the variation of the density of Brownian particles in a gravitational field in another way. Since the particles are independent, we can think of them as forming an ideal gas. By balancing the forces acting on an ideal gas under gravity, it is shown in many physical chemistry textbooks (e.g., Levine, 1995, Section 15.8) that

$$0 = k_B T\frac{d\rho_m}{dz} + mg\rho_m \qquad (9.38)$$

where k_B is the Boltzmann constant and T is the temperature of the fluid. [Equation (9.38) is identical to Eq. (8.54) obtained previously by using this method.] Equation (9.38) can be easily solved to give

$$\rho_m = \rho_m(0)\exp(-mgz/k_B T) \qquad (9.39)$$

Alternatively, knowing the principles of statistical mechanics, we can write Eq. (9.39) immediately by using the canonical distribution of equilibrium statistical mechanics (see Section 1.5.2). By comparing Eqs. (9.38) and (9.35), we obtain

$$D = k_B T/\gamma \qquad (9.40)$$

Equation (9.40) is another relation obtained by Einstein (Einstein, 1956, p. 12). This equation relates the diffusion constant, which is given by the mean-square fluctuations of z per unit time, to the friction coefficient, which is responsible for thermal dissipation. This equation is an example of a fluctuation-dissipation relation (see below).

Why are the two quantities, fluctuations and the dissipation coefficient, related at all? Both of these arise owing to the random force that the fluid particles exert on the Brownian particles. One may think of dissipation as the average effect of this random force and the fluctuations of the random force around this average as causing the spread or the diffusion of the particles from their original positions. Both ought to be related, because they are both rooted in the random bombardment of the Brownian particles by the fluid. Precisely this point of view was adopted by Langevin to modify the deterministic Newton's second law of motion to include the effects of the forces due to the fluid. We now turn to Langevin's equation.

9.5   THE LANGEVIN EQUATION

So far, we studied Brownian motion by deriving and solving equations for the probability density of observing a particular value of z at a time t. Another way of describing Brownian motion was suggested by Paul Langevin in the early part of this century. Following Langevin, we separate the total force on a Brownian particle into the average Stokes frictional force, which the particle experiences as it moves through the surrounding fluid, and a fluctuating force. We assume no other forces on the particle apart from the forces exerted by the fluid. In one dimension, this separation can be written as

$$m\frac{dv}{dt} = -\gamma v + R(t) \tag{9.41}$$

where v is the velocity of a Brownian particle of mass m, γ is the friction coefficient, and $R(t)$ is the stochastic force. This equation is the *Langevin equation* in one dimension. Our aim now is to study the velocity of the Brownian motion in detail.

Equation (9.41) for $v(t)$ can be solved directly by using the theory of first-order differential equations or by the Laplace transform method. In either case, we get

$$v(t) = v_0 \exp\left(\frac{-\gamma t}{m}\right) + \frac{1}{m} \exp\left(\frac{-\gamma t}{m}\right) \int_0^t \exp\left(\frac{\gamma t}{m}\right) R(\tau)\, d\tau \tag{9.42}$$

It can be verified by direct substitution that Eq. (9.42) satisfies Eq. (9.41). We next take an ensemble average of Eq. (9.42). This averaging can be done in two ways. Either one can consider a large number of independent Brownian particles and average over them, or one can take a time average of the motion of a single Brownian particle over widely separated intervals so as to ensure that motions over the time intervals are independent. Because there is no external force on the particle, its ensemble averaged velocity should be zero after a long time. This result can be ensured by assuming that

$$\langle R(t) \rangle = 0 \tag{9.43}$$

With this assumption, the ensemble average of the velocity at time t is obtained by taking the average of Eq. (9.42). We get

$$\langle v(t) \rangle = v_0 \exp(-\gamma t/m) \tag{9.44}$$

since the mean of $R(\tau)$ vanishes. Similarly, the mean-square velocity is

$$
\begin{aligned}
\langle v^2(t) \rangle = {}& v_0^2 \exp\left(\frac{-2\gamma t}{m}\right) \\
& + \frac{\exp(-2\gamma t/m)}{m^2} \int_0^t dt' \int_0^t \exp\left[\frac{\gamma(t' + t'')}{m}\right] \langle R(t')R(t'') \rangle \, dt''
\end{aligned}
\tag{9.45}
$$

where the cross-term involving $\langle R(t) \rangle$ vanishes.

At this point, we assume that the random force is a stationary Markovian process, so that

$$\langle R(t')R(t'') \rangle = \Gamma \delta(t' - t'') \tag{9.46}$$

where Γ is a time-independent constant and $\delta(t' - t'')$ is the Dirac delta function. Use of the delta function assumes that the stochastic force is impulsive, so that collisions are uncorrelated. The vanishing of the force autocorrelation function at different times ensures the Markovian nature of the stochastic process, which the random force represents. Insertion of Eq. (9.46) into Eq. (9.45) gives

$$\langle v^2(t) \rangle = v_0^2 \exp\left(\frac{-2\gamma t}{m}\right) + \frac{\Gamma}{2\gamma m}\left[1 - \exp\left(\frac{-2\gamma t}{m}\right)\right] \tag{9.47}$$

where the usual delta function property has been used.

At long times,

$$\langle v^2(\infty) \rangle = \frac{\Gamma}{2\gamma m} = \frac{k_B T}{m} \tag{9.48}$$

or

$$\Gamma = 2k_B T\gamma \tag{9.49}$$

where only one dimension is being considered. Equation (9.49), which relates the coefficient of friction to the fluctuating force autocorrelation function, is known as the *fluctuation–dissipation theorem*.

These results enable us to obtain a Fokker–Planck equation for the velocity of a Brownian particle. Assuming that the stochastic variable now represents the velocity of the particle, we have $\langle \xi \rangle = \langle \Delta v \rangle$ and $\langle \xi^2 \rangle = \langle (\Delta v)^2 \rangle$. By expanding Eq. (9.44) for short times,

$$\langle v(t) \rangle \approx v_0 \left(1 - \frac{\gamma t}{m} \right) \tag{9.50}$$

and

$$\langle \Delta v \rangle \approx -v_0 \frac{\gamma t}{m} \tag{9.51}$$

where only the linear terms have been kept because t is small. Now,

$$\langle (\Delta v)^2 \rangle = \langle (v - \langle v \rangle)^2 \rangle = \langle v^2 \rangle - \langle v \rangle^2 \tag{9.52}$$

By expanding Eq. (9.47) for short times and using the resulting expression and Eq. (9.50) in Eq. (9.52), we get

$$\langle (\Delta v)^2 \rangle = \frac{\Gamma t}{m^2} \tag{9.53}$$

For this case, the Fokker–Planck equation becomes

$$\frac{\partial P}{\partial t} = \frac{v_0 \gamma}{m} \frac{\partial P}{\partial v} + \frac{\Gamma}{2m^2} \frac{\partial^2 P}{\partial v^2} \tag{9.54}$$

Equation (9.54) is mathematically identical to Eq. (9.32). Hence, the solution is similar to that of Eq. (9.32).

9.6   BROWNIAN MOTION AS A TIME SERIES

A series of discrete numerical values of a stochastic variable is called a *time series* in the field of statistical analysis. Since Brownian motion consists of a series of values of the position coordinate at different times, it is possible to treat it by time series analysis. In fact, by using the procedure in this section, one can generate Brownian motion on a computer.

One starts with a random number generator that produces *uniform deviates*. These are random numbers that lie in a specified range such as 0 to 1 or -1 to $+1$. The distribution function for uniform deviates is the uniform probability distribution. Hence, at a given time all uniform deviates are equally probable. By taking the generation of uniform deviates, say y_i, at discrete times t_i, as a stochastic process, we obtain *uniform white noise*, in which the y_i are independent and thus uncorrelated so that

$$\langle y_i y_{i-1} \rangle = \langle y_i \rangle \langle y_{i-1} \rangle = 0 \tag{9.55}$$

The term white noise is used because the Fourier transform of the process with respect to time gives a uniform distribution in the resulting frequency space. In analogy to light, this implies a mixture of all colors, giving rise to the term white noise.

It is possible to generate white noise using distributions other than uniform. For example, one could use a Gaussian distribution characterized by a given mean and variance. Random number generators that use a Gaussian distribution with zero mean and unit variance produce *normal deviates*. The statistical distribution function in this case has the form

$$P(y_i) = \frac{2}{\sqrt{\pi}} \exp(-y_i^2) \tag{9.56}$$

A plot of y_i versus time gives *Gaussian white noise*. The variable y_i is called a standardized random variable.

By summing y_i up to time t, we can define a stochastic process $Z(t)$ as an observation $z(t)$.

$$z(t) = \sum_{i=0}^{t} y_i \tag{9.57}$$

A series of observations $z(1), z(2), \ldots, z(t)$ of the cumulative process $Z(t)$ represents a time series, which is a realization of the stochastic process. The y_i can be considered as a series of "shocks," which drives the process. We saw earlier that Brownian motion is a Gaussian process. If we represent this process by Eq. (9.57), we take z to be a displacement caused by the random shocks. The shocks represent the random fluctuations of the bath in which the Brownian particle moves.

It can be shown by using the *central limit theorem* (Feller, 1968, Vol. II, Chapter VIII) that the variance of $z(t)$ is proportional to t.

$$\text{Var}[z(t)] \propto t \tag{9.58}$$

Also, if $\Delta z(t + \tau) = z(t + \tau) - z(t)$,

$$\text{Var}[\Delta z(t + \tau)] \propto \tau \tag{9.59}$$

in agreement with Einstein's result [Eq. (9.29)]. The variance represents the "spread" in the Gaussian distribution function. Equation (9.58) indicates that this spread increases with time as the particle samples the entire space available to it.

Equation (9.57), which states that Brownian motion is a sum over Gaussian white noise, provides a method of generating Brownian motion on a computer. Shown in Figure 9.1 is a plot of y_i versus time for 100 normal deviates generated by the FORTRAN program GSSWTN (listed in Appendix D.4). The Brownian motion generated from this white noise is shown in Figure 9.2 as a plot of $z(t)$ versus time for 100 Brownian motion points, each point being a sum of five white noise points. This figure was generated by the FORTRAN program BRNMTN (listed in Appendix D.1).

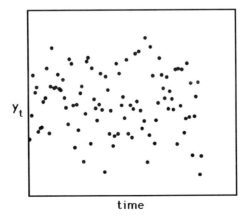

Figure 9.1   Gaussian white noise.

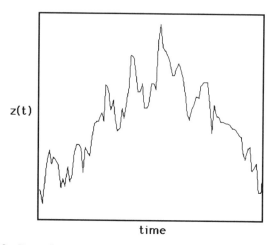

Figure 9.2 Brownian motion generated from the white noise of Figure 9.1.

9.7 MICROSCOPIC DERIVATION OF THE LANGEVIN EQUATION

Langevin proposed his Eq. (9.41) in 1908 on purely phenomenological grounds. Since then, there has been a great deal of progress in nonequilibrium statistical mechanics. Among other things, we have learned how to derive equations like Langevin's from a microscopic description of the many-body system. A seminal approach called the Zwanzig–Mori approach consists of the following steps. Start with a Hamiltonian of the many-body system. Separate the variables into two categories: important and unimportant. Solve the equations of motion of the unimportant variables and eliminate them by substituting the solutions into the equations of motion of the relevant or important variables. Take an ensemble average and, by making suitable assumptions about the distributions of the eliminated variables, obtain the stochastic equations for the relevant variables. This program can be carried out beautifully by using the projection operator method. We shall discuss this method in detail in Chapter 10. In this section, to illustrate this procedure, we derive the Langevin equation from a many-body Hamiltonian following Ford, Kac, Mazur (Ford et al., 1965) and Zwanzig (Zwanzig, 1973).

Consider a system of $N + 1$ particles, one of which, called the Brownian particle, has a mass m, coordinate x, and momentum p. The other N particles have masses m_i and coordinates and momenta labeled as x_i and p_i and constitute a system of independent harmonic oscillators. This N-particle system models the fluid surrounding the Brownian particle and is treated as a heat bath. The Brownian particle is *linearly* coupled to each of the N bath particles.

The Hamiltonian of the system is given by

$$\mathcal{H} = \frac{p^2}{2m} + \phi(x) + \sum_{i=1}^{N} \left(\frac{p_i^2}{2m_i} + \frac{1}{2} m_i \omega_i^2 \bar{x}_i^2 \right) \tag{9.60}$$

with

$$\bar{x}_i \equiv x_i - \frac{c_i}{m_i \omega_i^2} x \tag{9.61}$$

Here, $\phi(x)$ is the potential in which the Brownian particle is moving. The quantities ω_i are the angular frequencies of the harmonic oscillators, and the c_i are the coupling constants between the Brownian particle and each of the bath particles. Hamilton's equations of motion for the Brownian particle are

$$\frac{dx}{dt} = \frac{p}{m} \tag{9.62}$$

$$\frac{dp}{dt} = -\frac{d\phi}{dx} + \sum_{i=1}^{N} c_i \bar{x}_i \tag{9.63}$$

Hamilton's equations of motion of a bath particle are

$$\frac{dx_i}{dt} = \frac{p_i}{m_i} \tag{9.64}$$

$$\frac{dp_i}{dt} = -m_i \omega_i^2 \bar{x}_i \tag{9.65}$$

The momenta can be eliminated from these equations to give

$$m \frac{d^2 x}{dt^2} = -\frac{d\phi}{dx} + F_b(t) \tag{9.66}$$

where $F_b(t)$, the force on the Brownian particle due to the heat bath particles, is given by

$$F_b(t) = \sum_{i=1}^{N} c_i \bar{x}_i \tag{9.67}$$

and

$$m_i \frac{d^2 x_i}{dt^2} = m_i \omega_i^2 x_i - c_i x \tag{9.68}$$

Equations (9.66) and (9.68) are the equations of motion of forced oscillators and can be solved using standard techniques such as the Laplace transform. Our aim is to eliminate the bath particles. So, we solve Eq. (9.68) formally to get

$$x_i(t) = x_i(0) \cos \omega_i t + \frac{\dot{x}_i(0) \sin \omega_i t}{\omega_i} + \frac{c_i}{m_i \omega_i} \int_0^t x(t') \sin \omega_i(t - t') dt' \quad (9.69)$$

where $x_i(0)$ and $\dot{x}_i(0)$ are the initial values of the position and velocity of the bath particle labeled i. As usual, we have denoted a time derivative by a dot. The last term in Eq. (9.69) can be treated by a partial integration to yield

$$\frac{c_i}{m_i \omega_i^2} \left[x(t) - x(0) \cos \omega_i t - \int_0^t \dot{x}(t') \cos \omega_i(t - t') \right] dt' \quad (9.70)$$

Substituting this expression into Eq. (9.69) and then using the resulting expression in Eq. (9.67) and simplifying, we find that the expression for the bath force reduces to

$$F_b(t) = \sum_{i=1}^N \left[c_i \bar{x}_i(0) \cos \omega_i t + \frac{c_i}{\omega_i} \dot{x}_i(0) \sin \omega_i t \right.$$

$$\left. - \frac{c_i}{m_i \omega_i^2} \int_0^t dt' \dot{x}(t') \cos \omega_i(t - t') \right] \quad (9.71)$$

So far we have not made any assumptions about the heat bath. Now we assume that the whole system is in equilibrium at temperature T and time $t = 0$. From the Hamiltonian (9.60), the equilibrium hypothesis implies that at $t = 0$ the canonically conjugate variables

$$\bar{x}_i(0) \equiv x_i(0) - \frac{c_i}{m_i \omega_i^2} x(0) \quad (9.72)$$

and

$$p_i(0) \equiv m_i \dot{x}_i(0) \quad (9.73)$$

are distributed according to the canonical distribution

$$P[\bar{x}_i(0), \dot{x}_i(0)] \sim \exp\left(-\frac{\mathcal{H}}{k_B T}\right) = \prod_{i=1}^N \exp\left[-\frac{m_i \dot{x}_i^2(0)}{2k_B T} - \frac{m_i \omega_i^2 \bar{x}_i^2(0)}{2k_B T}\right]$$

$$(9.74)$$

This equation implies the following relations about the canonical ensemble averages:

$$\langle \bar{x}_i(0) \rangle = \langle \dot{x}_i(0) \rangle = 0$$

$$\langle \bar{x}_i(0)\bar{x}_j(0) \rangle = \delta_{ij} \frac{k_B T}{m_i \omega_i^2} \qquad \langle \dot{x}_i(0)\dot{x}_j(0) \rangle = \delta_{ij} \frac{k_B T}{m_i} \qquad (9.75)$$

where δ_{ij} is the Kronecker delta and all the other two-variable averages vanish identically. By using these relations in Eq. (9.71), it is found that the ensemble average of the bath force is given by

$$\langle F_b(t) \rangle = -m \int_0^t \dot{x}(t') \zeta(t - t')\, dt' \qquad \zeta(t - t') = \frac{1}{m} \sum_{i=1}^N \frac{c_i}{m_i \omega_i^2} \cos \omega_i(t - t')$$

$$(9.76)$$

This average force depends on the velocity of the Brownian particle. In this, it is similar to the hydrodynamic Stokes' law force, which is also proportional to the velocity. But the force of Eq. (9.76) at time t depends not only on the velocity at time t but on all the velocities the particle had at previous times. In other words, the force has a memory of the history of the motion of the Brownian particle. The quantity $\zeta(t)$ is called a *memory friction kernel*. Therefore, it is evident that the process described by the Hamiltonian (9.60) is not a Markovian process. We shall discuss such processes in detail in Chapter 10.

If we subtract the average force given by Eq. (9.76) from the total bath force given by Eq. (9.71), we obtain the stochastic or the random force. This force is given by

$$R(t) \equiv F_b(t) - \langle F_b(t) \rangle = \sum_{i=1}^N \left[c_i \bar{x}_i(0) \cos \omega_i t + \frac{c_i}{\omega_i} \dot{x}_i(0) \sin \omega_i t \right] \qquad (9.77)$$

Clearly,

$$\langle R(t) \rangle = 0 \qquad (9.78)$$

By using Eq. (9.71) and the properties of the thermal canonical distribution given by Eqs. (9.75), it can be shown that the correlation in the random force is given by

$$\langle R(t)R(t') \rangle = mk_B T \zeta(t - t') \qquad (9.79)$$

By combining Eqs. (9.66) and (9.71), it is seen that Eq. (9.66) is in the form of the Langevin equation with the random force given in Eq. (9.77). There is one important difference. The equation we have derived is the Langevin equation with memory or the *generalized Langevin equation* (GLE),

$$m\frac{d^2x}{dt^2} = -\frac{d\phi}{dx} - m\int_0^t \dot{x}(t')\zeta(t - t')\,dt' + R(t) \tag{9.80}$$

To reduce it to the usual free-particle Markovian Langevin equation (9.41), we assume $\phi(x) = 0$ and compare Eqs. (9.79) and (9.46) to get

$$\zeta(t) = (\Gamma/mk_{\mathrm{B}}T)\delta(t - t') = (2\gamma/m)\delta(t - t') \tag{9.81}$$

where the fluctuation–dissipation relation (9.49) has been used. Now, Eq. (9.80) reduces to Eq. (9.41).

9.8 BROWNIAN MOTION AND IRREVERSIBILITY

A Brownian particle undergoes irreversible, random movements. For this reason, theories of Brownian motion are intimately bound to theories of irreversible behavior. Approaches that explain Brownian motion can therefore be used to understand irreversible behavior in general.

All averages such as position and velocity can be rigorously calculated from the Liouville equation. However, this equation is generally difficult to formulate and to solve. The solutions of the Liouville equation are reversible, so that the equation must be modified in some way if it is to specifically show irreversibility. Some attempts at modification include coarse graining and time smoothing of a reduced Liouville equation (Isihara, 1971, Section 6.1). Another attempt, the Boltzmann equation, was studied in detail in Chapter 8. In the Boltzmann equation, irreversibility was introduced through Boltzmann's assumption of molecular chaos.

In this chapter, irreversibility has been introduced through the Markovian assumption. The Chapman–Kolmogorov equation is the basic equation of Markovian behavior. As a practical matter, variations of this equation that are more amenable to physical interpretation are used. These include the master (see Section 13.3.4) and Fokker–Planck equations.

The Langevin equation is not restricted to Markovian behavior, although the equation is simplified if Markovian behavior is introduced, as in Eq. (9.46). In fact, much effort has been devoted to studying the irreversible behavior of the non-Markovian Langevin equation (9.80), which we have derived here by starting from a microscopic Hamiltonian. Chapter 10 will present the derivation of the GLE from a very general point of view and explore methods for studying the GLE.

9.9 EXERCISES AND PROBLEMS

1. Calculate the root-mean-square displacement in the x direction after 1.00 h for a hemoglobin molecule in infinitely dilute aqueous solution at $25°C$ if the diffusion coefficient for hemoglobin under these conditions is $7.00 \times 10^{-11} m^2 s^{-1}$.

2. Find $\langle z \rangle$ using the probability density given by Eq. (9.23).

3. Relate ρ_m in Eq. (9.39) to the pressure of an ideal gas. Find the pressure of air (MW $= 29$) at 5000 ft. Take the acceleration due to gravity to be $9.81 m s^{-2}$.

4. Derive Eq. (9.47) from Eq. (9.45).

5. Show that Eq. (9.69) satisfies Eq. (9.68) by direct substitution. You will need the following theorem for differentiation under the integral sign

$$\frac{\partial}{\partial \alpha} \int_{a(\alpha)}^{b(\alpha)} f(x; \alpha) \, dx = \int_{a(\alpha)}^{b(\alpha)} \frac{\partial f(x; \alpha)}{\partial \alpha} \, dx - a' f(a; \alpha) + b' f(b; \alpha) \quad (9.82)$$

where the primes denote the derivative with respect to α.

6. Prove the relations (9.75) using the distribution given in Eq. (9.74).

7. Derive Eq. (9.79) using Eqs. (9.77) and (9.75).

10

ZWANZIG–MORI FORMALISM

10.1 INTRODUCTION

In Chapter 9, we treated systems close to equilibrium. We found that there are two general methods of treating such systems. In the first method, one deals with the probability distributions and tries to derive equations satisfied by the probability densities by making suitable assumptions. Examples are the Fokker–Planck equation and the master equation, which we did not discuss. The second method is to deal directly with the averages of the stochastic variables of interest. For example, we discussed the equations satisfied by the velocity of the Brownian particle. Such approaches are based on the Langevin equation and its generalizations.

Both of the above techniques originated in phenomenology; that is, they were based on empirical evidence and took into account the random motions in an ad hoc fashion. For a long time, many scientists struggled to obtain a microscopic basis for Fokker–Planck and Langevin equations. A dramatic advance was made by Zwanzig in 1960 (Zwanzig, 1960) when he presented a rigorous derivation of a generalized Fokker–Planck equation from first principles using the Liouville equation. Some years later, Mori (Mori, 1965) used the same techniques to derive a generalized Langevin equation. The Zwanzig–Mori formalism, as it is now called, liberated nonequilibrium statistical mechanics from phenomenology, unified the various empirical equations used to study systems not in equilibrium, and was responsible for a host of new techniques to deal with nonequilibrium phenomena.

We start with a brief discussion of *linear response theory*, which was formulated to handle systems near equilibrium. Then, we present the projection operator formalism of Zwanzig and Mori, which can be generalized to nonlinear systems. We shall see how the Markovian assumption, made in Chapters 8 and 9, can be removed in order to accommodate non-Markovian behavior, so that a generalized Langevin equation can be derived starting from the Liouville theorem. This derivation goes beyond the derivation given in Section 9.7, since it does not depend on a particular model of the heat bath. The power of the Zwanzig–Mori approach is illustrated by discussions of spectral line shapes and the Rouse–Zimm model of polymer dynamics.

10.2   LINEAR RESPONSE THEORY

If a system is in a nonequilibrium state, the variables describing it depend on space and sometimes time. For example, the concentration of some chemical in solution may depend on the region where it is measured. This imbalance gives rise to flows, the rate of which depends on the strength of the imbalance, which acts as a driving force. The flow gives rise to various transport processes described by certain transport coefficients. In the regime where the imbalance is not too large, or the system is close to equilibrium, the response of the system is linear. In such a case, empirical transport equations are of the form $J = \chi F$, where J is a flux, χ is a transport coefficient, and F is a generalized force. Some common transport equations (in one dimension) are the heat conduction equation

$$\frac{\partial Q}{\partial t} = -\kappa \frac{\partial T}{\partial x} \qquad \text{Fourier's law} \qquad (10.1)$$

for the transport of heat per unit area (Q) across a temperature gradient, the viscosity equation

$$\frac{\partial P}{\partial t} = -\eta \frac{\partial v_y}{\partial x} \qquad \text{Newton's law} \qquad (10.2)$$

for the transport of momentum per unit area (P) across a velocity gradient, the equation for diffusion

$$\frac{\partial N}{\partial t} = -D \frac{\partial c}{\partial x} \qquad \text{Fick's law} \qquad (10.3)$$

for the transport of moles per unit area (N) across a concentration gradient, and the electrical conductivity equation

$$\frac{\partial q}{\partial t} = -\sigma \frac{\partial \phi}{\partial x} \qquad \text{Ohm's law} \qquad (10.4)$$

for the transport of charge per unit area (q) across an electric potential gradient. The quantities κ, η, D, and σ are the transport coefficients. The gradients are set up by a fluctuation or perturbation, which moves the system away from equilibrium. The gradients in these equations then act as generalized forces that drive the system back to equilibrium.

Each of these transport equations has a corresponding equation of continuity, which expresses a conservation law. For example, for matter flow, we have the usual equation of continuity that expresses the conservation of mass. By combining the transport equation with its corresponding equation of continuity, one obtains a hydrodynamic equation of transport. For a typical dynamical variable $A(t)$, this equation is of the form

$$\frac{\partial \langle A \rangle}{\partial t} = \chi \nabla^2 \langle A \rangle \tag{10.5}$$

in three dimensions. An example of a hydrodynamic equation is Fick's second law,

$$\frac{\partial \rho_m}{\partial t} = D \nabla^2 \rho_m \tag{10.6}$$

where ρ_m is the mass density.

The phenomenological hydrodynamic transport equations can be derived from microscopic theory (Kubo, 1957) in an elegant form known as linear response theory. The main result of linear response theory is the relation between the frequency-dependent transport coefficient $\chi(\omega)$ and the time autocorrelation function of the dynamical variable. For a discussion of time correlation functions and linear response theory, see Berne and Harp (Berne and Harp, 1970). Briefly, linear response theory relates the dynamical variable $A(t)$ to a weak applied force $F(t)$ such that there is a linear relation between the two:

$$\langle A(t) \rangle = \int_0^t \Phi(\tau) F(t - \tau)\, d\tau = \int_0^t \Phi(t - \tau) F(\tau)\, d\tau \tag{10.7}$$

where $\Phi(t)$ has been called by Kubo the *response function* or the *after-effect function*. Equation (10.7) is of a form known as a convolution and can be solved by taking the Fourier transform of both sides and making use of the convolution theorem (Churchill, 1958, Section 13) to get

$$\langle \tilde{A}(\omega) \rangle = \chi(\omega) \tilde{F}(\omega) \tag{10.8}$$

where $\chi(\omega)$ is the Fourier transform of $\Phi(t)$. The complicated integral equation has been reduced to an algebraic equation and can be solved trivially. The transport coefficient $\chi(\omega)$ is called a generalized *susceptibility* (or *admittance*). It follows from linear response theory that a force of a given frequency ω can excite a response of only the same frequency. This statement is not true of nonlinear response, where other frequencies may arise, as they do in nonlinear optical phenomena.

Linear response theory then derives the properties of the susceptibility and relates the susceptibility to the response function and to the autocorrelation function of A. We do not pursue linear response further here, since we want to focus attention on an alternate and more fundamental approach to the treatment of time correlation functions, the projection operator approach. This approach, often called the *Zwanzig–Mori formalism*, was originally applied to linear response, although efforts have been made to extend the formalism to nonlinear situations [see, e.g., Nordholm and Zwanzig (Nordholm and Zwanzig, 1975)]. Not only does the Zwanzig–Mori approach provide a rigorous basis for linear transport theory, it also suggests new ways of calculating the correlation functions. The linear transport theory, on the other hand, shows how to relate the correlation functions to experimentally observable quantities, without saying how correlation functions may be evaluated for a given system.

10.3   ZWANZIG KINETIC EQUATION

10.3.1   Hilbert Space Formulation of Dynamics

In this section we discuss the classical case. Later, we shall indicate briefly how to deal with the quantum case. We know from Chapter 1 that at the foundation of classical statistical mechanics is the Liouville equation (1.130):

$$\frac{\partial \rho}{\partial t} = [\mathcal{H}, \rho] \tag{10.9}$$

where $[\mathcal{H}, \rho]$ is the Poisson bracket of the Γ-space statistical distribution function (which may be called the probability density in Γ space) with the Hamiltonian. This equation is rewritten in the form of an operator equation,

$$\frac{\partial \rho}{\partial t} = -i\widehat{L}\rho \tag{10.10}$$

where the Liouvillian operator is defined as

$$\widehat{L} = i[\mathcal{H}, \ ] \tag{10.11}$$

In discussing equilibrium statistical mechanics, we focused on distribution functions for which the Poisson bracket vanishes. In the present case, one has to solve the Liouville equation for the probability density and then calculate the averages of the dynamical variables for comparison with experiment. For a given dynamical variable A, the average, which is time-dependent in general, is given by

$$\langle A \rangle_t = \int A(q, p, 0)\rho(q, p, t)\, dp\, dq \tag{10.12}$$

This average may be compared with the ensemble average given by Eq. (1.56). One may take the opposite view in which the probability density is fixed but the observable is moving in time according to the equation

$$\frac{\partial A(q, p, t)}{\partial t} = i\widehat{L}A(q, p, t) \tag{10.13}$$

in which case the average is given by

$$\langle A \rangle_t = \int A(q, p, t)\rho(q, p, 0)\, dp\, dq \tag{10.14}$$

On looking at Eqs. (10.12) and (10.14), one is reminded of the Schrödinger and the Heisenberg pictures of quantum mechanics. In fact, the analogy can be used to construct a Hilbert space of dynamical variables and probability densities in order to exploit the mathematical machinery of Hilbert space and thus simplify the derivations. We note that both definitions of the average resemble the definition of a scalar product in Hilbert space. Thus, the vectors corresponding to the dynamical variables may be written as $|A_1\rangle, |A_2\rangle, |A_3\rangle$, and so on, and the probability densities may be written as $|\rho\rangle$. The average may be written simply as

$$\langle A \rangle_t = \langle A(0)|\rho(t)\rangle = \langle A(t)|\rho(0)\rangle \tag{10.15}$$

in the Dirac bra–ket notation; we have omitted the arguments q, p for brevity. In the Hilbert space of dynamics, the Liouville equation can be written as

$$\frac{\partial|\rho\rangle}{\partial t} = -i\widehat{L}|\rho\rangle \tag{10.16}$$

where, from now on, we will use Dirac's bra and ket notation. The equation of motion of an observable vector is written as

$$\frac{\partial|A\rangle}{\partial t} = i\widehat{L}|A\rangle \tag{10.17}$$

where the sign difference in the right-hand sides of Eqs. (10.16) and (10.17) is a reflection of the usual duality of the Schrödinger and the Heisenberg pictures.

The quantum case proceeds similarly. The Liouvillian is defined as

$$\widehat{\mathcal{L}} = \frac{1}{i\hbar}[\widehat{\mathcal{H}},\ ] \tag{10.18}$$

and Liouville's equation can be written as

$$\frac{\partial \widehat{\rho}}{\partial t} = -i\widehat{\mathcal{L}}\widehat{\rho} \tag{10.19}$$

From now on we shall work with the classical case exclusively.

10.3.2 Projection Operators

Zwanzig's approach consists in focusing on only those aspects of the problem that are physically important and omitting irrelevant details. In most of the cases, it eliminates the fast degrees of freedom and concentrates on the relatively slow variables to obtain their equations of motion. The method used to eliminate the irrelevant variables in Zwanzig's scheme is the formalism of projection operators, which were introduced in Chapter 2. There the projection operators were operators in the Hilbert space of quantum mechanics; here they are operators in the Hilbert space of dynamics, whether it is classical or quantal in nature. The projection operator method is successful, because the projection operator corresponding to a given vector, say $|A_1\rangle$, projects out (or selects) the component corresponding to $|A_1\rangle$ from a general vector.

We define a *projection operator* for the ith vector as

$$\widehat{P}_i = |A_i\rangle\langle A_i| \tag{10.20}$$

On applying this operator to the probability density vector, we get

$$\widehat{P}_i|\rho(t)\rangle = |A_i\rangle\langle A_i|\rho(t)\rangle \tag{10.21}$$

This equation gives a new vector directed along $|A_i\rangle$ of length $\langle A_i|\rho(t)\rangle$. The vector $|\rho(t)\rangle$ is said to be projected onto the A_i axis. Figure 10.1 shows a projection onto the A_1 axis. Another projection operator can be defined as $\widehat{Q}_i = 1 - \widehat{P}_i$. This operator projects onto a subspace orthogonal to the initial subspace A_i. The sum $\widehat{P}_i + \widehat{Q}_i$ is the identity. It can be shown that both $\widehat{P}_i$ and $\widehat{Q}_i$ are Hermitian operators satisfying $\widehat{P}_i^2 = \widehat{P}_i$ and $\widehat{Q}_i^2 = \widehat{Q}_i$.

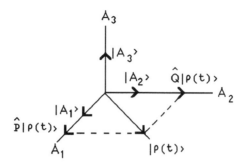

Figure 10.1   Projections of the probability density vector.

10.3.3 Derivation of the Kinetic Equation

Zwanzig (Zwanzig, 1960) divides Hilbert space into a relevant part, say A_1, that contains only those dynamical variables of immediate interest to the problem at hand and into an irrelevant part, A_2, consisting of all other dynamical variables of the system. This separation is much like a separation of a system into a subsystem and a bath (reservoir). The derivation of a kinetic equation for the probability density of the relevant part begins with the Liouville equation.

$$\frac{\partial|\rho(t)\rangle}{\partial t} = -i\widehat{L}|\rho(t)\rangle = -i\widehat{L}(\widehat{P} + \widehat{Q})|\rho(t)\rangle = -i\widehat{L}[|\rho_1(t)\rangle + |\rho_2(t)\rangle] \quad (10.22)$$

Equation (10.22) has split the probability density into a relevant probability density $|\rho_1(t)\rangle$ and an irrelevant density $|\rho_2(t)\rangle$. We now operate on this equation with $\widehat{P}$.

$$\widehat{P}\frac{\partial|\rho(t)\rangle}{\partial t} = -i\widehat{P}\widehat{L}[|\rho_1(t)\rangle + |\rho_2(t)\rangle] \quad (10.23)$$

$$\frac{\partial\widehat{P}|\rho(t)\rangle}{\partial t} = \frac{|\partial\rho_1(t)\rangle}{\partial t} = -i\widehat{P}\widehat{L}[|\rho_1(t)\rangle + |\rho_2(t)\rangle] \quad (10.24)$$

Because $\widehat{P}$ and $\widehat{L}$ are noncommuting operators, it is necessary to preserve the order in which these operators appear, even in the classical case. Operating with $\widehat{Q}$ in a like manner gives

$$\frac{\partial|\rho_2(t)\rangle}{\partial t} = -i\widehat{Q}\widehat{L}[|\rho_1(t)\rangle + |\rho_2(t)\rangle] \quad (10.25)$$

Laplace transforms can be used to solve Eq. (10.25) for $|\rho_2(t)\rangle$.

$$|\rho_2(t)\rangle = \exp(-i\widehat{QL}t)|\rho_2(0)\rangle - i\int_0^t \exp(-i\widehat{QL}\tau)\widehat{QL}|\rho_1(t-\tau)\rangle\,d\tau \quad (10.26)$$

By substituting Eq. (10.26) into Eq. (10.24)

$$\frac{\partial|\rho_1(t)\rangle}{\partial t} = -i\widehat{PL}|\rho_1(t)\rangle - i\widehat{PL}\exp(-i\widehat{QL}t)|\rho_2(0)\rangle$$

$$- \int_0^t \widehat{PL}\exp(-i\widehat{QL}\tau)\widehat{QL}|\rho_1(t-\tau)\rangle\,d\tau \quad (10.27)$$

Equation (10.27) is Zwanzig's kinetic equation.

10.3.4 Discussion of Zwanzig's Equation

Because the probability density contains all the statistical information known about a given system, Eq. (10.26) can be looked upon as representing the information that was originally in the A_2 subspace at $t = 0$ plus the information in the A_1 subspace that "leaked" into A_2 in the time t. Each term of Eq. (10.27) will be discussed separately (see Nordholm and Zwanzig, 1975):

1. The term $i\widehat{PL}|\rho_1(t)\rangle$ is the part of $|\rho(t)\rangle$ that is relevant to the A_1 subspace and remains in the A_1 subspace. It is the Markovian part of the time rate of change of $|\rho_1\rangle$.
2. The term $i\widehat{PL}\exp(-i\widehat{QL}t)|\rho_2(0)\rangle$ represents information that was originally in A_2, has propagated in A_2 [$\exp(-i\widehat{QL}t)$ is called a *propagator*], and whose time rate of change is influencing the A_1 subspace at time t. Because this effect depends on the value of $|\rho_2\rangle$ at $t = 0$, it is called an inhomogeneous effect. If $|\rho_2(0)\rangle = 0$, there is no inhomogeneity.
3. The term $\int_0^t \widehat{PL}\exp(-i\widehat{QL}\tau)\widehat{QL}|\rho_1(t-\tau)\rangle\,d\tau$ represents the time rate of change of information that was in A_1 at time $t - \tau$, leaked into A_2, was propagated in time along A_2, and then leaked back into A_1. The integration sums over earlier times. This term represents memory effects and is therefore non-Markovian.

The first term on the right-hand side of Eq. (10.27) often vanishes. To see this, write

$$-i\widehat{PL}\widehat{P}|\rho(t)\rangle = -\widehat{P}(i\widehat{L})\widehat{P}|\rho(t)\rangle$$

$$= -|A_1\rangle\langle A_1|i\widehat{L}|A_1\rangle\langle A_1|\rho(t)\rangle$$

$$= -|A_1\rangle\langle A_1|\dot{A}_1\rangle\langle A_1|\rho(t)\rangle \quad (10.28)$$

which vanishes whenever $|A_1\rangle$ and $|\dot{A}_1\rangle$ are orthogonal. As usual, the dot means the time derivative. If we ignore the first term in Eq. (10.27) and take $|\rho_2(0)\rangle = 0$, we are left with

$$\frac{\partial|\rho_1(t)\rangle}{\partial t} = -\int_0^t \widehat{P}\widehat{L}\exp(-i\widehat{Q}\widehat{L}\tau)\widehat{Q}\widehat{L}|\rho_1(t-\tau)\rangle\,d\tau \qquad (10.29)$$

This equation is the starting point for the derivation of a master equation (Zwanzig, 1964) or a Fokker–Planck equation (Zwanzig, 1961). Equation (10.27) involves no approximations and is therefore reversible. The "irreversibility" comes about as information leaks out of the relevant subspace.

10.4   GENERALIZED LANGEVIN EQUATION

10.4.1   Mori's Approach

Mori's work (Mori, 1965) has to do with the time development of a dynamical variable $A(t)$ in the A_1 subspace. In order to derive a generalized Langevin equation (GLE), we work in a Hilbert space of dynamical variables. The average value of the dynamical variable $A(t)$ is

$$\langle A(t)\rangle = \langle A|\rho_1(t)\rangle \qquad (10.30)$$

In time, $\langle A(t)\rangle$ evolves according to

$$\frac{\partial\langle A(t)\rangle}{\partial t} = \left\langle A\frac{\partial}{\partial t}|\rho_1(t)\right\rangle \qquad (10.31)$$

We now substitute Eq. (10.27) into Eq. (10.31). First we remark about the Mori definition of the projection operator, which is different from the Zwanzig definition. Mori defines the projection operator for the vector $|A\rangle$ to be

$$\widehat{P} = \frac{|A\rangle\langle A|}{\langle A^2\rangle} \qquad (10.32)$$

This definition has the advantage that the projection operators become dimensionless, making some of the later interpretations easier. For notational convenience, we shall continue to use $\widehat{P} = |A\rangle\langle A|$ but shall indicate those instances when the Mori definition affects the mathematical expressions. Each of the resulting terms will be treated in turn now:

1. Since $\langle A| \equiv \langle A|\widehat{P}$

$$
\begin{aligned}
\langle A|\widehat{P}(-i\widehat{L})\widehat{P}|\rho(t)\rangle &= \langle A|A\rangle\langle A| - i\widehat{L}|A\rangle\langle A|\rho(t)\rangle \\
&= -\langle A|\dot{A}\rangle\langle A|\rho(t)\rangle \\
&= \langle \dot{A}|A\rangle\langle A(t)\rangle \\
&= \Omega\langle A(t)\rangle
\end{aligned}
\tag{10.33}
$$

where $\langle \dot{A}|A\rangle = -\langle A|\dot{A}\rangle$ and $\Omega = \langle i\widehat{L}A|A\rangle$ has dimensions of frequency using the Mori definition of the projection operator. If $|A\rangle$ is orthogonal to $|\dot{A}\rangle$, $\Omega = 0$.

2. Expansion of the second term gives $\langle A|\widehat{P}(-i\widehat{L})\exp(-i\widehat{Q}\widehat{L}t)|\rho_2(0)\rangle$. This expansion can be written

$$
\begin{aligned}
\langle A|\widehat{P}(-i\widehat{L})\exp(-i\widehat{Q}\widehat{L}t)\widehat{Q}|\rho(0)\rangle &= \langle i\widehat{L}A|\exp(-i\widehat{Q}\widehat{L}t)\widehat{Q}|\rho(0)\rangle \\
&= \langle \exp(i\widehat{Q}\widehat{L}t)\widehat{Q}i\widehat{L}A|\rho(0)\rangle \\
&= \langle F(t)|\rho(0)\rangle \\
&= \langle F(t)\rangle
\end{aligned}
\tag{10.34}
$$

where $|F(t)\rangle = |\exp(-i\widehat{Q}\widehat{L}t)\widehat{Q}(-i\widehat{L})A\rangle$ is a vector in a subspace orthogonal to $|A\rangle$. In the steps leading to Eq. (10.34), we have used the fact that the operators $\widehat{L}$ and $\widehat{Q}$ are Hermitian and i is anti-Hermitian. From the definition of $|F(t)\rangle$, it follows that $\langle A|F(t)\rangle = 0$.

3. Expansion of the integrand in the third term gives

$$
\begin{aligned}
\langle A|\widehat{P}(i\widehat{L})\exp(-i\widehat{Q}\widehat{L}\tau)\widehat{Q}(i\widehat{L})\widehat{P}|\rho(t - \tau)\rangle & \\
= \langle A|(i\widehat{L})\exp(-i\widehat{Q}\widehat{L}\tau)\widehat{Q}(i\widehat{L})|A\rangle\langle A(t - \tau)\rangle & \\
= \langle i\widehat{L}A|F(\tau)\rangle\langle A(t - \tau)\rangle & \\
= \langle \widehat{Q}i\widehat{L}A|F(\tau)\rangle\langle A(t - \tau)\rangle & \\
= \langle F(0)|F(\tau)\rangle\langle A(t - \tau)\rangle &
\end{aligned}
\tag{10.35}
$$

where use has been made of $|F(t)\rangle \equiv \widehat{Q}|F(t)\rangle$.

Equation (10.31) can now be written as

$$
\frac{\partial\langle A(t)\rangle}{\partial t} = \Omega\langle A(t)\rangle + \langle F(t)\rangle - \int_0^t \langle F(\tau)|F(0)\rangle\langle A(t - \tau)\rangle\, d\tau
\tag{10.36}
$$

Here we have used the fact that $\langle F(0)|F(t)\rangle = -\langle F(t)|F(0)\rangle$. By introducing a weight function in the definition of the scalar product and choosing it to be a delta function (Nordholm and Zwanzig, 1975), one can show that the final equation is (in the Heisenberg picture)

$$\frac{\partial A(t)}{\partial t} = \Omega A(t) - \int_0^t K(\tau)A(t-\tau)\,d\tau + F(t) \qquad (10.37)$$

where

$$K(\tau) = \langle F(\tau)F(0)\rangle \qquad (10.38)$$

is called the *memory function*. The memory function term is a generalization of the hydrodynamic Stokes' law friction to the case where friction has memory. The last term is the random noise term. Equation (10.37) is called the *generalized Langevin equation*.

The present GLE may be compared with the GLE derived in Chapter 9 for the special case of the velocity of the Brownian particle, Eq. (9.80). There is no term corresponding to the Ω term in the Brownian velocity GLE. On the other hand, the present equation does not have the term corresponding to the external potential of the Brownian GLE. The other two terms, the memory function and the random noise terms, are rather similar.

The present derivation is quite general, using the Liouville equation as a basis, and does not depend on the particular nature of the heat bath. The previous derivation is specific to the bilinear coupling between the system particle and the harmonic oscillator heat bath and is very difficult to generalize to other systems.

10.4.2   Correlation Functions

An equation of motion for time correlation functions $\langle A(t)A(0)\rangle$ can be obtained by multiplying Eq. (10.37) on the right by $A(0)$ and ensemble averaging:

$$\frac{\partial\langle A(t)A(0)\rangle}{\partial t} = \Omega\langle A(t)A(0)\rangle - \int_0^t K(\tau)\langle A(t-\tau)A(0)\rangle\,d\tau + \langle F(t)A(0)\rangle$$
$$(10.39)$$

or

$$\frac{\partial C(t)}{\partial t} = \Omega C(t) - \int_0^t K(\tau)C(t-\tau)\,d\tau \qquad (10.40)$$

where $\langle F(t)A(0)\rangle = 0$, since $|F\rangle$ and $|A\rangle$ are orthogonal. The autocorrelation function $C(t) = \langle A(t)A(0)\rangle$ represents the information that is available about A at time t. If we have complete information about A at $t = 0$, $C(t)$ represents the fraction of this information available at time t.

The memory function represents the decreasing effect of the fluctuating forces with time. These are the forces that control the system's behavior and thus determine how rapidly the system loses its memory of past events. If the memory function $K(\tau) = 2\gamma\delta(\tau)/m$ is used in Eq. (10.40), this equation can be solved for the velocity autocorrelation function of a Brownian particle. We get

$$\langle v(t)v(0)\rangle = \langle v^2\rangle \exp(-\gamma t/m) \tag{10.41}$$

in agreement with Section 9.5. Setting the memory function equal to a delta function makes the process Markovian. Because the process is also Gaussian, the process is said to be Gaussian–Markovian. Equation (10.41) is an example of *Doob's theorem* (Doob, 1942), which states that if the autocorrelation function of a Gaussian process is exponential, the process must be Markovian.

10.4.3   Volterra Equations and the Continued Fraction

In the Mori formalism, the autocorrelation function $C(t)$ belongs to the A_1 subspace, while the memory function $K(t)$ is a correlation function in the A_2 subspace. There is a hierarchy of subspaces A_3, A_4, and so on. The A_1 subspace is the space of the slow variables, and the A_2 subspace is the space of the faster variables. Consequently, $C(t)$ has a longer relaxation time than $K(t)$. Accordingly, it is possible to write a hierarchy of GLEs (neglecting Ω).

$$\frac{\partial C(t)}{\partial t} = -\int_0^t K_1(\tau)C(t-\tau)\,d\tau$$

$$\frac{\partial K_1(t)}{\partial t} = -\int_0^t K_2(\tau)K_1(t-\tau)\,d\tau$$

$$\vdots$$

$$\frac{\partial K_n(t)}{\partial t} = -\int_0^t K_{n+1}(\tau)K_n(t-\tau)\,d\tau \tag{10.42}$$

Each of the Eqs. (10.42) is an integral equation known as the *Volterra equation*. These equations are coupled equations, where each unknown depends on the next equation. Laplace transformation of these equations gives *Mori's continued fraction*:

$$\tilde{C}(s) = \cfrac{C(0)}{s + \cfrac{K_1(0)}{s + \cfrac{K_2(0)}{s + \cdots \cfrac{K_n(0)}{s + \tilde{K}_{n+1}(s)}}}} \tag{10.43}$$

where s is the Laplace transform variable. The solutions of equations such as Eq. (10.42) are discussed by Berne and Harp (Berne and Harp, 1970). We now consider some examples of the application of Zwanzig–Mori formalism.

10.5   EXAMPLE: SPECTRAL LINE SHAPE ANALYSIS

10.5.1   Dipolar Absorption of Radiation

Molecular absorption spectroscopy is a well-established field of study (Levine, 1995, Section 21.2). A molecular spectrum results when molecules absorb radiation corresponding to the frequencies of the various vibrational and rotational degrees of freedom. A spectrum is composed of absorption bands, the shapes of the bands being dependent on the nature of the sample (gas, liquid, or solid) and on the fluctuations of the intermolecular force field. As discussed by Levine, the probability of absorption depends on the dipole moment operator. We now show how this occurs and how the band shapes are functions of the dipole moment autocorrelation function.

By using Fermi's golden rule of time-dependent perturbation theory (see Appendix C), an absorption line shape $I(\omega)$ can be written (Gordon, 1965)

$$I(\omega) = 3 \sum_{m,n} \rho_m |\langle m|\widehat{M}|n\rangle|^2 \delta(\omega_n - \omega_m - \omega) \tag{10.44}$$

where ρ_m is the probability of finding the system in the state $|m\rangle$ and $\widehat{M}$ is the electric-dipole moment operator. By taking the definition of the delta function as in Appendix C and converting to the Heisenberg picture,

$$I(\omega) = \frac{3}{2\pi} \sum_{m,n} \rho_m \langle m|\widehat{M}|n\rangle\langle n|\widehat{M}|m\rangle \int_{-\infty}^{\infty} \exp[i(\omega_n - \omega_m - \omega)t]\, dt$$

$$= \frac{3}{2\pi} \int_{-\infty}^{\infty} \sum_{m,n} \rho_m \langle m|\widehat{M}|n\rangle\langle n| \exp(iE_n t/\hbar)\widehat{M} \exp(-iE_m t/\hbar)|m\rangle \exp(-i\omega t)\, dt$$

$$= \frac{3}{2\pi} \int_{-\infty}^{\infty} \sum_{m,n} \rho_m \langle m|\widehat{M}|n\rangle \langle n|\widehat{M}(t)|m\rangle \exp(-i\omega t)\,dt$$

$$= \frac{3}{2\pi} \int_{-\infty}^{\infty} \sum_{m} \rho_m \langle m|\widehat{M}(0)\widehat{M}(t)|m\rangle \exp(-i\omega t)\,dt$$

$$= \frac{3}{2\pi} \int_{\infty}^{\infty} \langle \widehat{M}(0)\widehat{M}(t)\rangle \exp(-i\omega t)\,dt$$

$$= \frac{3}{2\pi} \int_{-\infty}^{\infty} \langle \widehat{M}(t)\widehat{M}(0)\rangle \exp(i\omega t)\,dt \tag{10.45}$$

In going from the second to the third line, the Heisenberg rate of change of $\widehat{M}$ has been used [see Eq. (2.20)].

If we assume that the total dipole moment is a sum of individual-molecule moments $\widehat{\mu}_i$ (i.e., no collision-induced effects), then $\widehat{M}(t) = \sum_i \widehat{\mu}_i(t)$. We also assume no cross-correlation because either the sample is dilute or the cross-correlation is intrinsically negligible. With these assumptions,

$$I(\omega) = \frac{3N}{2\pi} \int_{-\infty}^{\infty} \langle \widehat{\mu}(t)\widehat{\mu}(0)\rangle \exp(i\omega t)\,dt \tag{10.46}$$

where $\langle \widehat{\mu}(t)\widehat{\mu}(0)\rangle$ is a single-molecule correlation function. All constant coefficients can be eliminated by using normalized intensities $\bar{I}(\omega)$.

$$\bar{I}(\omega) = \frac{I(\omega)}{\int_b I(\omega)\,d\omega} \tag{10.47}$$

Here, the subscript b stands for the range of frequencies making up the band. Taking the inverse Fourier transform of Eq. (10.47) gives

$$\langle \widehat{\mu}(t)\widehat{\mu}(0)\rangle = \int_b \bar{I}(\omega)\exp(-i\omega t)\,d\omega \tag{10.48}$$

Therefore, by studying the absorption line shape experimentally, one can learn about the dipole moment correlation function.

10.5.2 Pure Rotational Spectrum

The pure rotational spectrum of a molecule occurs in the far infrared or microwave regions. For these spectra, $\widehat{\mu}$ is a vector that lies along the electric dipole of the molecule. The dipole autocorrelation function

$$\langle \widehat{\mu}(t)\widehat{\mu}(0)\rangle / \langle \widehat{\mu}(0)\rangle^2$$

abbreviated $C(t)$, is obtained by Fourier transforming the rotational band shape according to Eq. (10.48). The memory function of the dipolar autocorrelation function can be found from Eq. (10.40). Typical rotational correlation and memory functions are shown in Figure 10.2. The rotational correlation function can be analyzed by expanding it in a Taylor's series about $t = 0$:

$$C(t) = 1 + t \left(\frac{\partial C(t)}{\partial t} \right)_0 + \frac{1}{2} t^2 \left(\frac{\partial^2 C(t)}{\partial t^2} \right)_0 + \cdots$$

$$= 1 + M_1(-it) + \frac{1}{2} M_2(-it)^2 + \cdots \tag{10.49}$$

where M_n is the nth spectral moment defined by

$$\left(\frac{\partial^n C(t)}{\partial t^n} \right)_0 = \int_b (-i\omega)^n \bar{I}(\omega) \, d\omega = (-i)^n M_n \tag{10.50}$$

For symmetric bands, all odd moments vanish, because the integrals are of odd functions over symmetric intervals, and

$$C(t) = 1 - \frac{1}{2} M_2 t^2 + \frac{1}{4!} M_4 t^4 + \cdots \tag{10.51}$$

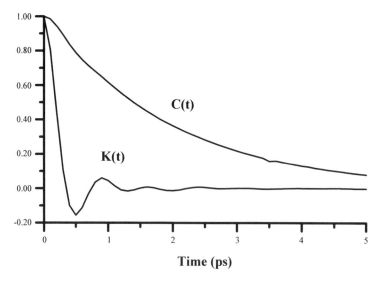

Figure 10.2   The rotational correlation function $C(t)$ and memory function $K(t)$.

We assume the molecule to be a rigid rotator,

$$M_2 \equiv \langle \omega^2 \rangle = \langle L^2 \rangle / I^2 = 2k_B T / I \tag{10.52}$$

In Eq. (10.52), we have used the facts that the angular momentum is $L = I\omega$, the average of the kinetic energy is $\langle I\omega^2/2 \rangle = \langle L^2/2I \rangle = k_B T$, and I is the moment of inertia of the molecule. Consequently, the rotational correlation function can be written as

$$\begin{aligned} C(t) &= 1 - [\langle L^2 \rangle / (2I^2)]t^2 + \cdots \\ &= 1 - (k_B T / I)t^2 + \cdots \end{aligned} \tag{10.53}$$

At short times, we may write the correlation function as

$$C(t) \approx 1 - (k_B T / I)t^2 \approx \exp(-k_B T t^2 / I) \tag{10.54}$$

Even in condensed phases, molecules behave as free rotators on a short time scale. Hence, Eq. (10.54) represents the correlation function of a free rotator. Because Eq. (10.54) is Gaussian, Fourier transformation of the free-rotator correlation function will give a Gaussian band shape. At longer times, the mean-square torque on a molecule due to other molecules causes the line shape to deviate from Gaussian. We now consider the condensed phase, where the molecule is no longer a free rotator and where its band shape is no longer Gaussian.

Several rotational models have been used to study molecular rotation in condensed phases. These models can be Markovian or non-Markovian. A very simple approximation was suggested by Peter Debye and is known as *Debye diffusion*. The memory function for this model is a delta function, $K(t) = 4D_r\delta(t)$, where D_r is the rotational diffusion coefficient. Use of Eq. (10.40) and neglect of Ω gives, for Debye diffusion,

$$\frac{\partial C(t)}{\partial t} = -\int_0^t 4D_r\delta(\tau)C(t-\tau)\,d\tau = -2D_r C(t) \tag{10.55}$$

Solving Eq. (10.55) for $C(t)$ gives $C(t) = C(0)\exp(-2D_r t)$. Fourier transformation of this exponential function gives a Lorentzian band shape. This band shape contrasts with the free-rotator model, which predicts a Gaussian band shape. Actual band shapes are between Gaussian and Lorentzian. They are often Lorentzian in the center and Gaussian on the wings. As mentioned above, according to Doob's theorem (Doob, 1942), a Gaussian process is Markovian if its autocorrelation function is exponential. Accordingly, the Debye rotational diffusion process is Markovian by Doob's theorem.

10.6   EXAMPLE: HYDRODYNAMICS OF FLEXIBLE POLYMERS

10.6.1   A Polymer Model

It is known that the addition of polymer molecules to viscous fluids can drastically alter the flow properties of the fluids. For example, the addition of high-molecular-weight polymers to water can dramatically reduce turbulent flow. It is therefore of some interest to devise a model of the hydrodynamic properties of solutions of flexible polymer molecules. A tractable model has been formulated by Rouse (Rouse, 1953) and Zimm (Zimm, 1956). It is known as the "bead-spring model" and is represented in Figure 10.3. The model is composed of N identical submolecule segments that join $N + 1$ identical "beads," which are the terminal atoms of the submolecule segments and thus serve to join the submolecule segments. Each segment is a portion of the polymer chain that is just long enough for the distance between its terminal atoms (beads) to obey a Gaussian distribution. The segments have complete flexibility and are restricted only in that they obey Gaussian statistics. The chain segments exert forces on the beads. The requirements of a Gaussian distribution of bead distances and a Boltzmann distribution of the canonical equilibrium distribution function allows us to treat the chain segments as harmonic-oscillator springs.

10.6.2   Fluid Dynamics

It is assumed that the fluid obeys the Navier–Stokes equation for an incompressible fluid:

$$\rho_m \frac{\partial \mathbf{v}}{\partial t} + \rho_m (\mathbf{v} \cdot \nabla)\mathbf{v} = -\nabla P + \eta \nabla^2 \mathbf{v} \tag{10.56}$$

where ρ_m is the mass density of the fluid, $\mathbf{v}$ is its velocity at a given point in the fluid, P is the hydrostatic pressure, η is the coefficient of viscosity, and ∇ is the del operator. No external force is assumed to be present. The left-hand side of Eq. (10.56) is the so-called inertial force per unit volume, $\rho_m d\mathbf{v}/dt$,

Figure 10.3   The "bead-spring" model.

which must be equal to the gradient of the stress on the right-hand side. The hydrostatic pressure represents the isotropic part of the stress tensor, while the coefficient of viscosity is associated with the anisotropic (shear) part of the stress tensor (Landau and Lifshitz, 1975).

When an object is immersed in a viscous fluid, there will be developed a frictional force, which retards the flow of the fluid relative to the object. The effect of the object on the fluid flow depends on the shape and dimensions of the object and on the velocity of the fluid. For large velocities, the nonlinear term in Eq. (10.56) becomes large, resulting in turbulent flow. For this reason, we assume laminar flow and set $\rho_m(\mathbf{v} \cdot \nabla)\mathbf{v} = 0$. For steady-state flow, $\partial \mathbf{v}/\partial t = 0$ and Eq. (10.56) becomes

$$\nabla P = \eta \nabla^2 \mathbf{v} \tag{10.57}$$

10.6.3   The Rouse–Zimm Model

We immerse the bead-spring polymer model in a viscous fluid of velocity $\mathbf{v}$. The interaction of the fluid with the polymer is assumed to take place at the beads only. The hydrodynamic frictional force imparted by the jth bead to the fluid is given by Stokes' law, $-\gamma(\mathbf{v}_j - \mathbf{u}_j)$, where γ is the coefficient of friction, $\mathbf{v}_j$ is the velocity the fluid would have at the position of the jth bead in the absence of the bead, and $\mathbf{u}_j = \dot{\mathbf{r}}_j$ is the velocity of the jth bead. The velocity $\mathbf{v}_j$ takes into account the hydrodynamic interaction between the fluid and the polymer, which results in a velocity gradient in the fluid. The hydrodynamic frictional force imparted by the fluid to the bead is

$$\mathbf{F}_j(\text{hydro}) = -\gamma(\mathbf{u}_j - \mathbf{v}_j) \tag{10.58}$$

The effect of the springs must be transmitted to the fluid via the beads. This transmission gives rise to a mechanical force. To calculate this force, we first transform from orthogonal Cartesian coordinates to a set of $3N - 6$ normal coordinates q_j for the N harmonic oscillators (Levine, 1995, p. 706). The potential energy then can be written

$$U = \frac{b}{2} \sum_j q_j^2 \tag{10.59}$$

where b is the force constant. The mechanical force on the jth bead due to the harmonic potential of the $(j - 1)$th spring is

$$\mathbf{F}_j(\text{mech}) = -\nabla_j U \tag{10.60}$$

The flexibility in the polymer chain means that we are dealing with viscoelasticity. For a viscoelastic Langevin equation, the inertial force is equal to the frictional force, which is made up of the hydrodynamic force, the mechanical force on the beads, and finally a random force. The random force arises from the thermal motions of the beads, which changes the polymer configuration. The polymer, which has accessible to it a Gaussian distribution of configurations, is assumed to undergo a Brownian diffusion among its possible conformers. As the beads diffuse, they move through the velocity gradient of the fluid, which in turn affects the polymer configuration. The net result is a nonequilibrium distribution of bead positions. Under these conditions, it is impossible to know the potential energy between the segments of the polymer chain. As in Section 4.4.2, we define a potential of mean force w, which is a function of the normal coordinates and the time. The canonical statistical distribution function for the polymer is

$$f(t) = C \exp(-bw) \tag{10.61}$$

where C is a constant. The random force on the fluid due to the jth bead is $-\nabla_j w$, and the random force on the jth bead is

$$\mathbf{F}_j(\text{random}) = -k_\text{B} T \nabla_j \ln f \tag{10.62}$$

The equation of motion for the polymer is

$$m\frac{d\mathbf{u}_j}{dt} = \mathbf{F}_j(\text{hydro}) + \mathbf{F}_j(\text{random}) + \mathbf{F}_j(\text{mech})$$

$$= -\gamma(\mathbf{u}_j - \mathbf{v}_j) - k_\text{B} T \nabla_j \ln f - \nabla_j U \tag{10.63}$$

Under conditions of steady, laminar flow, the inertial force vanishes and Eq. (10.63) becomes

$$\gamma(\mathbf{u}_j - \mathbf{v}_j) = -k_\text{B} T \nabla_j \ln f - \nabla_j U \tag{10.64}$$

It is possible to solve for the velocity of the jth bead by using hydrodynamics (see Bixon, 1973, p. 1460):

$$\mathbf{u}_j = -\sum_k \mathcal{D}_{jk} \cdot (\nabla_k \ln f + \nabla_k U/k_\text{B} T) \tag{10.65}$$

where $\mathcal{D}_{jk}$ is a diffusion tensor. The behavior of f with time can be found by substituting Eq. (10.65) into the equation of continuity, Eq. (1.124), which becomes, in the present case,

$$\frac{\partial f}{\partial t} = -\sum_j \nabla_j \cdot (f\mathbf{u}_j) \tag{10.66}$$

Equation (10.66) is in the form of a Fokker–Planck equation [cf. Eq. (9.16)]. The solution of Eq. (10.66) for the distribution function has been worked out by Zimm (Zimm, 1956), and this solution has been used by him to calculate the hydrodynamic properties of polymer solutions such as viscosity. We are interested in the solution of Eq. (10.66) in order to obtain an expression for the equilibrium ensemble-averaged motion of the normal coordinates. For this, we take the Zwanzig–Mori projection operator approach.

10.6.4   Optimized Rouse–Zimm Model

The right-hand side of Eq. (10.66) can be treated as a diffusion operator $\widehat{L}$ in the spirit of Eq. (10.10).

$$\frac{\partial f}{\partial t} = -\widehat{L}f \tag{10.67}$$

We seek an expression for the equilibrium ensemble-averaged motion of the normal coordinates, $\langle q_i(t) \rangle$. The normal coordinates reside in the space of the slow variables, the A_1 subspace. Equation (10.36) is directly applicable.

$$\frac{\partial \langle q_i(t) \rangle}{\partial t} = -\Omega_i \langle q_i(t) \rangle + \int_0^t K_i(\tau) \langle q_i(t - \tau) \rangle \, d\tau \tag{10.68}$$

where

$$\Omega_i = \langle \widehat{L} q_i | q_i \rangle, \tag{10.69}$$

$$K_i(\tau) = \langle \widehat{Q}\widehat{L} q_i | \exp(-\widehat{Q}\widehat{L}\tau)(\widehat{Q}\widehat{L}) | q_i \rangle \tag{10.70}$$

and

$$\langle F(t) \rangle = \langle q_i | \widehat{P}(-\widehat{L}) \exp(\widehat{Q}\widehat{L}t)\widehat{Q} | f \rangle = 0 \tag{10.71}$$

The change of signs between Eq. (10.68) and Eq. (10.37) is due to the missing i in Eq. (10.67) compared to Eq. (10.10). Equation (10.71) represents an inhomogeneous term (see Section 10.3.4) and vanishes, since we assume no initial inhomogeneity. The quantity Ω_i in Eq. (10.69) has dimensions of frequency. In this case it represents diffusion of the beads, and in the Rouse–Zimm model it is related to the diagonal element of the averaged diffusion tensor:

$$\Omega_i = \frac{\langle D_i \rangle}{\langle q_i | q_i \rangle} \tag{10.72}$$

where $\langle q_i | q_i \rangle$ comes from the Mori definition of the projection operator.

The solution of Eq. (10.68) for the normal coordinates and their corresponding correlation functions gives the polymer dynamics and is the solution of our problem. Bixon (Bixon, 1973) and Zwanzig (Zwanzig, 1974) have discussed this solution in detail and have obtained an optimized Rouse–Zimm model. It is pointed out by these authors that Eq. (10.68) is quite general within linear response theory. If a Gaussian equilibrium distribution function and an averaged diffusion tensor are used, then it is shown by these authors that the memory function $K_i(\tau)$ vanishes for the Rouse–Zimm model. If a different distribution function is selected, the spring force constants are changed, and the memory function does not vanish. In this case, the memory function provides a correction to the Rouse–Zimm model. This general approach is an optimized Rouse–Zimm model in the following sense. Even if the distribution function is not Gaussian and even if there is no preaveraging of the diffusion tensor, the memory function introduces only second-order effects, which are much smaller than the $\Omega_i \langle q_i(t) \rangle$ term.

In summary, for the Rouse–Zimm model, Eq. (10.68) is

$$\frac{\partial \langle q_i(t) \rangle}{\partial t} = -\Omega_i \langle q_i(t) \rangle \tag{10.73}$$

with the solution

$$\langle q_i(t) \rangle = \langle q_i(0) \rangle \exp(-\Omega_i t) \tag{10.74}$$

Likewise, the correlation function $C_i(t) \equiv \langle q_i(t) q_i(0) \rangle$ is found from the solution of the equation

$$\frac{\partial C_i(t)}{\partial t} = -\Omega_i C_i(t) \tag{10.75}$$

With the correlation functions in hand, it is possible to obtain the transport coefficients within the linear response theory. For example, Bixon (Bixon, 1973) obtained the coefficient of viscosity. However, the neglect of nonlinear terms in Eq. (10.56) precludes high frequencies in the viscosity $\eta(\omega)$ of the polymer solution. If the nonlinear terms are retained, linear response is inappropriate and the Zwanzig–Mori formalism, as presented above, cannot be used. For low-frequency viscosity, for which the steady, laminar flow is not disrupted, the memory function does not vanish, but it may be neglected. For systems far from equilibrium, where turbulent flow exists, new approaches have been developed. These approaches are discussed in Chapters 12 and 13.

10.7   EXERCISES AND PROBLEMS

1. Show that the projection operator $\widehat{P}$ is (a) Hermitian and (b) idempotent (i.e., $\widehat{P} = \widehat{P}^2$).

2. From the definition of $|F(t)\rangle$, show that, if A has dimensions of velocity, $|F(t)\rangle$ has dimensions of force per unit mass.

3. By substituting Eq. (10.26) into Eq. (10.25) show that the former is a solution of the latter. You will need Eq. (9.82).

4. Derive Eq. (10.41) from Eq. (10.40).

5. The relaxation time for rotational diffusion is taken to be the time required for the autocorrelation function $C(t)$ to reach $1/e$.

    (a) Using this definition, find the free rotational relaxation time for the HCl molecule at $25°C$.

    (b) Determine the rotational diffusion coefficient for a colloidal particle whose rotational relaxation time is $50\,\mu s$ in aqueous solution.

6. In the bead-spring model, let the distance between any two consecutive beads be $\mathbf{r}$. If the springs are harmonic oscillators, show that the Boltzmann distribution is also a Gaussian function of $\mathbf{r}$. Take the average oscillator energy to be $3k_B T/2$.

11

ACTIVATED BARRIER-CROSSING PROBLEM

11.1 INTRODUCTION

The steady state is a nonequilibrium state in which the macroscopic properties of the system have become independent of time but may still depend on position. An important steady state arises in chemical reactions. To understand this steady state, consider the following very popular model, which uses a one-dimensional potential shown in Figure 11.1. It is assumed that, in the many-body problem of reacting species, there is one special degree of freedom called the *reaction coordinate*. The motion of the species along this coordinate on the potential energy curve shown in Figure 11.1 results in the chemical reaction. In the spirit of the reduced description, which results in the

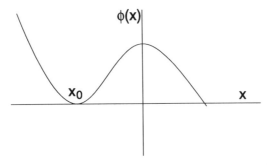

Figure 11.1 One-dimensional potential used in the barrier-crossing problem.

contraction of the many degrees of freedom into just the reaction coordinate, the potential should be thought of as the potential of mean force. The reactant species start in the well, cross-over the barrier and form products on the other side of the barrier. Near the top of the barrier is a special state called the *transition state*, which corresponds to the *activated complex* formed from the reactants. If one starts with the reactants in the well with a given energy, how do they climb the barrier? In quantum mechanics, one may consider tunnelling, which indeed becomes important at low temperatures. But there is another effect that is present at all temperatures. The degrees of freedom of the system, excluding the reaction coordinate, but including the solvent, if any, form a heat bath. Interaction with the heat bath provides fluctuations by which the reactants are activated to the top of the barrier. Once they reach the top, the reactants are in a transition state (or form an activated complex), which is deactivated by the heat bath to form products, which in turn are removed from the system. A steady state is reached after some time if the number of reactants is maintained by an unending supply in the well. The central problem of chemical reaction theory is to calculate the reaction rate and to study its dependence on the properties of the potential and the heat bath to determine universal features, if any. For a review of chemical reaction theory, see Hynes (Hynes, 1985).

The model that we have been discussing is well known in physics, where it is used to study the decay of metastable states in the presence of a heat bath. It is known as the *activated barrier crossing* (ABC) problem and occurs in such diverse fields as diffusion in solids, surface diffusion, transport across cell membranes, chemical reactions, tunnelling through domain walls in ferromagnets, nucleation of vortices in superfluid helium, decay of the zero-voltage state in Josephson superconducting junctions, electron transport in semiconductors, and transitions in nonlinear optics. For a recent review of the ABC problem discussing applications in physics, see Hänggi et al. (Hänggi et al., 1990).

The mathematics of the ABC problem is the same in all the above applications, although the language used to describe the various phenomena may be different. We shall use the language of a physical particle escaping the metastable state and the language of chemical reaction theory interchangeably. A correspondence between the two is given in Table 11.1, which is based on Table 1 of Singh and Robinson (Singh and Robinson, 1995). Some of the terms occurring in this table will be explained later.

We start with a derivation of the first universal property, the *Arrhenius law* and go on to describe the *Kramers theory*, in which one finds another universal property, the proportionality of the chemical reaction rate to the inverse of the solvent viscosity. The seminal work of Grote and Hynes (GH) (Grote and Hynes, 1980), which takes into account the memory effects of the heat bath

TABLE 11.1   Correspondence of Terminology between the ABC Problem and the Chemical Reaction Theory

ABC Problem	Chemical Reaction Theory
Position Coordinate	Reaction Coordinate
Heat bath	Solvent
Particle in the well	Reactant
Escaped particle	Product
Potential	Potential of mean force
Barrier height	Activation energy
Low-damping limit	Energy-diffusion limit
High-damping limit	Spatial-diffusion limit
Memory friction kernel	Solvent friction
Memory relaxation time	Solvent response time
Coupling strength	Solvent–solute coupling strength
Strong coupling regime	Solvent caging regime
Weak coupling regime	Nonadiabatic solvation regime
Critical coupling	Dividing coupling
Escape rate	Chemical reaction rate
Reduced rate coefficient	Dynamical transmission coefficient

friction is described next. Then, the scaling and universality properties of the critical point in the GH rate expression are described. We end with a few remarks about applications of the theory in other directions.

11.2   TRANSITION STATE THEORY

11.2.1   Arrhenius Law

The first important result in the ABC problem was the famous Arrhenius law (see Levine, 1995, Section 17.7), which states that, for a wide variety of chemical reactions, the *rate constant* satisfies the equation

$$k = \nu \exp(-E_b/k_B T) \tag{11.1}$$

where ν is a prefactor and E_b is the energy of activation (the barrier height). This relation has been used to create Arrhenius plots, the plots of the logarithm of the rate versus $1/T$, for generations. It should be noted that sometimes there are deviations from Eq. (11.1), and the reaction is said to exhibit a non-Arrhenius behavior. We restrict our attention to systems obeying Eq. (11.1). The Boltzmann factor in Eq. (11.1) simply expresses the fraction of collisions between reacting species that have an energy in excess of the activation energy, E_b. A mathematical derivation of the rate constant is given by the *transition state theory* (TST), to which we now turn.

11.2.2 Assumptions

We shall discuss the TST in terms of the model potential in Figure 11.1. The first assumption of the ABC approach is that $E_b \gg k_B T$. This assumption, which is called the *metastability assumption*, ensures the following. Since the barrier is high, very few particles are escaping over the barrier. Therefore, most of the particles are in the well and the distribution function of the system differs little from the equilibrium Maxwell–Boltzmann distribution. Clearly, it is a near-equilibrium phenomenon in terms of the classification introduced in Chapter 8. Mathematically, this assumption simplifies the calculations very much, because it implies that there is a very slow process, the escape process, which has the longest time associated with it. This assumption makes it justifiable to use the contracted description of the Langevin or the Fokker–Planck equations, which are themselves derived from the fast-variable-elimination approach of Zwanzig and Mori. Without this assumption, one would have to solve the complicated nonequilibrium problem in which even a steady state may not exist.

The next two assumptions are specific to the TST. The first one is the *equilibrium assumption*. It states that thermodynamic equilibrium must be applicable to both reactant and activated species, and any deviations from the Maxwell–Boltzmann distribution can be neglected. This assumption makes it possible to use the equilibrium ensemble methods in the TST. The second assumption is the *no-recrossing assumption*. It states that any particle that has crossed over to the other side by climbing over the barrier is gone forever and will not recross.

11.2.3 Derivation of the TST Rate Constant

We consider a system of a fixed number N of particles. They are being bombarded by the particles of the heat bath in which they are immersed. In chemistry, one says that the reactants are colliding with the solvent molecules. Since the Maxwell–Boltzmann distribution is applicable, we can calculate the flux, or the number of particles crossing the barrier per second, which in chemical kinetics is called the *rate of conversion*. The rate of conversion divided by the number of reactant particles is called the rate constant (or the escape rate coefficient) and is an example of the *flux-over-population* method used in the ABC problem.

We now restrict our discussion to a unimolecular decomposition, where each particle interacts only with the barrier potential and the heat bath. The individual particles do not interact with one another, and we can focus on one particle at a time. The partition function of one particle is

(see Section 1.6.3)

$$Z_1 = \frac{1}{h} \int_{-\infty}^{\infty} \int_{-\infty}^{\infty} \exp[-\beta p^2/(2m) - \beta\phi(x)] \, dp \, dx \qquad (11.2)$$

where m is the mass of each particle and $\phi(x)$ is the potential in which the particles are moving. For the sake of concreteness, we take the potential to be a piecewise parabolic potential given by

$$\phi(x) = \begin{cases} \frac{1}{2}m\omega_0^2(x - x_0)^2 & x < x_1 \\ E_b - \frac{1}{2}m\omega_b x^2 & x > x_1 \end{cases} \qquad (11.3)$$

where ω_0 is the well frequency, E_b is the barrier height, and ω_b is the parameter describing the shape of the parabolic barrier, often called the imaginary barrier "frequency." The barrier top is at the origin, the well minimum is at the point $x = x_0 < 0$, and the well and the barrier are assumed to be joined at the "seam" $x = x_1 > x_0$.

The classical distribution function is given by

$$f(p, x) = \frac{N}{hZ_1} \exp\left[-\beta p^2/(2m) - \beta\phi(x)\right] \qquad (11.4)$$

which is normalized for convenience such that

$$\int_{-\infty}^{\infty} \int_{-\infty}^{\infty} f(p, x) \, dp \, dx = N \qquad (11.5)$$

The flux over the top of the barrier is given by

$$j = \int_0^{\infty} \left(\frac{p}{m}\right) f(p, 0) \, dp \qquad (11.6)$$

where the no-recrossing assumption has been used by restricting the momentum integral to positive momenta, the top of the barrier is at $x = 0$ and $f(p, 0) = (N/hZ_1)\exp[-\beta E_b - \beta p^2/(2m)]$. The flux integral can be evaluated to give

$$j = \frac{N}{hZ_1\beta} \exp(-\beta E_b) \qquad (11.7)$$

The partition function given by Eq. (11.2) can be evaluated as follows. The momentum integral is a Gaussian integral and can be evaluated easily. Because of the metastability assumption, most of the particles are in the well region, so that, in the integral over x, the potential in the barrier region may be neglected and the range of integration extended over all space without much error. This assumption yields

$$Z_1 \approx \sqrt{\frac{2m\pi}{\beta h^2}} \int_{-\infty}^{\infty} \exp[-\beta m\omega_0^2(x - x_0)^2/2] \qquad (11.8)$$

which gives

$$Z_1 = \frac{2\pi}{\beta h \omega_0} \qquad (11.9)$$

The rate constant now follows by using this result in Eq. (11.7) and is given by

$$k_{\text{TST}} = \frac{j}{N} = \frac{\omega_0}{2\pi} \exp(-\beta E_b) \qquad (11.10)$$

which is the well-known TST rate constant for our potential. Sometimes it is written in the alternative form

$$k_{\text{TST}} = \frac{k_B T}{h} \frac{1}{Z_1} \exp(-\beta E_b) \qquad (11.11)$$

This equation is almost the form one finds in chemistry textbooks, wherein the numerator has an extra factor $Z^{\#}$ called the partition function of the transition state. For our potential, this quantity happens to be 1. From this, one can get other forms of the TST rate constant. For example, for an isobaric-isothermal reaction, one gets another form familiar to chemists,

$$k_{\text{TST}} = \frac{k_B T}{h} \exp(-\beta \Delta G) \qquad (11.12)$$

where ΔG is the excess of the Gibbs free energy of the transition state over the ground state. For an interesting account of the notation # for the transition state, see Eyring (Eyring, 1980).

11.2.4   Discussion of the TST

The transition state theory provides a simple physical interpretation of the Arrhenius equation. It is simply the fraction of particles at the top of the barrier. The prefactor in the TST rate constant has the interpretation of an attempt frequency, the frequency at which the particles try to escape, or, in the chemical kinetics language, the rate at which the reactants attempt to form the activated complex.

However, the TST is based on contradictory assumptions. The point is that equilibrium cannot be valid everywhere, because then there is no net flux of particles anywhere owing to the fact that the Maxwell–Boltzmann distribution is symmetric in velocities. The assumption of equilibrium is abandoned at the top of the barrier by supposing the existence of a Maxwell demon, who lets the particles go only in the positive direction. This deliberate breaking of the symmetry in the classical distribution is what gives a nonzero rate of conversion. Because of these considerations, the TST is best thought of as a limiting case of more logical theories, such as the Kramers theory.

In the Kramers theory, the heat bath does more than provide the barrier height E_b and the temperature T. The motion of the escaping particles is taken into account in more detail by using the hydrodynamic approach, whereby the viscosity of the solvent is introduced as an explicit parameter. The TST can be seen as a limiting case of the Kramers theory as follows. Because the equilibrium between the reactant particles and those of the heat bath is always and everywhere maintained, one says that the viscosity is assumed to be infinite in the TST. On the other hand, by letting the particles stream through to the other side of the barrier without any interruptions from the heat bath, the TST assumes a zero viscosity. We see a contradiction again, which is resolved as follows. The TST is properly derived from the Kramers theory with a viscosity that is different in the well and the barrier regions. The TST assumes an infinite viscosity in the well region to ensure instantaneous equilibrium in that region and assumes a zero viscosity in the barrier region to ensure a safe unhindered passage in the barrier region. Interpreted in this way, the TST appears as a logical limiting case of the Kramers theory. We turn now to the Kramers theory.

11.3   KRAMERS THEORY

11.3.1   Introduction

In a landmark paper (Kramers, 1940), Kramers removed the assumptions of the TST in order to derive more general expressions for the rate. This paper had far-reaching consequences in many branches of physics, chemistry, and biology. Kramers used the Fokker–Planck approach, which is equivalent to the Langevin equation, in order to take into account the frictional effects of the heat bath explicitly. The following picture of the escape problem emerges from his study. The escape of the particles over a barrier can be thought of as a two-step process. In the first process, the particles gain energy by means of spontaneously occurring fluctuations. After they reach the barrier region, the second process takes over. Here the particles perform spatial diffusion or Brownian motion, random walking back and forth over the barrier top, eventually diffusing away. The total escape rate depends on both of these processes, which in turn depend on the viscosity of the solvent.

For very low viscosities, the equilibrium in the well region is disturbed only slightly and the particles move with almost fixed energy, which changes slightly whenever there are collisions with the heat bath particles. Eventually, they reach the barrier top, from where they stream across without disturbance because of low viscosity. Because the energy of the particles fluctuates around the mean value, a diffusion equation in energy can be written and solved to give the rate. The climb from the well to the barrier top is the rate-limiting step. The particles are said to be in the *energy diffusion* regime, where the Kramers rate is found to be proportional to viscosity.

The above situation is no longer applicable for high viscosities. In this case, the energy of the particles changes rapidly in order to bring them into velocity equilibrium with the solvent in such a way that the Maxwell distribution of velocities is established rapidly. The particles find themselves very quickly near the barrier and the diffusion across the barrier becomes the rate-limiting step. In this case, called the *spatial diffusion* regime, the second assumption of the TST breaks down as the particles recross the boundary between the reactant and the product side many times before settling into the other side of the barrier. In contrast, the TST assumes that the particles escape with 100% efficiency. The Kramers rate is found to be proportional to the *inverse* of the viscosity in this case. The system is said to be in the *Smoluchowski diffusion* limit.

Kramers derived the rate expression in both regions and also in an extended case where the viscosity is moderate to large, the so-called *intermediate viscosity* regime. Later authors proposed that a result for the rate for the whole viscosity range (low to high) may be obtained by adding the rates in parallel. We shall discuss this below.

11.3.2 High-Damping Limit

By using Einstein's method (see Chapter 9), Kramers first derived the following Fokker–Planck equation in the phase space for the distribution function $f(p, x, t)$:

$$\frac{\partial f}{\partial t} = -\frac{p}{m}\frac{\partial f}{\partial x} + \gamma'\frac{\partial(pf)}{\partial p} + \frac{\partial}{\partial p}\left(f\frac{\partial\phi}{\partial x}\right) + \frac{\partial\phi}{\partial x}\frac{\partial f}{\partial p} + m\gamma'k_\mathrm{B}T\frac{\partial^2 f}{\partial p^2} \quad (11.13)$$

where $\gamma' = \gamma/m$, and γ is the coefficient of friction (or the damping parameter) introduced in Chapter 9 and is related to the diffusion constant by the relation $D = k_B T/\gamma$. Equation (11.13) was first derived by Klein (Klein, 1922) and is called the Klein–Kramers equation. The solution of this equation for all values of damping is still an unsolved problem. Kramers solved it in the high- and low-damping cases by ingenious arguments. We discuss the high-damping case first.

We see from Eq. (9.44) that the velocity of a Brownian particle reaches its equilibrium value in a time of the order of $1/\gamma'$. For high damping this time is very short. Therefore, in this case the velocity equilibrium is reached rapidly. Kramers showed that Eq. (11.13) can be linearized around the equilibrium distribution to give a simplified equation,

$$\frac{\partial \overline{f}}{\partial t} = D \frac{\partial}{\partial x} \left(\beta \overline{f} \frac{\partial \phi}{\partial x} + \frac{\partial \overline{f}}{\partial x} \right) \tag{11.14}$$

where the relation $\gamma = 1/(\beta D)$ has been used and the bar over f means an integration over momentum. This equation was also obtained by Smoluchowski while studying Brownian motion and is known by his name. This equation implies a flux j given by

$$j = -D \left(\beta \overline{f} \frac{\partial \phi}{\partial x} + \frac{\partial \overline{f}}{\partial x} \right) \tag{11.15}$$

which can be written in the form,

$$j = -D \exp(-\beta \phi) \frac{\partial}{\partial x} \left[\overline{f} \exp(\beta \phi) \right] \tag{11.16}$$

In the limit $\gamma \to \infty$, the system changes from being in a steady state to one in equilibrium, the diffusion constant is zero, and the flux $j = 0$ as expected. In the TST, the equilibrium is artificially broken with the no crossing rule, which makes it possible to count only the particles traveling to the right, giving the nonzero flux as in Eq. (11.7). No such assumption is made in the Kramers theory. We do know, however, that the equilibrium distribution is obtained at the bottom of the well and the distribution is essentially zero at a point $x = x_2 > 0$ far away from the barrier top in the region of positive x. Keeping this in mind, we integrate Eq. (11.16) from $x = x_0$ to $x = x_2$ to get

$$j \int_{x_0}^{x_2} \exp(\beta \phi) \, dx = -D \left[\overline{f} \exp(\beta \phi) \right]_{x_0}^{x_2} \tag{11.17}$$

In the integral on the left-hand side of Eq. (11.17), the biggest contribution comes from the region where the potential ϕ is largest, that is, the barrier region. Replacing the potential by its form in the barrier region and extending the limits from $-\infty$ to ∞, gives a Gaussian integral, which can be evaluated easily. The integral and the result are

$$\int_{-\infty}^{\infty} \exp(\beta E_b) \exp\left(\frac{-\beta m \omega_b^2 x^2}{2}\right) dx = \sqrt{\frac{2\pi}{\beta m \omega_b^2}} \exp(\beta E_b) \qquad (11.18)$$

On the right-hand side of Eq. (11.17), the distribution function $\overline{f}$ is practically zero at the upper limit, and what remains is the lower limit. The potential is zero at the bottom of the well, and the distribution function at the bottom of the well can be obtained from Eq. (11.4) by integrating it with respect to momentum and using the value of Z_1 from Eq. (11.9). The result is

$$\overline{f}(x_0) = N\omega_0 \sqrt{\frac{\beta m}{2\pi}} \qquad (11.19)$$

By using this result and Eq. (11.18) in Eq. (11.17) we get the Kramers rate coefficient:

$$k_{Kh} = \frac{1}{\gamma'} \frac{\omega_0 \omega_b}{2\pi} \exp(-\beta E_b) \qquad (11.20)$$

where the subscript K implies Kramers and h stands for the high-damping limit. Sometimes it is convenient to introduce a reduced rate coefficient given by the ratio of the Kramers rate coefficient to the TST rate constant. In the present case, it is given by

$$k_{Kh}^* = \frac{1}{\gamma^*} \qquad (11.21)$$

where the dimensionless damping parameter has been introduced by the relation $\gamma^* = \gamma'/\omega_b$. This high-damping, inverse-viscosity dependence of the rate is found in a large number of physical and chemical problems. Kramers also solved the rate problem in a more general case, where the damping is not too large. The result is

$$k_K^* = \left(1 + \frac{1}{4}\gamma^{*2}\right)^{1/2} - \frac{1}{2}\gamma^* \qquad (11.22)$$

It can be seen that, in the limit of large γ^* ($\gamma^* \gg 1$), this result reduces to Eq. (11.21), as it should. But in the limit of small γ^*, it reduces to the TST rate constant. This last result is of course not correct. The real story of how the TST is violated when the friction is small is discussed next.

11.3.3   Low-Damping Limit

In this case, the damping parameter γ^* is small, so that the particle moves with constant energy, occasionally experiencing random fluctuations. Kramers showed, by a suitable reduction of Eq. (11.13), that the particle performs a Brownian motion with the energy as a variable and an energy diffusion constant given by

$$D_E = k_B T \gamma' J(E) \tag{11.23}$$

where the quantity $J(E)$ is the energy-dependent action integral,

$$J(E) = \oint p \, dx \tag{11.24}$$

In this equation, the integral is over a whole cycle of almost periodic motion. The reduced rate coefficient is given by

$$k_{Kl}^* = \beta \gamma' J(E_b) \tag{11.25}$$

where the subscript l on the reduced rate indicates the low-friction limit.

Since the Kramers problem cannot be solved for a general value of γ, many attempts have been made to approximate the exact result for the rate coefficient. One formula for the full Kramers rate k_{Kf}^* is

$$\frac{1}{k_{Kf}^*} = \frac{1}{k_K^*} + \frac{1}{k_{Kl}^*} \tag{11.26}$$

with the understanding that, the escape being a two-step process, the times of escape ought to add. This form seems to fit computer simulations done in the regime where the Kramers theory is likely to be valid. A plot of the Kramers rate coefficient versus the reduced damping parameter must give a linear curve in the region of low damping ($\gamma^* \ll 1$) and an inverse relationship in the region of high damping ($\gamma^* \gg 1$) in agreement with Eqs. (11.25) and (11.21). Consequently, there must be a maximum in the curve for intermediate damping ($\gamma^* \simeq 1$). This maximum is called the *Kramers turnover*. Rate studies as a function of solvent viscosity have indeed shown an increase in rate as the solvent viscosity is lowered. However, the observation of a Kramers turnover is rare. Experimentally, it is difficult to reduce the solvent viscosity enough to see the turnover.

The inverse viscosity dependence of the rate in the presence of solvents was the experimental norm till fractional dependence of the rate on viscosity was seen by Fleming and co-workers (Fleming, 1986, pp. 185–192). They were able to fit the reduced rate coefficient for many chemical reactions to the form

$$k^* \propto \eta^{-\theta} \tag{11.27}$$

with the exponent θ between 0 and 1. This result cannot be explained by the Kramers theory. One needs to go to Grote–Hynes theory, which is discussed in Section 11.4.

11.4   GROTE–HYNES THEORY

11.4.1   Introduction

Kramers theory goes beyond the TST by using the Brownian motion Markovian theory as its basis. The long-time hydrodynamic response of the system is taken into account by using the Stokes' law friction. This situation is ideal if the escaping particle can be assumed to be a massive Brownian particle moving slowly in a solvent. Then, on the time scale of the motion of the escaping particle, the solvent particles appear to be moving very fast and their response can be represented by a Markovian friction. On the other hand, if the reactants and the solvent particles are about equal in mass and size, this picture becomes questionable. A simple way to improve the picture is to use a non-Markovian friction, which leads naturally to the use of the GLE and the associated nonequilibrium theory. The GLE was first used by Grote and Hynes (Grote and Hynes, 1980) to obtain the famous GH rate equation, a cornerstone result in chemical reaction theory. This equation is the generalization of the Kramers moderate damping Markovian result, Eq. (11.22). Later, Grote and Hynes (Grote and Hynes, 1982) and Carmeli and Nitzan (Carmeli and Nitzan, 1982) generalized the Kramers low-damping energy-diffusion result, Eq. (11.25), to the case of a non-Markovian solvent friction.

The GH theory is embodied in the above-mentioned two results, which are based on the *stable states picture* using time correlation functions. For a discussion of this method, see the Hynes review mentioned above. The theory has since been reexamined from many different viewpoints. For references, see the Hänggi review. A particularly appealing derivation of the moderate damping result was given by Pollak (Pollak, 1986). Our discussion is based on Pollak, Grabert, and Hänggi (PGH) (Pollak et al., 1989), which is closer to our discussion of the TST and the Kramers theories.

11.4.2   Memory Friction Kernel

In Section 9.7, we showed how the GLE can be obtained from a microscopic approach based on a particularly simple model of the heat bath. In the present problem, the Brownian particle of that section becomes the escaping particle, and the heat bath represents the solvent, whose memory friction is given by the second part of Eq. (9.76). We rewrite this equation as

$$\zeta(t) = \frac{1}{m} \sum_{i=1}^{\mathcal{N}} \frac{c_i}{m_i \omega_i^2} \cos \omega_i(t) \tag{11.28}$$

It should be remembered that $\mathcal{N}$ is the number of bath oscillators and should not be confused with N, which is the number of escaping particles. In the following sections, we shall always choose $N = 1$ without loss of generality. By a suitable choice of the solute–solvent coupling constants c_i, one can get a given friction in the continuum limit $\mathcal{N} \to \infty$. When a particular model of friction is needed, we shall use the exponential friction as the memory friction kernel (MFK)

$$\zeta(t) = \frac{1}{\alpha} \exp\left(\frac{-t}{\alpha \gamma'}\right) \tag{11.29}$$

first introduced by Straub et al. (Straub et al., 1985, 1986). The integral of the MFK over time is γ', and, in the limit $\alpha \to 0$, it reduces to the Markovian friction

$$\zeta(t) = 2\gamma'\delta(t) \tag{11.30}$$

which indicates that the parameter γ' is the same γ' that is proportional to the damping parameter γ in earlier sections.

The non-Markovian MFK depends on two parameters, γ and α. The parameter $1/\alpha$ is the value of the MFK at $t = 0$ and can be thought of as the strength of friction or the strength of the coupling between the escaping particle and the heat bath. The Markovian case then corresponds to the strong coupling limit. For a fixed α, the parameter γ is the relaxation time of the

MFK, because it governs how fast the MFK goes to zero at long times. A large damping parameter γ indicates a slow decay and vice versa.

The MFK is related to the autocorrelation of the random force by the fluctuation–dissipation theorem, Eq. (9.79). In the exponential friction case,

$$\langle R(t)R(0) \rangle = \frac{mk_B T}{\alpha} \exp\left[\frac{-t}{\alpha\gamma'}\right] \tag{11.31}$$

from which we conclude that the parameters α and γ determine the inverse strength and the decay time, respectively, of the autocorrelations of the random force of the solvent. Since small α approximates the Markovian case, we expect to see large deviations from the Kramers results for large α or weak coupling. The strong coupling and the weak coupling that arise for different values of α should not be confused with the cases of high damping and low damping that arise depending on the value of γ. The two are quite distinct and lead to independent regimes for the behavior of the rate.

11.4.3   The Case $\mathcal{N} = 1$

The escape rate is obtained by diagonalizing the microscopic Hamiltonian and studying the resulting normal modes. We illustrate the procedure for the case of a single bath particle, in which case the Hamiltonian (9.60) becomes

$$\mathcal{H}_1 = \frac{p^2}{2m} + \phi(x) + \frac{p_1^2}{2m_1} + \frac{1}{2}m_1\omega_1^2\left(x_1 - \frac{c_1}{m_1\omega_1^2}x\right)^2 \tag{11.32}$$

The first step in diagonalizing this Hamiltonian is to introduce the mass-weighted coordinates and momenta defined by

$$x' = \sqrt{m}x \qquad x_1' = \sqrt{m_1}x_1 \tag{11.33}$$

$$p' = p/\sqrt{m} \qquad p_1' = p_1/\sqrt{m_1} \tag{11.34}$$

in which case the Hamiltonian becomes

$$\mathcal{H}_1 = \frac{1}{2}p'^2 + \phi(x') + \frac{1}{2}p_1'^2 + \frac{1}{2}\omega_1^2\left(x_1' - \frac{c_1}{\sqrt{mm_1}\omega_1^2}x'\right)^2 \tag{11.35}$$

In the barrier region,

$$\phi(x') = E_b - \frac{1}{2}\omega_b^2 x'^2 \tag{11.36}$$

so that the symmetric matrix corresponding to the coordinates x' and x'_1 is given by

$$
V = \begin{pmatrix} -\omega_b^2 + c_1^2/(mm_1\omega_1^2) & -c_1/(\sqrt{mm_1}) \\ -c_1/(\sqrt{mm_1}) & \omega_1^2 \end{pmatrix}
\tag{11.37}
$$

In the absence of coupling, $c_1 = 0$, and the matrix is already diagonal, with the diagonal elements given by the uncoupled squared frequencies $-\omega_b^2$ and ω_1^2, corresponding to the barrier frequency and the single bath frequency. In the presence of coupling, both of these will be changed to new frequencies, which are given by the eigenvalues of the matrix V. The coordinates of the particles, bath, and the system are uncoupled for zero coupling but are coupled when c_1 is not zero.

The 2×2 matrix V can be diagonalized by a linear transformation, which in this case corresponds to just a rotation of the axes. Let the normal coordinates be denoted by ρ and ρ_1. Denoting the eigenvalues by Ω^2, we get the equation

$$
\Omega^4 - \omega_1^2\Omega^2 \left(1 - \frac{\omega_b^2}{\omega_1^2} + \frac{c_1^2}{mm_1\omega_1^2} \right) - \omega_1^2\omega_b^2 = 0
\tag{11.38}
$$

This equation gives two squared frequencies. By focusing on the eigenvalue corresponding to the new barrier frequency and calling it $\lambda_\rho^2 = -\Omega^2$, one can show by manipulation of Eq. (11.38) that λ_ρ satisfies the relation

$$
\lambda_\rho^2 = \frac{\omega_b^2}{1 + c_1^2[mm_1\omega_1^2(\omega_1^2 + \lambda_\rho^2)]^{-1}}
\tag{11.39}
$$

The other eigenvalue is the changed bath frequency, is denoted by λ_1^2, and is given by a similar equation. One can show directly from Eq. (11.38) or by using the fact that the determinant of a symmetric matrix is invariant under diagonalization that

$$
\lambda_\rho\lambda_1 = \omega_b\omega_1
\tag{11.40}
$$

The kinetic energy part of the Hamiltonian is proportional to a unit matrix and can be rewritten in terms of the new momenta or velocities. The diagonalized Hamiltonian is given by

$$
\mathcal{H}_1 = \tfrac{1}{2}\dot\rho^2 + \tfrac{1}{2}\dot\rho_1^2 + E_b + \tfrac{1}{2}(\lambda_1^2\rho_1^2 - \lambda_\rho^2\rho^2)
\tag{11.41}
$$

where the dots indicate time derivatives as usual.

A similar diagonalization can be performed in the well region where the potential is given by

$$\phi(x) = \frac{1}{2}m\omega_0^2(x - x_0)^2 \tag{11.42}$$

First, the constant x_0 is eliminated by changing to a shifted x and then the mass-weighted coordinates are introduced as before. By assuming that the eigenvalues are given by λ_0, which corresponds to the new well frequency, and λ_1', which is the new bath oscillator frequency, we may derive a relation similar to Eq. (11.40). It is found to be

$$\lambda_0\lambda_1' = \omega_0\omega_1 \tag{11.43}$$

11.4.4   Derivation of the GH Rate Coefficient

The diagonalization for the case of $\mathcal{N}$ bath oscillators proceeds similarly by performing an orthogonal transformation in $\mathcal{N} + 1$ dimensions. The details are given by Pollak (Pollak, 1986) and PGH (Pollak et al., 1989). The results can be seen as a generalization of the equations of Section 11.4.3. The normal mode Hamiltonian in the barrier region is found to be

$$\mathcal{H} = \tfrac{1}{2}\dot{\rho}^2 + \tfrac{1}{2}\sum_i \dot{\rho}_i^2 + E_b + \tfrac{1}{2}\sum_i \lambda_i^2\rho_i^2 - \tfrac{1}{2}\lambda_\rho^2\rho^2 \tag{11.44}$$

We see from this Hamiltonian that the normal system mode and the normal bath modes are decoupled; indeed that was the purpose of the diagonalization. There is a barrier in the new ρ space. It occurs at $\rho = 0$, and the value of the potential there is E_b, whereas the new (imaginary) barrier frequency is λ_ρ. In honor of GH, this frequency is called the *Grote–Hynes* frequency. Equation (11.40) now becomes

$$\lambda_\rho \prod_i \lambda_i = \omega_b \prod_i \omega_i \tag{11.45}$$

A similar diagonalization can be made in the well region, and the relation corresponding to Eq. (11.43) is given by

$$\lambda_0 \prod_i \lambda_i' = \omega_0 \prod_i \omega_i \tag{11.46}$$

From the last two equations, it follows that

$$\prod_i (\lambda_i'/\lambda_i) = \omega_0\lambda_\rho/(\omega_b\lambda_0) \tag{11.47}$$

a relation that will be used later.

The beauty of the normal mode Hamiltonian is that the normal modes are decoupled, even though the original coordinates were coupled. If one considers the normal modes corresponding to the original escaping particle, there is no recrossing by the normal mode ρ near the barrier, because ρ does not interact with the normal bath modes in the barrier region; there are no terms in the normal-mode Hamiltonian corresponding to terms containing c_i in the original Hamiltonian. Therefore, we can use the second assumption of the TST to calculate the escape rate for the ρ mode with impunity.

How about the first assumption of the TST? It is not satisfied everywhere. The reason is that as one travels away from the barrier region, the ρ mode and the bath modes do couple and equilibrium is disturbed. In fact, one can write down the coupled equations of these modes in the well region. These equations have not been solved exactly yet, although some progress has been made to solve them iteratively in the weak-coupling limit (see the PGH paper). However, if one assumes equilibrium in the normal modes, the GH rate equation results. Let us now derive this equation by using the technique used in the derivation of the TST rate constant.

Assuming equilibrium, the multidimensional distribution function normalized to one particle is given in the barrier region by

$$f(\rho, \dot{\rho}, \rho_i, \dot{\rho}_i)$$
$$= \frac{1}{hZ} \exp(-\beta E_b) \exp\left[\frac{-\beta(\dot{\rho}^2 - \lambda_\rho^2\rho^2)}{2}\right] \prod_i \exp\left[\frac{-\beta(\dot{\rho}_i^2 + \lambda_i^2\rho_i^2)}{2}\right] \tag{11.48}$$

The flux at the top of the barrier is obtained from this equation by putting $\rho = 0$, integrating it over all the bath modes, multiplying it by $\dot{\rho}$, the velocity of the normal escaping mode, and integrating over only positive velocities. The integrals are easy to do, and we get

$$j = \frac{1}{hZ\beta} \left(\frac{2\pi}{\beta}\right)^{\mathcal{N}} \exp(-\beta E_b) \prod_i \lambda_i^{-1} \tag{11.49}$$

The partition function can be evaluated as follows. By using the metastability assumption, the main contribution to the partition function comes from the well region. In terms of the normal modes, the partition function is given by

$$Z_1 = \frac{1}{h} \int \int \exp\left[\frac{-\beta(\dot{\rho}'^2 + \lambda_0^2 \rho'^2)}{2}\right] d\rho \, d\dot{\rho}$$

$$\prod_i \exp[-(\beta/2)(\dot{\rho}_i'^2 + \lambda_i'^2 \rho_i'^2)] \, d\rho_i' \, d\dot{\rho}_i' \tag{11.50}$$

where ρ_i' and $\dot{\rho}_i'$ are the normal modes and their velocities in the well region. All the integrals are Gaussian, and the result is

$$Z_1 = \frac{1}{h\lambda_0} \left(\frac{2\pi}{\beta}\right)^{\mathcal{N}+1} \prod_i \lambda_i'^{-1} \tag{11.51}$$

By combining this equation with the expression for j, Eq. (11.49), and using Eq. (11.47), we get the GH rate coefficient

$$k_{GH} = \frac{\omega_0 \lambda_\rho}{2\pi\omega_b} \exp(-\beta E_b) \tag{11.52}$$

from which the reduced GH rate coefficient is found to be

$$k_{GH}^* = \frac{\lambda_\rho}{\omega_b} \tag{11.53}$$

The GH frequency λ_ρ can be shown to satisfy the following relation, which generalizes Eq. (11.39),

$$\lambda_\rho^2 = \frac{\omega_b^2}{1 + \sum_i c_i^2 [mm_i \omega_i^2 (\omega_i^2 + \lambda_\rho^2)]^{-1}} \tag{11.54}$$

We still need to express this relation in terms of the MFK given by Eq. (11.28), which is done as follows. Take the Laplace transform of Eq. (11.28) to give

$$\hat{\zeta}(s) = \frac{1}{m} \sum_i \frac{c_i}{m_i \omega_i^2} \left(\frac{s}{s^2 + \omega_i^2}\right) \tag{11.55}$$

Compare this Laplace transform with the sum occurring in the denominator of Eq. (11.54) to see that the latter is related to the value of the former at $s = \lambda_\rho$. Use this relation to obtain

$$\lambda_\rho^2 = \frac{\omega_b^2}{1 + \hat{\zeta}(\lambda_\rho)/\lambda_\rho} \tag{11.56}$$

which is the celebrated GH equation. By treating this equation as a quadratic in λ_ρ, it is possible to rewrite it in another form,

$$\lambda_\rho = \omega_b \left(1 + \frac{\widehat{\zeta}(\lambda_\rho)}{4\omega_b^2} \right)^{1/2} - \frac{\widehat{\zeta}(\lambda_\rho)}{2} \tag{11.57}$$

from which the reduced GH rate coefficient may be rewritten as

$$k_{GH}^* = \left(1 + \frac{\widehat{\zeta}(\lambda_\rho)}{4\omega_b^2} \right)^{1/2} - \frac{\widehat{\zeta}(\lambda_\rho)}{2\omega_b} \tag{11.58}$$

11.4.5 Discussion of the GH Rate Coefficient

The GH rate coefficient reduces to the Kramers result for the Markovian case, as expected. For the Markovian MFK,

$$\zeta(t) = 2\gamma'\delta(t) \tag{11.59}$$

whose Laplace transform is given by

$$\widehat{\zeta}(s) = \gamma' \tag{11.60}$$

Substituting this into Eq. (11.58) reduces it to the Kramers reduced rate coefficient, which is given by Eq. (11.22).

The GH rate coefficient depends on the value of the MFK at the GH frequency, which is usually of the order of the barrier frequency ω_b. On the other hand, the Kramers rate coefficient depends on the long-time (hydrodynamic) behavior of the MFK, which corresponds to small frequencies. Since the friction is likely to have a very different behavior at short and long frequencies or at long- and at short-time scales, the GH theory will give results quite different from the Kramers theory.

By choosing suitable forms for the MFK, the GH rate coefficient was examined as a model for dipole isomerization in the presence of polar solvents by van der Zwan and Hynes (van der Zwan and Hynes, 1983, 1984). They found a rich behavior in different cases depending on the coupling strength $1/\alpha$ and the damping parameter γ'. In various regimes, the GH rate coefficient depends on simple powers of γ'. In addition, van der Zwan and Hynes discovered the existence of a dividing coupling strength at which the behavior of the rate coefficient changes qualitatively. The critical behavior of the rate is discussed in detail in Section 11.5.

11.5   CRITICAL PHENOMENA IN THE GH THEORY

11.5.1   Singularity in the GH Theory

The nature of the GH theory around its singularity has been studied recently by Singh et al. (Singh et al., 1992, 1994) (SKR) by mapping the singularity to a critical point. To discuss the singularity, we choose the exponential friction model, Eq. (11.29), whose Laplace transform is

$$\widehat{\zeta}(s) = \frac{\gamma'}{1 + s\alpha\gamma'} \tag{11.61}$$

By substituting this result into Eq. (11.57) and simplifying, we get the the following cubic equation in the reduced GH frequency:

$$r^{*3} + \frac{r^{*2}}{\alpha^*\gamma^*} - r^*\left(1 - \frac{1}{\alpha^*}\right) - \frac{1}{\alpha^*\gamma^*} = 0 \tag{11.62}$$

where we have introduced the convenient notations

$$\alpha^* = \alpha\omega_b^2 \qquad r^* = \lambda_\rho/\omega_b \tag{11.63}$$

The quantity α^* is the reduced inverse coupling strength. The quantity r^* is the reduced GH frequency. Equation (11.62) will always have a real and positive root, which is identified with the reduced GH escape rate. Figure 11.2 shows the reduced GH frequency calculated from Eq. (11.62) for a range of parameters.

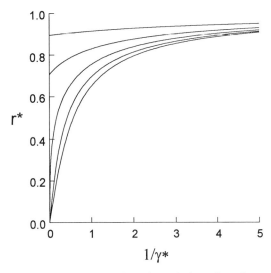

Figure 11.2   Reduced GH frequency. The values of α^* are from the top $5, 2, 1, 0.2, 0.5$.

For infinite damping, $\gamma^* \to \infty$, and Eq. (11.62) reduces to

$$r^{*3} - r^*(1 - 1/\alpha^*) = 0 \tag{11.64}$$

which has the real, positive solutions

$$r^* = (1 - 1/\alpha^*)^{1/2}, \qquad 0 \tag{11.65}$$

The first solution is applicable for $\alpha^* > 1$, and the second one for $\alpha^* \leq 1$. Summarizing, for infinite γ^*,

$$r^* = \begin{cases} (1 - 1/\alpha^*)^{1/2} & \alpha^* > 1 \\ 0 & \alpha^* \leq 1 \end{cases} \tag{11.66}$$

We see that for infinite γ^*, the rate has a singular behavior at the point $\alpha^* = 1$. At this point, the rate is continuous, having the value 0, but the derivative of the rate is infinite for $\alpha^* > 1$ and zero otherwise. Similarly, by studying Eq. (11.62) for large but finite γ^*, one can show that the following behavior of the rate is obtained:

$$r^* \approx \begin{cases} 1 - 1/\alpha^* & \alpha^* > 1 \\ 1/[(1 - \alpha^*)\gamma^*] & \alpha^* < 1 \\ 1/\gamma^{*1/3} & \alpha^* = 1 \end{cases} \tag{11.67}$$

This equation shows the singularity at α^* even more clearly. The behavior of the reduced GH frequency for an exponential MFK was obtained by PGH. The correspondence of this singular behavior to critical phenomena was made by SKR, who exploited the relationship to show that the GH rate satisfies the properties of scaling and universality.

11.5.2   Scaling and Universality in the GH Theory

By comparing Eq. (11.67) with Eqs. (3.33) and (3.38), we see that the following correspondence can be made (ignoring nonuniversal amplitudes):

$$M \iff r^* \qquad T \iff 1/\alpha^* \qquad H \iff 1/\gamma^* \tag{11.68}$$

With this correspondence, one might say that the critical point in the GH theory, with an exponential MFK, is at $1/\gamma^* = 0$ and $\alpha^* = 1$ and that the critical exponents β and δ have the values 0.5 and 3, respectively, just as in the Weiss mean-field theory.

To discover the scaling behavior of the rate near the critical point, we expand Eq. (11.62) near the critical point, that is, for small $1/\gamma^*$ and small

$\Delta = 1 - \alpha_c^*/\alpha^*$, where the critical reduced coupling strength α_c^* is unity for the exponential MFK. Exploiting the critical analogy, we define a scaled rate f and a scaled distance from the critical strength, y, by the relations

$$f = r^*\gamma^{*1/3} \qquad y = \Delta\gamma^{*b} \tag{11.69}$$

where the exponent b is unknown at this point. We do know that $b > 0$ because, in the scaling limit, $\gamma^* \to \infty$ and $\Delta \to 0$ such that y remains finite. Note that we have already used the fact that the rate scales as $1/\gamma^{*1/3}$, which is known from Eq. (11.67). Now we substitute for r^* and Δ in terms of f and y into Eq. (11.62) to get

$$\frac{f^3}{\gamma^*} + \frac{f^2}{\gamma^{*5/3}}\left(1 - \frac{y}{\gamma^{*2b}}\right) - \frac{yf}{\gamma^{*(b+1/3)}} - \frac{1}{\gamma^*}\left(1 - \frac{y}{\gamma^{*b}}\right) = 0 \tag{11.70}$$

In the scaling limit, as $\gamma^* \to \infty$, the second term in Eq. (11.70) is of higher order than the first one and can be dropped. Also, in the same limit, the last term reduces to $1/\gamma^*$ because $b > 0$. Therefore, we get

$$\frac{f^3}{\gamma^*} - \frac{yf}{\gamma^{*(b+1/3)}} - \frac{1}{\gamma^*} \approx 0 \tag{11.71}$$

Clearly, we get scaling if we choose $b = \frac{2}{3}$. With this choice of b and taking the scaling limit, Eq. (11.71) gives the following cubic for the scaling function:

$$f^3 - yf - 1 = 0 \tag{11.72}$$

Consequently, the reduced GH frequency obeys the scaling equation

$$r^* \approx \frac{1}{\gamma^{*1/3}}f(y) \qquad y = \Delta\gamma^{*2/3} \tag{11.73}$$

with the scaling function given by Eq. (11.72). A plot of the scaling function f with respect to y is given in Figure 11.3.

Thus, Eq. (11.73) is interpreted as follows. For any given values of α^* and γ^*, the reduced GH frequency is obtained from Eq. (11.62) as the real and positive solution. Thus the reduced GH frequency, in general, depends on two variables. But in the neighborhood of the singularity or the critical point where $\gamma^* = \infty$, $\alpha_c^* = 1$, the scaled rate, $r^*\gamma^{*1/3} = f$, depends on just the one variable y, which is a combination of Δ, the distance from the critical point, and a suitable power of γ^*, in this case two-thirds. This behavior is similar to

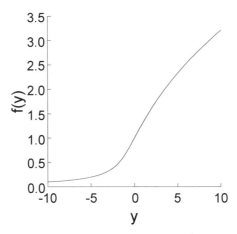

Figure 11.3 Plot of the scaling function.

the scaling of the magnetization in a ferromagnet or of the density of a fluid near their respective critical points.

The cubic equation for the scaling function can be solved by standard methods. The solution is (see Singh and Robinson, 1994)

$$
f(y) = \begin{cases}
\left(\dfrac{4y}{3}\right)^{1/2} \cosh\left[\dfrac{1}{3}\cosh^{-1}\left(\dfrac{27}{4y^3}\right)^{1/2}\right] & y > 0 \\[3mm]
\left(\dfrac{4|y|}{3}\right)^{1/2} \sinh\left[\dfrac{1}{3}\sinh^{-1}\left(\dfrac{27}{4|y|^3}\right)^{1/2}\right] & y < 0
\end{cases}
\tag{11.74}
$$

Note that for $y > (27/4)^{1/3}$ the argument of the cosh and $\cosh^{-1}$ in the first of these cases is less than 1, so that the cosh changes into a cosine. These equations may be studied to obtain the behavior of the scaling function in the various regimes, or the behavior may be obtained directly from Eq. (11.72). The resulting behavior of f can be substituted in Eq. (11.73) to obtain the behavior of the reduced GH frequency in that regime. For example, in the strong coupling regime, $\alpha^* \lesssim 1$ and γ^* becomes very large. From Eq. (11.73), this makes y negative and very large. From Eq. (11.72) one sees that in this regime $f \approx 1/|y|$, so that from Eq. (11.73) the rate behaves as $r^* \approx |\Delta|^{-1}\gamma^{*-1}$. From Eq. (11.74) one sees that the second line is now applicable. In this case, the argument of the sinh becomes small so that one may use successively $\sinh^{-1} x \approx x$ and $\sinh x \approx x$ to obtain once again the relation $f \approx 1/|y|$. This strong coupling regime is called the solvent caging regime in chemical reaction theory (see van der Zwan and Hynes, 1983, 1984). The other cases may be dealt with similarly. The results are gathered in Table 11.2.

TABLE 11.2   The Behavior of the GH Scaling Function and Rate in Various Regimes

ABC	Chemical Reactions	Parameter Conditions	y Behavior	$f(y)$	r^*
Strong	Solvent caging	$\alpha^* \lesssim 1, \gamma^* \to \infty$	$y \to -\infty$	$\|y\|^{-1}$	$\gamma^{*-1}\|\Delta\|^{-1}$
Critical	Dividing coupling	$\alpha^* = 1, \gamma^* \to \infty$	$y = 0$	1	$\gamma^{*-1/3}$
Weak	Nonadiabatic solvation	$\alpha^* \gtrsim 1, \gamma^* \to \infty$	$y \to +\infty$	$y^{1/2}$	$\Delta^{1/2}$

It is seen from this table that the critical exponents do not depend on the quantities like the barrier height or the well frequency. This behavior is an indication of the universality of critical exponents. The scaling function f is also universal as seen from Eq. (11.74). It turns out that the values of the exponents in Table 11.2 are specific to the exponential MFK. The work of SKR considered general MFKs of the form

$$\zeta(t) = \frac{A}{\alpha} \exp(-B|t/\alpha\gamma'|^p) \tag{11.75}$$

and showed that the critical exponents and the scaling function are independent of A and B but depend on p, again indicating universality.

11.6   CONCLUDING REMARKS

Various universal phenomena have been discovered in chemical reaction theory over the years. It started with the Arrhenius law and the Smoluchowski $1/\gamma^*$ decay of the rate. Since the GH theory admits of a memory relaxation time, it was only to be expected that when the relaxation time of the solvent autocorrelation function becomes large, critical phenomena should emerge. Thus, the recent discovery of a singularity in the GH rate is in line with expectations, and it has brought the theory of chemical reaction rates within the realm of critical phenomena, where the heavy artillery of this field in the form of RG and MCRG is bound to yield new insights into this old subject.

In Chapters 12 and 13, we concentrate on phenomena far from equilibrium, where many exciting new and unusual effects can be found.

11.7   EXERCISES AND PROBLEMS

1. Calculate the TST rate if the potential in the well region is a square well of width a,

$$\phi(x) = \begin{cases} 0 & x_1 - a < x < x_1 \\ \infty & x \le x_1 - a \end{cases} \tag{11.76}$$

2. Consider the potential

$$\phi(x) = -Ax^2 + Bx^4, \qquad (11.77)$$

where A and B are positive. Show that this potential has two minima and a maximum. Calculate the barrier height, the well and the barrier frequencies, respectively. Use all this information to calculate the TST rate for the escape from the left well over the barrier.

3. Calculate the rate coefficient in the Kramers high-damping limit for the potentials in the above two problems. Show that the reduced rate coefficient is the same in both cases.

4. Derive Eq. (11.49).

5. Derive Eq. (11.51).

6. Derive Eqs. (11.61) and (11.62).

7. Derive the entries in the last two columns of Table 11.2.

12

OSCILLATING CHEMICAL REACTIONS AND CHAOS

12.1 INTRODUCTION

It is well known that, upon mixing the reactant chemicals, a typical reaction will proceed by well-established rate laws till equilibrium is attained. Again, typically, such systems may be assumed to be approximately isolated, so that the second law of thermodynamics dictates that they will proceed to equilibrium with an increase in entropy. In nature, there exist many oscillating chemical reactions that are characterized by being far from equilibrium. In these systems, the long-time behavior is not one of equilibrium, but rather one of oscillating behavior. These systems appear to be in conflict with the second law of thermodynamics, and, for this reason, such reactions were treated as laboratory curiosities for many years after their discoveries. However, in biological systems it had become apparent (see Schrödinger, 1944; Nicolis and Prigogine, 1977) that self-organization required open systems, which were far from equilibrium. Biological systems show periodic oscillations, examples of which include heartbeats, menstrual cycles, glycolysis, and biosynthesis of some proteins. Hence, it is reasoned that chemical systems, capable of oscillating, must also be far from equilibrium and must be open systems if the oscillations are to be sustainable. Accordingly, the apparent contradiction with the second law of thermodynamics was resolved, because, in nonisolated systems, the entropy can increase, decrease, or remain unchanged depending on the interactions of the system with the surroundings, thus implying that the ultimate fate of the system may not be the usual equilibrium state.

In this chapter, we first briefly sketch the history of oscillating chemical reactions. Then we look at some model systems, which have been devised to help understand the experimental observations. We shall see that these systems are all described by nonlinear ordinary differential equations (ODEs) that employ feedback through autocatalysis steps. The procedures for solving the ODEs will be discussed in detail. The model systems are chosen to illustrate various aspects of oscillating reactions. These systems include the Lotka–Volterra, Brusselator, and Oregonator models. In addition, we discuss difference equations and their relation to ODEs, and we use the one-dimensional logistic difference equation to illustrate many features of oscillating reactions, including the phenomenon of chaos.

We next focus our attention on the Belousov–Zhabotinskii (BZ) reaction, since a very large amount of work has been done on this system, and, as a result, the behavior of the BZ reaction is becoming well understood. The Oregonator model will be discussed as a model of the BZ reaction. The experimental results of the BZ reaction, such as bifurcations, period doublings, and chaos, will be discussed and analyzed with the help of statistical mechanical concepts. We shall show that concepts of phase space, distribution functions, probability densities, ergodic behavior, correlation functions, and time series are essential to properly understand oscillating systems.

12.2 HISTORY OF OSCILLATING CHEMICAL REACTIONS

12.2.1 History to 1970

In 1921 Bray (Bray, 1921) studied the iodate catalyzed decomposition of hydrogen peroxide and observed that, in certain narrow concentration regions, the concentrations of the oxygen and iodine intermediates oscillated in a periodic fashion. More work was done by Bray and Liebhafsky (Bray and Liebhafsky, 1931) on what is now known as the Bray–Liebhafsky (BL) oscillator. Since 1931, several investigations of the BL oscillator have shown that the reaction is very complex, but no definitive mechanism has resulted (Furrow, 1985, p. 171).

In 1920, Lotka published a model that provided for periodic oscillations based on an autocatalytic mechanism. Later, in 1926, Volterra used the Lotka equations to model predator–prey populations of fish. The mechanism is often referred to as the Lotka–Volterra (LV) model. We discuss it in detail in Section 12.3.

Even though the LV model provided a mechanism for a periodic reaction and the BL reaction clearly demonstrated oscillations, very little work was done on oscillating reactions till 1961 when Zhabotinskii took up the study of the chemical oscillator, which was discovered in 1951 by Belousov (Zhabotinskii, 1964; Field and Burger, 1985, p. 1). Belousov had studied the cerium catalyzed oxidation of citric acid by bromate ion. He observed color changes that oscillated between the yellow Ce^{4+} and the colorless Ce^{3+}. Zhabotinskii did extensive studies of the system, substituting malonic acid for the citric acid substrate. It was shown that not only were periodic oscillations observed for well-stirred systems, but that nonstirred systems showed spatial patterns of propagating waves. Zhabotinskii and co-workers continued their study through the 1960s and proposed possible mechanisms.

12.2.2   History after 1970

The last 25 years have seen considerable progress in the understanding of the BZ reaction and in the discovery of many other oscillating reactions. A breakthrough occurred in the early 1970s when Noyes and Field at the University of Oregon collaborated with the Hungarian, Körös, to provide a mechanism for the BZ reaction. This mechanism, referred to as the FKN mechanism, was able to explain much of what was known about the BZ reaction. The FKN mechanism consists of some 11 elementary reactions involving 15 chemical species, 7 of which are intermediates (Field et al., 1972).

The success of the FKN mechanism in explaining the BZ reaction stimulated searches for other oscillating reactions. Consequently, many bromate-driven oscillators were discovered by replacing the cerium ion with ferroin and manganese, by using other organic substrates, by the addition of various halide ions, and by changing the conditions of the reaction. A major breakthrough by researchers was made by changing from a stirred closed system to a continuous-flow stirred tank reactor (CSTR) in the mid-1970s. Only in this way could a systematic study of periodic oscillations and chaos be done. Such temporal oscillations in CSTRs are the main focus of this chapter.

The FKN mechanism is too complex for a detailed mathematical study, because the kinetics of the FKN mechanism gives seven coupled nonlinear first-order ODEs. Field and Noyes in 1974 (Field and Noyes, 1974) found a way to reduce the FKN mechanism to a manageable set of elementary reactions, which model the salient parts of the BZ reaction. This model is known as the Oregonator and will be discussed in Section 12.6.1.

Besides bromate oscillators, chlorite and iodate oscillators (e.g., the BL reaction) have been discovered and studied. However, the BZ reaction remains the most studied and best understood of all the oscillator reactions, since it presents the experimenter with a wide diversity of behavior and complexity unequaled so far in any other reaction.

12.3   THE LOTKA–VOLTERRA MODEL

12.3.1   Solution of the Lotka–Volterra Equations

The LV model is a mechanism for producing sustained oscillations in a reaction involving two intermediates. The essential part of the mechanism is feedback involving the intermediates, which is necessary for nonlinearity and for oscillations. Hence, only those steps that directly affect the kinetics of the intermediates are included in the model mechanism. For an overall reaction A $\longrightarrow$ P with intermediates X and Y, the mechanism is

$$A + X \xrightarrow{k_1} 2X \tag{LV 1}$$

$$X + Y \xrightarrow{k_2} 2Y \tag{LV 2}$$

$$Y \xrightarrow{k_3} P \tag{LV 3}$$

with rate constants k_1, k_2, and k_3, respectively. Reactions (LV 1) and (LV 2) are autocatalytic and provide feedback. The rate equations are

$$\frac{d[X]}{dt} = k_1[A][X] - k_2[X][Y] \tag{12.1a}$$

$$\frac{d[Y]}{dt} = k_2[X][Y] - k_3[Y] \tag{12.1b}$$

To get these into more familiar notation, we let $a = k_1[A], b = k_2$, and $c = k_3$. Equations (12.1) now become

$$\frac{dx}{dt} = x(a - by) \tag{12.2a}$$

$$\frac{dy}{dt} = y(bx - c) \tag{12.2b}$$

The time t is the independent variable, the concentrations x and y are the dependent variables, and a, b, c are positive constants. The reactant concentration [A] must be fixed either by making huge quantities available or replenishing it continuously in a CSTR, making the LV system a nonlinear open system.

Our aim is to solve Eqs. (12.2), to study the time dependence of x and y, and to learn about the final states of the systems and how they are approached. In general, the solution of nonlinear differential equations is very difficult. For most nonlinear ODEs no analytic solutions have been found, so we are forced to use numerical methods of solution. Fortunately, computers are available to help us, and there are available algorithms, which give accurate solutions. Before we turn to the computer, however, there is much that can be done to understand the nature of the solutions analytically.

The first step is to look at the steady states, which are given by the points $dx/dt = 0$ and $dy/dt = 0$. These are often called *critical points* or *fixed points*. Simple algebra shows that the critical points for Eqs. (12.2) are at $x_c = 0$, $y_c = 0$ and $x_c = c/b$, $y_c = a/b$. Next, we ask about the behavior of the equations in the immediate vicinities of the critical points or the behavior of the system not too far from the steady states. For this purpose, we can assume linear behavior of the ODEs in these regions. Therefore, if we write Eqs. (12.2) as $dx/dt = F_1(x, y)$ and $dy/dt = F_2(x, y)$, expand each $F_i(x, y)$ $(i = 1, 2)$ in a Taylor's series about the critical point (x_c, y_c), and truncate at the linear term, we get

$$F_i(x, y) \approx F_i(x_c, y_c) + (x - x_c)\left(\frac{\partial F_i}{\partial x}\right)_c + (y - y_c)\left(\frac{\partial F_i}{\partial y}\right)_c \qquad (12.3)$$

At the point $(x_c, y_c) = (0, 0)$, we get $F_1 = ax$ and $F_2 = -cy$.

We now have the linearized forms of Eqs. (12.2). In the vicinity of $(0,0)$ we have

$$\frac{dx}{dt} = ax \qquad (12.4a)$$

$$\frac{dy}{dt} = -cy \qquad (12.4b)$$

We can now use the methods of linear ODEs to get analytical solutions for Eqs. (12.4). A simple integration gives $x = x_0 \exp(at)$ and $y = y_0 \exp(-ct)$, where x_0 and y_0 are now the initial concentrations, which, however, must be in the vicinity of the critical point. If we start the system at the point $x_0 = 0$, $y_0 = 0$, the solution is $x = 0$ and $y = 0$. Therefore, let us choose $x_0 \neq 0$, $y_0 \neq 0$ in the immediate vicinity of the critical point. We see that, as t becomes large, $\exp(at)$ becomes exponentially large and x moves away from the critical point. Likewise, as t becomes large, $\exp(-ct)$ becomes very small and y moves toward the critical point. Critical points of this type are called *saddle points* and the steady state in this case is said to be unstable. Consequently, the point $(x_c, y_c) = (0, 0)$ is a point of an unstable steady state. Instability has the usual meaning that the system moves rapidly away from the critical point if disturbed slightly.

For the point $(c/b, a/b)$ we get $F_1 = -c(y - a/b)$ and $F_2 = a(x - c/b)$. Therefore, Eqs. (12.2), when linearized in the vicinity of $(c/b, a/b)$, become

$$\frac{dx}{dt} = -c\left(y - \frac{a}{b}\right) \qquad (12.5a)$$

$$\frac{dy}{dt} = a\left(x - \frac{c}{b}\right) \tag{12.5b}$$

The solution of Eqs. (12.5) is not as simple as that of Eqs. (12.4), since Eqs. (12.5) are a system of linear coupled equations. However, these equations can be simplified by a transformation of coordinates to a new set of coordinates X and Y given by a translation of axes: $X = x - c/b$ and $Y = y - a/b$. The critical point is now at $X_c = 0$ and $Y_c = 0$, and Eqs. (12.5) become

$$\frac{dX}{dt} = -cY \tag{12.6a}$$

$$\frac{dY}{dt} = aX \tag{12.6b}$$

If we divide Eq. (12.6a) by Eq. (12.6b) we get

$$\frac{dX}{dY} = -\frac{cY}{aX} \tag{12.7}$$

or

$$aX\, dX + cY\, dY = 0 \tag{12.8}$$

The solution of this equation follows by inspection and is

$$\tfrac{1}{2}aX^2 + \tfrac{1}{2}cY^2 = C_1 \tag{12.9}$$

where C_1 is an integration constant. In general, equations like this are solved as follows. Equation (12.6) is in the form $M(X, Y)\, dX + N(X, Y)\, dY = 0$. We use the test for exactness, which says that this type of equation is exact if $\partial M/\partial Y \equiv \partial N/\partial X$. It is seen immediately that Eq. (12.6) is exact. Being exact, this equation has a solution

$$\int_0^X aX\, dX + \int_0^Y cY\, dY = C_1 \tag{12.10}$$

from which Eq. (12.9) again follows. By comparing Eq. (12.9) with the standard equation of an ellipse

$$\frac{X^2}{A^2} + \frac{Y^2}{B^2} = 1 \tag{12.11}$$

we see that Eq. (12.9) describes an ellipse whose center is at $X = 0, Y = 0$ and whose semiaxes are $\sqrt{2C_1/a}$ and $\sqrt{2C_1/c}$.

The graph of X versus Y is called a *phase portrait*. It gives the orbit or trajectory that is traced out by a given solution for given boundary conditions. In the present case, the phase portrait consists of an ellipse. The above procedure does not give the time dependence of the variables X and Y. The time dependence is obtained as follows.

Equations (12.6), being linear, can be solved by the methods of linear algebra. Just as in the case of Eqs. (12.4), we assume solutions of the form $X = \alpha_1 \exp(\lambda t)$ and $Y = \alpha_2 \exp(\lambda t)$. If these solutions are substituted back into Eqs. (12.6), we get

$$-\lambda\,\alpha_1 - c\,\alpha_2 = 0 \tag{12.12a}$$

$$a\,\alpha_1 - \lambda\,\alpha_2 = 0 \tag{12.12b}$$

We wish to solve these two linear equations for the unknown coefficients α_1 and α_2. As is well known from linear algebra, a set of simultaneous homogeneous linear equations of this sort have a nontrivial solution if and only if the determinant of the coefficients vanishes. The resulting secular determinant is

$$\begin{vmatrix} -\lambda & -c \\ a & -\lambda \end{vmatrix} = 0 \tag{12.13}$$

or, in matrix notation,

$$|J - \lambda I| = 0 \tag{12.14}$$

where λ is the eigenvalue, I is the 2×2 unit matrix, and J is the Jacobian matrix,

$$J = \begin{pmatrix} \partial F_1/\partial X & \partial F_1/\partial Y \\ \partial F_2/\partial X & \partial F_2/\partial Y \end{pmatrix} \tag{12.15}$$

Solving the determinantal equation (12.13) for λ gives $\lambda = \pm i\sqrt{ac}$, which is purely imaginary. Critical points with purely imaginary eigenvalues are called *centers*. No limiting behavior can be determined as $t \to \infty$ for imaginary eigenvalues, and orbits associated with a center are said to be marginally stable. We see that, without solving the nonlinear equations (12.2), it has been possible to determine that these equations have two critical points, one of which is an unstable saddle point and the other is a marginally stable center.

The constants in Eqs. (12.12) cannot be determined uniquely, but their ratios can be found from initial conditions. As is well known, the general solutions are sums of the exponentials or sines and cosines owing to the principle of superposition. We may write

$$X = X_0 \cos(\sqrt{act}) + X_1 \sin(\sqrt{act}) \tag{12.16a}$$

$$Y = Y_0 \sin(\sqrt{act}) + Y_1 \cos(\sqrt{act}) \tag{12.16b}$$

The four constants in these equations are not all independent. We choose $X_1 = 0$, which gives

$$X = X_0 \cos(\sqrt{act}) \tag{12.17}$$

By substituting this solution in Eq. (12.6a), we get

$$Y = X_0 \sqrt{(a/c)} \sin(\sqrt{act}) \tag{12.18}$$

These are periodic solutions, the periods T being given by $T = 2\pi/\sqrt{ac}$.

The solution given by Eqs. (12.17), (12.18), and the phase portrait for $a = 5, c = 3$, and $X_0 = 0.1$ are shown in Figure 12.1. For other initial conditions, other ellipses will be obtained in Figure 12.1c. The common center of the ellipses is marginally stable, because a perturbation of the system neither attracts nor repells the orbit to or from the center.

Having disposed of the linear case, we turn now to the nonlinear Eqs. (12.2). It is possible to get the phase portrait for Eqs. (12.2) without solving for the time dependency of x and y. By dividing Eq. (12.2a) by Eq. (12.2b), we can obtain an exact equation:

$$\frac{bx - c}{x} dx - \frac{a - by}{y} dy = 0 \tag{12.19}$$

which has the solution

$$\int_{x_0}^{x} \frac{bx - c}{x} dx - \int_{y_0}^{y} \frac{a - by}{y} dy = 0 \tag{12.20}$$

or

$$b(x + y) - c \ln x - a \ln y - b(x_0 + y_0) + c \ln x_0 + a \ln y_0 = 0 \tag{12.21}$$

Just as Eq. (12.9) gives the phase portrait, Figure 12.1(c), for the linearized LV equation, Eq. (12.21) gives the phase portrait for the nonlinear LV equation.

The time dependence of x and y can be found from Eqs. (12.2) using a Runge–Kutta (RK) method. This method is available with some mathematics software packages. A simple RK FORTRAN program for this problem is listed in Appendix D.9 under the title LOTKA. The graphical solutions of

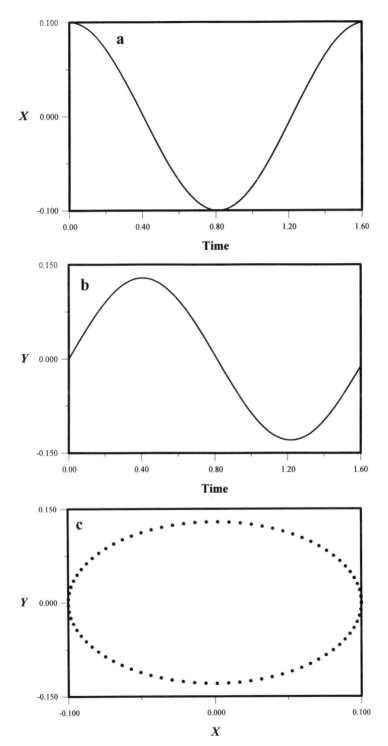

Figure 12.1 Linearized solutions for the Lotka–Volterra equation evaluated at the critical point, which is a center. (a) Equation (12.17). (b) Equation (12.18). (c) Phase portrait of the marginally stable center given by Eqs. (12.16).

Eqs. (12.2) are shown in Figure 12.2 for $a = 25$, $b = 0.3$, $c = 9$, $x_0 = 10$, and $y_0 = 15$. The phase portrait of the nonlinear LV equation depends on the initial conditions. Once the initial conditions are fixed, the system proceeds along the orbit containing the initial points. Equations (12.2) show that, when x is large and y is small, dx/dt and dy/dt are both positive for the parameters of Figure 12.2. The derivatives are both negative for small x and large y. Consequently, the system proceeds in a counterclockwise direction and oscillates with a period $T = 0.5635$ as shown in the time series of Figures 12.2(a and b). The center at $x = 30$ and $y = 250/3$ attracts the orbit, but as the saddle point is approached the orbit is repelled; it is the competition between the stable center and the unstable saddle point that gives the orbit of Figure 12.2(c).

Physically, what Figure 12.2 tells us is that, if species X is originally in excess, it will reach a maximum according to Reaction (LV 1). Then the species Y concentration rapidly increases according to Reaction (LV 2) till it begins to exhaust species X, at which point the species Y concentration falls, and the process repeats. For the predator–prey relationship, X is the prey and Y is the predator. In this case, the prey population increases because there are few predators. However, with increasing prey available, the predator population increases. The increase of predators continues till the prey population is almost exhausted, at which time the predators nearly all starve to death, and the process repeats. Clearly, the system, being far from equilibrium and open, does not tend simply to a stable steady state but oscillates.

12.3.2 Statistical Mechanics and the Lotka–Volterra Model

Statistical mechanics is concerned with the evaluation of macroscopic properties by the averaging of these properties over large numbers of molecules. The averages are ensemble averages, which are equal to time averages by the ergodic hypothesis. The space in which the averages are calculated is phase space. For systems at equilibrium or at a steady state, the volume in phase space is conserved according to Liouville's theorem. Phase space is conserved in the LV model, where the initial conditions determine the orbit of the system. Even in the absence of mechanical and thermal fluctuations, mass fluctuations can disturb the steady state and lead to a new steady state. Liouville's theorem is still valid provided that all accessible regions of phase space (for all possible initial conditions) are included. Even so, the LV system is far from reaction equilibrium. It is only when the reactant A is being depleted that the second law applies, and the system irreversibly approaches equilibrium.

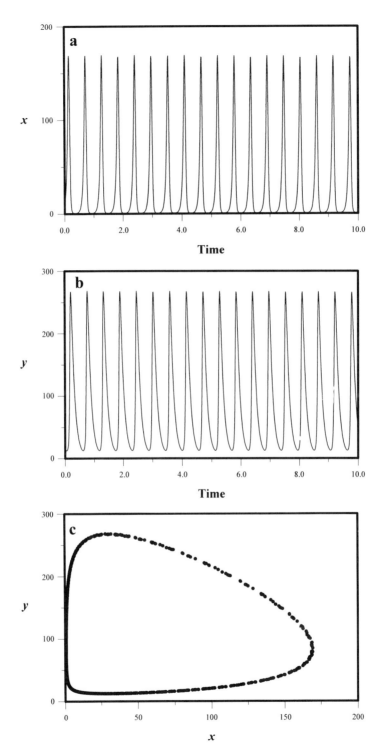

Figure 12.2 Solution of the Lotka–Volterra equation for $a = 25$, $b = 0.3$, and $c = 9$. (a) Time series for species X. (b) Time series for species Y. (c) Phase portrait in the x, y plane.

It is possible to apply many of the concepts of equilibrium statistical mechanics to systems that are far from equilibrium. Thus, statistical concepts such as ensemble averages, correlation functions, time series, and power spectra cannot only be invoked but, indeed, are necessary to interpret and understand data from oscillating chemical reactions. Figures 12.2(a and b) are examples of time series. It is the interplay of these two time series that determines the rate at which the species X and Y oscillate. The Y species time series, slightly lagging the X species time series, causes the Y concentration to build up very rapidly once the rate of X formation has peaked. From these time series can be obtained time correlation functions and power spectra (Nicolis, 1971). The use of these concepts will be illustrated below.

12.4   THE BRUSSELATOR MODEL

12.4.1   Solution of the Brusselator Equations

As we have seen, the trajectory of the LV model depends on the initial conditions and is only marginally stable. A simple model that is independent of initial conditions, after transients have died out, and has stable solutions was proposed by Prigogine and his group at Brussels. This model is called the Brusselator. The overall reaction is $A + B \rightarrow D + E$. The mechanism is

$$A \xrightarrow{k_1} X \tag{B 1}$$

$$B + X \xrightarrow{k_2} Y + D \tag{B 2}$$

$$2X + Y \xrightarrow{k_3} 3X \tag{B 3}$$

$$X \xrightarrow{k_4} E \tag{B 4}$$

with rate constants k_1, k_2, k_3, and k_4, respectively. As in the case of the LV model, X and Y are intermediates. The autocatalytic step is (B 3). The rate equations are

$$\frac{d[X]}{dt} = k_1[A] - (k_2[B] + k_4)[X] + k_3[X]^2[Y] \tag{12.22a}$$

$$\frac{d[Y]}{dt} = k_2[B][X] - k_3[X]^2[Y] \tag{12.22b}$$

To get these equations into more convenient notation, let $a = k_1[A]$ and $b = k_2[B]$. The Brusselator equations are obtained by taking k_3 and k_4 equal to 1. We note that both a and b must be positive. Equations (12.22) now become, in analogy to Eqs. (12.2),

$$\frac{dx}{dt} = a - (b + 1)x + x^2y \qquad (12.23a)$$

$$\frac{dy}{dt} = bx - x^2y \qquad (12.23b)$$

As shown above, these reactions are assumed to occur in a CSTR, and we characterize them by the stability of their critical points. Inspection of Eqs. (12.23) shows that for positive a and b there is only one critical point, which occurs at $x_c = a$, $y_c = b/a$. Unlike the coefficients in the LV equations, the values of the coefficients a and b can affect the nature of the solutions. However, for now we assume fixed values for a and b. By writing Eqs. (12.23) in the form $dx/dt = F_1(x, y)$ and $dy/dt = F_2(x, y)$ and expanding the $F_i(x, y)$ in a Taylor's series about (x_c, y_c) as in Eq. (12.3), we get, in the linear approximation,

$$\frac{dx}{dt} = (b - 1)x + a^2y + a(1 - 2b) \qquad (12.24a)$$

$$\frac{dy}{dt} = -bx - a^2y + 2ab \qquad (12.24b)$$

These linearized ODEs can be solved by using the eigenvalue method discussed above. In order to get a set of homogeneous linear equations, it is necessary to transform to a new coordinate system defined by $X = x - a$ and $Y = y - b/a$. The secular determinant is

$$\begin{vmatrix} b - 1 - \lambda & a^2 \\ -b & -a^2 - \lambda \end{vmatrix} = 0 \qquad (12.25)$$

The secular equation arising from Eq. (12.25) can be written

$$\lambda^2 - \lambda \operatorname{tr} J + |J| = 0 \qquad (12.26)$$

where $\operatorname{tr} J$ is the trace of the Jacobian matrix and $|J|$ is the determinant of the Jacobian matrix, equal to $(b - 1 - a^2)$ and a^2, respectively. The solution of the secular equation gives two eigenvalues:

$$\lambda = \frac{\operatorname{tr} J \pm \sqrt{(\operatorname{tr} J)^2 - 4|J|}}{2} \qquad (12.27)$$

where the discriminant may be written as $D = (b - 1 - a^2)^2 - 4a^2 = [b - (1 + a)^2][b - (1 - a)^2]$. The eigenvalues depend on the relative signs of tr J, $|J|$ and the discriminant; several possibilities arise.

1. tr $J > 0$, $|J| > 0$, and $D > 0$ give the results that both the eigenvalues are positive ($\lambda_1 > 0$ and $\lambda_2 > 0$). These conditions for the Brusselator are $b - 1 - a^2 > 0$, $a^2 > 0$, and $(b - 1 - a^2)^2 > 4a^2$. These conditions are met if $b > (1 + a)^2$. Solutions exist for $a = 1$ and $b > 4$, $a = 2$ and $b > 9$, $a = 3$ and $b > 16$, $a = 4$ and $b > 25$, and so on. The solutions $X = \alpha_1 \exp(\lambda t)$ and $Y = \alpha_2 \exp(\lambda t)$ for positive values of λ give critical points that are called *repellers* in that all solutions are repelled from the critical points and, as a result, are unstable.

2. tr $J < 0$, $|J| > 0$, and $D > 0$ give the results that both the eigenvalues are negative ($\lambda_1 < 0$ and $\lambda_2 < 0$). For the Brusselator, these conditions are $b - 1 - a^2 < 0$, $a^2 > 0$, and $(b - 1 - a^2)^2 > 4a^2$. These are met for $b < (1 - a)^2$. Solutions exist for $a = 2$ and $b < 1$, $a = 3$ and $b < 4$, $a = 4$ and $b < 9$, and so on. Because both eigenvalues are negative, the critical points are called *attractors* and are stable.

3. $|J| < 0$ gives the results that one eigenvalue is positive and the other is negative, and we choose $\lambda_1 > 0$ and $\lambda_2 < 0$. As seen in Section 12.3.1, this is a saddle point. Although mathematically possible, this solution does not arise physically, since it requires that $a^2 < 0$. Accordingly, the Brusselator has no saddle points.

4. $|J| > 0$ and $D < 0$ give the complex result of the form $\lambda = \alpha \pm i\beta$. For the Brusselator, these conditions are $a^2 > 0$ and $(b - 1 - a^2)^2 < 4a^2$, which happens if $(1 - a)^2 < b < (1 + a)^2$. Solutions exist for $a = 1$ and $0 < b < 4$, $a = 2$ and $1 < b < 9$, $a = 3$ and $4 < b < 16$, $a = 4$ and $9 < b < 25$, and so on. A negative value of α gives an attractor and a stable critical point, whereas a positive value of α gives a repeller and an unstable critical point. Attractors exist for $b < 1 + a^2$ and repellers exist for $b > 1 + a^2$. Consequently, the nature of the solution depends on the relative values of a and b . When $\alpha = 0$, the solution is pure imaginary and is a center (see Section 12.3.1). A center exists when tr $J = 0$ or, $b = 1 + a^2$, at which point the solution changes from attractor to repeller; this has important consequences for the nonlinear Brusselator equations.

The above behavior can be summarized in the parameter space of a and b, shown in Figure 12.3. The positive plane is divided into four regions separated by the three parabolas, $b = 1 + a^2$, $b = (1 + a)^2$, and $b = (1 - a)^2$. For values of a and b lying in regions I and II the system has repelling or unstable critical

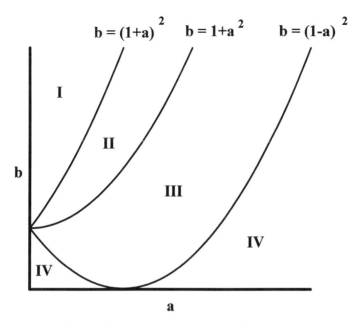

Figure 12.3   Parameter space of the Brusselator.

points, while in regions III and IV, the critical points are stable or attractors. In addition, in regions II and III the eigenvalues λ have nonzero imaginary parts. Summarizing, the linear analysis shows that there is one steady state and it is stable, unstable, or a center depending on the values of the parameters a and b.

In the full nonlinear case, the Brusselator equations (12.23) must be solved numerically. This solution has been accomplished using the FORTRAN program BRSLTR in Appendix D.2 for the last possibility above, which gives the complex solution $\alpha \pm i\beta$. This result is shown in Figures 12.4–12.7 for an attractor, two repellers, and a bifurcation point to be discussed below. Figure 12.4 shows the trajectory for a typical attractor. Starting at $x_0 = 5$ and $y_0 = 5$, all trajectories are attracted to the critical point, which then remains stable. Figures 12.5 and 12.6 show trajectories for typical repellers; the trajectory in Figure 12.5(c) starts from a point $x_0 = 2$ and $y_0 = 2$ inside the closed orbit, and Figure 12.6(c) shows a trajectory that starts from a point $x_0 = 4$ and $y_0 = 4$ outside the closed orbit. It is seen that the initial trajectories asymptotically approach trajectories that move around the critical points. In this case the critical points themselves are unstable; the orbital trajectories of the system are called *limit cycles*. The limit cycles are the attractors of the system's dynamical motion. The limit cycle is a one-dimensional curve describing a periodic orbit. Consequently, for stable critical points the attractors

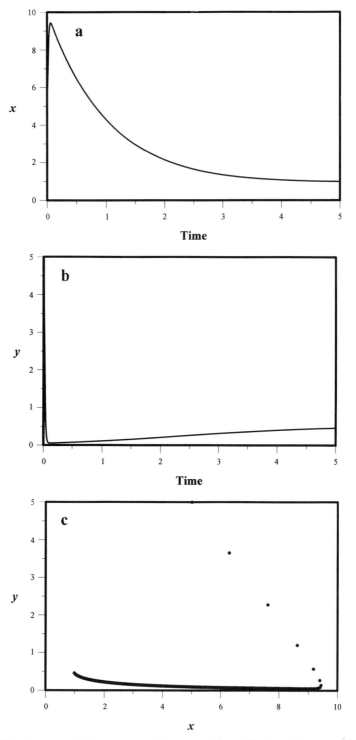

Figure 12.4   Attractor for Brusselator with $a = 1$ and $b = 0.5$. The critical point is at $x = 1$ and $y = 0.5$. (a) Time series for X. (b) Time series for Y. (c) Phase portrait.

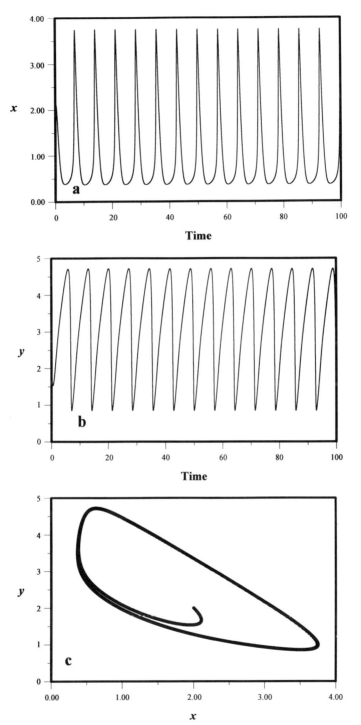

Figure 12.5 Limit cycle for Brusselator with $a = 1$ and $b = 3$. The trajectory is attracted to the limit cycle. The critical point is at $x = 1$ and $y = 3$. (a) Time series for X. (b) Time series for Y. (c) Phase portrait.

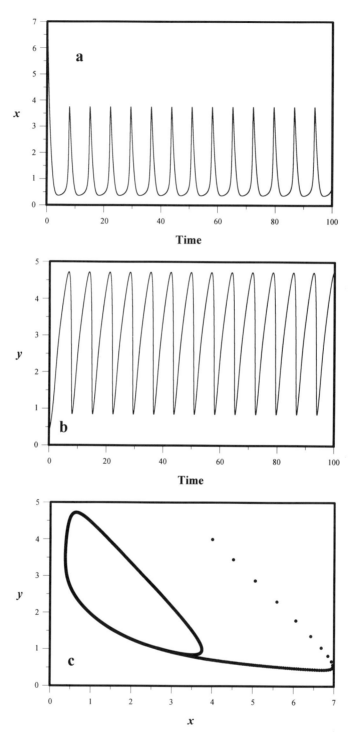

Figure 12.6 Limit cycle for Brusselator with $a = 1$ and $b = 3$. The trajectory is attracted to the limit cycle. The critical point is at $x = 1$ and $y = 3$. (a) Time series for X. (b) Time series for Y. (c) Phase portrait.

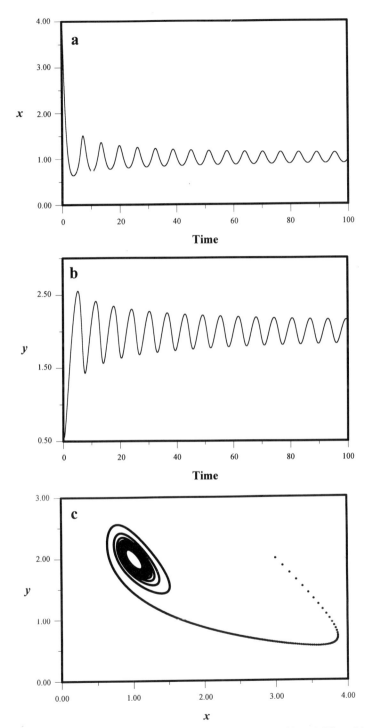

Figure 12.7 Hopf bifurcation point for Brusselator with $a = 1$ and $b = 2$. The critical point, at $x = 1$ and $y = 2$, can never be reached. (a) Time series for X. (b) Time series for Y. (c) Phase portrait.

are the critical points, while for the unstable critical points the attractors are the limit cycles.

In between the attractors and repellers are the centers that occur when $b = 1 + a^2$. This situation is shown in Figure 12.7. In this case, the trajectory moves endlessly around the critical point, coming continuously and infinitesimally closer to the critical point but is never able to reach the critical point. This situation is one between the stable critical point and the limit cycle. It signifies a change in the nature of the attractor. The point where the solution of the Brusselator equations changes from a stable steady state to an oscillation is called a *bifurcation*. This particular bifurcation is called a *Hopf bifurcation*.

All this can be studied with the help of Figure 12.3. For values of a and b lying on the middle parabola that separates regions II and III, we get bifurcation points. For fixed a, on changing b, as the system passes through the bifurcation point, the nature of the solution changes qualitatively. This behavior is very much analogous to what happens at a critical point in the theory of phase transition as discussed in Chapter 3. In a zero field in a ferromagnet, on lowering the temperature, the paramagnetic state becomes unstable and the system develops spontaneous magnetization at the critical temperature. The system literally bifurcates into two thermodynamic phases that must coexist below the Curie point or the critical temperature. Clearly, the Curie point in a ferromagnet is the analogue of the Hopf bifurcation point in the Brusselator.

A Hopf bifurcation occurs for a nonlinear ODE when the eigenvalues are pure imaginary. If the attractor changes from a critical point to a limit cycle as the bifurcation parameter is increased, the Hopf bifurcation is said to be *primary*. There is another type of Hopf bifurcation called a *secondary bifurcation*. A secondary Hopf bifurcation requires at least three variables and gives rise to an attractor that is the surface of a torus. The limit cycle of the primary Hopf bifurcation can be considered to be a torus with zero cross-sectional radius.

All trajectories of the y versus x phase portraits of the Brusselator equations move clockwise toward the attractor, which is either the stable critical point or the limit cycle. In the latter case, after the transients have died out, the system is in the limit cycle and remains in this steady state unless disturbed. The disturbance can take the form of a fluctuation of external parameters, such as temperature or flow rate, or a fluctuation of internal variables, such as concentrations of reactants. If we follow the details of the time series of either Figures 12.5(a and b) or 12.6(a and b), we see that, when the concentration of species X is a maximum, the concentration of species Y is a minimum. At this point, Reaction (B 2) dominates, and the Y species concentration slowly increases, while the X species concentration decreases rapidly to a small value. When the Y concentration reaches a maximum, Reaction (B 3) becomes dominant, and the Y concentration rapidly decreases, while the autocatalytic nature of the reaction causes X to form rapidly till its concentration is a maximum. The process then repeats.

12.4.2   Statistical Characterization of the Brusselator Model

For actual chemical reactions, such as the BZ reaction, the basic experimental data reside in a time series, which is the measurement of the concentration of a given species as a function of time. The notion of a time series as a series of observations $z(1), z(2), \ldots, z(t)$ was introduced in Section 9.6. There are two types of time series, *deterministic* and *statistical*. A time series is deterministic if its values $z(t)$ are exactly determined by a mathematical function. In Chapter 9, we saw that $z(t)$ can represent a statistical time series if the process $Z(t)$ is stochastic in that there exists an underlying probability mechanism. In this case, the "shocks" $z(t)$ arise from a stochastic process.

In the case of the Brusselator, the time series is deterministic in that it is exactly determined by nonlinear ODEs. Time series can be characterized by either correlation functions or power spectra. The *power spectrum* (or *spectral density*) $S(\omega)$ is defined as

$$S(\omega) = \left| \int_{-\infty}^{\infty} z(t) \exp(i\omega t)\, dt \right|^2 \tag{12.28}$$

The power spectrum, which represents a transformation from the time domain to the frequency domain, is so named because of its use by electrical engineers for the average power dissipated in a resistor. In Section 10.5, the power spectrum was represented by $I(\omega)$, where it was referred to as the spectral line shape. As was shown in Eq. (10.46), the autocorrelation function and power spectrum are related through a Fourier transform. This relation is referred to as the Wiener–Khinchin theorem. Because of this theorem, a time series can be characterized by either its power spectrum or its autocorrelation function. We shall use the power spectrum, because the nature of motion can be judged very quickly by looking at the power spectrum. We shall see examples of this later.

Power spectra can be obtained using the FORTRAN program PWRSPR in Appendix D.12. This program was used to get the power spectra of the time series of Figures 12.5(a and b). The power spectra are shown in Figures 12.8 and 12.9, respectively. We see that there are peaks in the power spectra at certain frequencies, which indicate periodic oscillations. The peaks in Figures 12.8 and 12.9 are spaced equidistant at an average frequency ν_{ps} of 0.14000. These power spectra represent the fundamental and harmonic frequencies of the respective time series and show that the oscillations are periodic. In this case, the analysis is especially simple in that the frequency

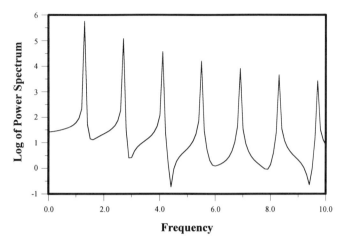

Figure 12.8   A portion of the power spectrum of the time series in Figure 12.5(a).

of the power spectra is close to the average frequency ν_{ts} of the peaks in the time series, which is 0.13978. Another advantage of using power spectra is as follows. In general, only the time series of one variable needs to be characterized, since the power spectra are identical for all the variables of a given dynamical process (e.g., cf. Figures 12.8 and 12.9). The power spectrum for the time series of Figure 12.7(a) is shown in Figure 12.10. In this case, the power spectrum damps out rapidly and the structure is lost. We can see at a glance that this power spectrum characterizes a nonperiodic process.

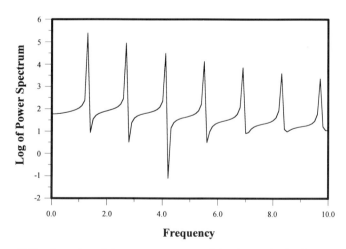

Figure 12.9   A portion of the power spectrum of the time series in Figure 12.5(b).

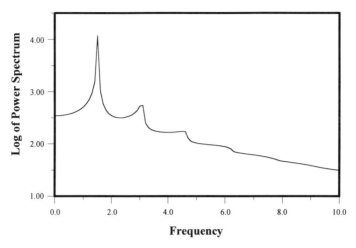

Figure 12.10   Power spectrum of the Brusselator at the bifurcation point.

For a system with three variables, more complexity is present than is observed in the Brusselator model. This complexity includes multiple periodic oscillations, multiple bifurcations, and sometimes chaotic behavior. Such complex behavior has been observed in the BZ reaction discussed in Section 12.6. In order to understand this more complex behavior, we can treat a multivariable ODE such as the Oregonator. However, the solution of three and higher variable ODEs is difficult because the equations are often "stiff." This means that one variable gets very large while another variable simultaneously gets very small. This behavior presents difficulties in the numerical solution of the ODEs. We shall briefly consider the Oregonator model later on. Fortunately, it is possible to study most of the effects observed in multivariable ODEs, including chaos, by considering a one-dimensional difference equation. The equation we have in mind is known as the *logistic equation.*

12.5   THE APPROACH TO CHAOS IN DYNAMICAL SYSTEMS

12.5.1   Nonlinear Dynamics

Nonlinear dynamics occurs in biology, chemistry, ecology, economics, engineering, and physics. Dynamics simply means that there are equations that tell us how to go from one state of the system to the next in time. Nonlinear means that the equations of motion are nonlinear. The central task in dynamics is to predict the ultimate behavior or evolution of the system and to study the stability properties of these ultimate states.

For linear systems, there is a standard method called linear stability analysis, which we used to study chemical oscillations in the previous sections. The method can also be applied to nonlinear systems near their critical points by expanding linearly around these points and performing linear stability analysis. The hope is that even in the general nonlinear case, the solutions will not look too different near the critical points. For the nonlinear case, there are no standard methods except to use computer simulation, which has led to many discoveries. Much can be gained by looking at the phase portrait, time series, and the power spectrum of the system.

One popular tool for studying these systems is the *Poincaré section*, which is defined as follows. Create a phase portrait but not for all times. Restrict the times to a discrete series as is done in a stroboscopic set of pictures. By choosing the time lapse between the strobes suitably, a particularly simple picture emerges. This picture is called a section, because we are watching how the dynamic flow cuts through a section of the whole space.

We shall come back to chemical oscillations, but first we use a simple model to introduce basic concepts such as chaos.

12.5.2 Difference Equations

It is possible to approximate a differential equation by a difference equation. For example, Eqs. (12.2) can be written

$$\frac{x_{n+1} - x_n}{\Delta t} \approx x_n(a - by_n) \tag{12.29a}$$

$$\frac{y_{n+1} - y_n}{\Delta t} \approx y_n(bx_n - c) \tag{12.29b}$$

or

$$x_{n+1} \approx x_n \Delta t(a - by_n) + x_n \tag{12.30a}$$

$$y_{n+1} \approx y_n \Delta t(bx_n - c) + y_n \tag{12.30b}$$

Equations (12.30) are difference equations. Insertion of values for x_n and y_n gives values for x_{n+1} and y_{n+1}. Therefore, these can be considered to be dynamical equations for a discrete time. Repeated iteration, replacing x_n by x_{n+1} and y_n by y_{n+1}, allows all values of x_n and y_n to be determined. Of course, Eq. (12.30) is only an approximation to the differential equations (12.2), as can be seen by comparing the solution of $dx/dt = ax$ with the difference equation $x_{n+1} = a\Delta tx_n + x_n$, where the higher order terms in the Taylor's series expansion of $x_0 \exp(at)$ are lost in the difference equation. Apart from being approximate solutions to differential equations, difference equations are of interest in their own right.

A plot of x_{n+1} versus x_n is called an *iterated map*. In particular, this is a first-iterate map. A plot of x_{n+2} versus x_n is called a second-iterate map. In general, a plot of x_{n+k} versus x_n is a kth iterate map. If the difference equation is $x_{n+1} = f(x_n)$, the various iterated maps are denoted by $f(x_n), f^2(x_n), \ldots, f^k(x_n)$. Each of these maps can have points, called *fixed points*, that are equivalent to the critical points of differential equations. The fixed points x_c of the kth iterate of a difference equation occur when $x_c = f^k(x_c)$. In other words, the fixed points map onto themselves. It is much easier to obtain maps of difference equations than it is to obtain the solutions of ODEs. For this reason, we now consider a simple one-dimensional difference equation that shows all the complexity of a multidimensional differential equation, namely, the logistic equation.

12.5.3 The Logistic Equation

The logistic equation, which was originally used to model population growth, in differential form is

$$\frac{dx}{dt} = ax(1 - x) \tag{12.31}$$

where x is the fraction of the population and $a > 0$. This equation is solved easily by the method of partial fractions to obtain the solution

$$x(t) = \frac{x_0 \exp(at)}{1 - x_0 + x_0 \exp(at)} \tag{12.32}$$

where x_0 is the initial value of x. This solution has a very simple behavior. The long-time value of x is 1. For short times, the population increases or decreases according to whether $x_0 < 1$ or $x_0 > 1$. For $x_0 = 1$ the solution is $x = 1$ at all times. If one thinks that the behavior of the corresponding difference equation is going to be similar, one is in for a big shock. The logistic difference equation is one of the most complex systems, as we shall see below.

The difference form of the logistic equation is

$$x_{n+1} = ax_n(1 - x_n) \tag{12.33}$$

where a is again a positive parameter. As with ODEs, we first locate the fixed points, which are distinguished by the relation $x_{n+1} - x_n = 0$. If x_c is a fixed point,

$$ax_c(1 - x_c) - x_c = 0 \tag{12.34}$$

and the fixed points are at $x_c = 0, (a-1)/a$. The stability of these fixed points can be examined by considering a fluctuation $x'_n = x_n - x_c$ about each fixed point. The linear stability analysis can be done analytically. For the given fluctuation, Eq. (12.33) gives

$$x'_{n+1} + x_c = a(x'_n + x_c)(1 - x'_n - x_c) = ax_c(1 - x_c) + ax'_n(1 - 2x_c) + x'^2_n \quad (12.35)$$

Equation (12.35) can be linearized by neglecting the quadratic term. For the fixed point $x_c = 0$, the linearized equation is

$$x'_{n+1} = ax'_n \quad (12.36)$$

For the fixed point to be stable, the ratio $x'_{n+1}/x'_n < 1$, so that upon iteration the new fluctuation will be smaller and will approach the fixed point. Thus, $a < 1$ if the fixed point $x_c = 0$ is to be stable.

For the fixed point $x_c = (a - 1)/a$, the linearized equation is

$$x'_{n+1} = (2 - a)x'_n \quad (12.37)$$

The condition for this fixed point to be stable is that $(2 - a) < 1$ or that $1 < a < 3$. The stable solutions will move to the fixed points. However, what happens to the unstable solutions? The behavior of the unstable solutions depends on the value of the parameter a. Figure 12.11 shows the first-iterate maps of the logistic equation, for various values of a, obtained from the FORTRAN program LOGIS1 in Appendix D.7. The fixed points occur where

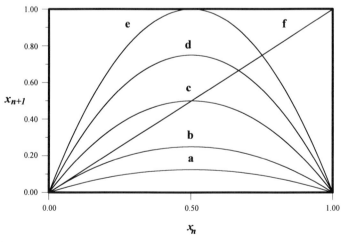

Figure 12.11 Logistic map for various values of the parameter a. In curve a, $a = 0.5$. In curve b, $a = 1$. In curve c, $a = 2$. In curve d, $a = 3$. In curve e, $a = 4$. Curve f is the fixed-point line $x_{n+1} = x_n$.

curves a–e intersect the diagonal f. Curve a intersects curve f only at $x_n = 0$. All the other curves intersect f at $x_n > 0$ as well as at $x_n = 0$. In order to see the nature of the fixed points, it is necessary to select an initial point and follow successive iterations of this point, as shown in Figure 12.12 for $a = 2.5, 3.35$, and 3.5; the dashed lines show successive iterations. Because $x_{n+1} = x_n$ along the diagonal, the successive iterations are conveniently shown by extending the dotted lines from the map to the diagonal line rather than showing their movement beginning each time from the abscissa. Whether the fixed point is stable or not can be seen by inspection. It is stable if the magnitude of the slope of the curve is less than 1 and unstable otherwise. We can see that the fixed point is stable in Figure 12.12(a) but is unstable in Figures 12.12(b and c).

If the fixed point is unstable, it is possible for it to have stable fixed points for higher order iterates, which is shown in Figure 12.12(b), where the first-iterate fixed point is unstable and the system moves to an attractor with two stable fixed points. In Figure 12.12(c), the system moves to an attractor with four stable fixed points. An attractor with two stable fixed points is said to be a period 2 cycle, and an attractor with four stable fixed points is said to be a period 4 cycle. In general, it is possible to have a period k cycle. Figure 12.13 shows the logistic second and fourth iterate maps for $a = 3.35$ and 3.5, respectively. The square dots (■) on the figures indicate the positions of the stable fixed points. The fixed points occur at the intersection of the diagonal line with the map. The intersections without the dots are the unstable fixed points.

It is evident from the above discussion that the number of stable fixed points depends on the value of the parameter a. As a increases, the number of fixed points doubles, which is called *period doubling*, and is a phenomenon observed in many oscillating reactions. As shown in Figure 12.12, the system either goes directly to the stable fixed point or it flips rapidly back and forth among the higher order fixed points. Each period doubling is a bifurcation, called a *flip bifurcation* for obvious reasons. The question arises as to how many period doublings a system can undergo. A period doubling occurs when the system becomes unstable for the new value of the bifurcation parameter and a new stability sets in. Period doubling could conceivably go on indefinitely, but for the logistic equation, as well as for physical systems, the system eventually becomes so unstable that it becomes chaotic, which is why period doubling is described as one of the routes to chaos. For chaos, stable fixed points cease to exist and the system moves throughout its accessible phase space, never visiting the same point twice. Chaotic systems lack periodicity. The signature of chaos is sensitivity to initial conditions. Thus, no matter how close two initial points are, after a while they will exponentially diverge. Sensitivity to initial conditions has become known as the butterfly effect; the

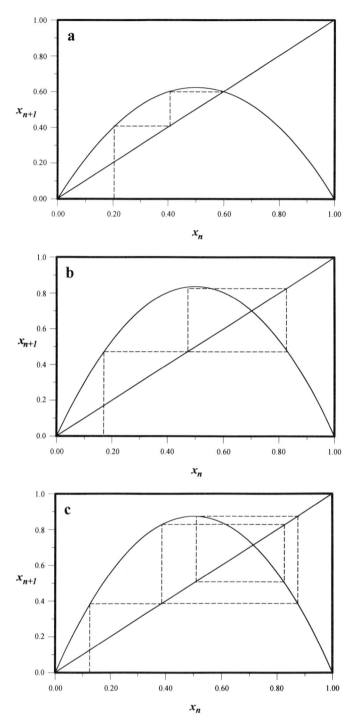

Figure 12.12 Logistic maps. (a) Successive iterations to a fixed point for $a = 2.5$. (b) Successive iterations to a period 2 iterate for $a = 3.35$. (c) Successive iterations to a period 4 iterate for $a = 3.5$.

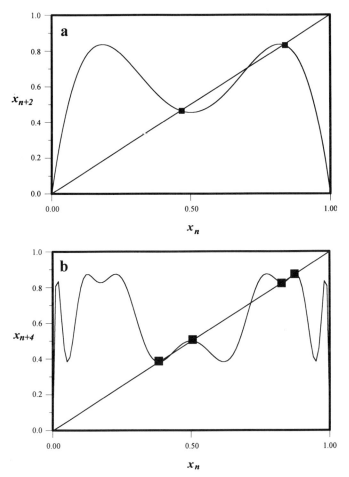

Figure 12.13   Logistic maps. Curve a is the second-iterate map for $a = 3.35$. Curve b is the fourth-iterate map for $a = 3.5$.

flutter of a butterfly's wings in the Brazilian jungle can produce a tornado on the Kansas plains!

Surprisingly, it is possible to increase the value of the parameter above that required for chaos, and the system can again exhibit stable fixed points before once again undergoing period doubling followed by chaos. This behavior can be seen in the bifurcation diagram of the logistic equation shown in Figure 12.14. This diagram represents the values obtained by the repeated use of the FORTRAN program LOGIS2 (Appendix D.8) for values of a ranging from 3.0 to 4.0 after removal of the transients. Computer programs, in both C and Basic, for generating this figure directly on a monitor can be found in the books by Tufillaro et al. and Drazin, respectively (Tufillaro et al.,

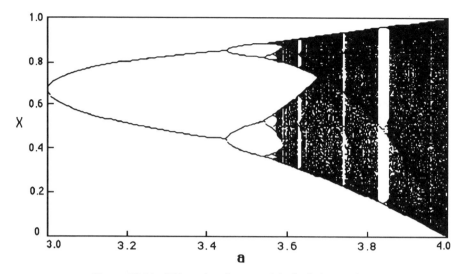

Figure 12.14 Bifurcation diagram of the logistic equation.

1992, pp. 77–78; Drazin, 1993, p. 95). Each point on this diagram represents a fixed point for a given value of a. At $a = 3.0$ a period doubling occurs. At $a = 3.45$ the system again undergoes period doubling, and doubling continues till chaos ensues. This diagram is very complex, and includes both even and odd periods that show several chaotic regions. The odd periods require what is known as a *tangent bifurcation* (also called *saddle-node bifurcation*). An example of a tangent bifurcation is shown in Figure 12.15 for a third-iterate

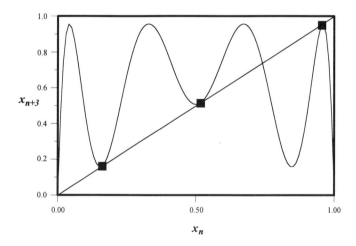

Figure 12.15 Third-iterate map of the logistic equation for $a = 3.8284$.

map of the logistic equation. The diagonal line intersects this map at five points. Two of these points are unstable period 1 cycles. The three points, indicated by square (■) dots, are the degenerate roots of a stable period 3 cycle. As the bifurcation parameter is increased, the degeneracy is lost and six period 3 roots are obtained, only three of which are stable. The splitting of degenerate roots into a stable branch and an unstable branch is the essence of the tangent bifurcation. As we shall see in Section 12.6, the BZ reaction shows complexity very similar to that of the logistic equation. There are many references that discuss the logistic equation in more detail (see, e.g., Drazin, 1993). Our task now is to characterize the behavior shown by the logistic equation.

12.5.4 Characterization of the Logistic Equation

The logistic equation can be represented by a time series, just as in the case of nonlinear ODEs. Also, as in the case of ODEs, the time series can be characterized by its power spectrum. The time series of the logistic equation for $a = 3.5$, 3.84, and 3.9 are shown in Figure 12.16. The lines between the square dots are a guide to the eye. These time series were calculated using the FORTRAN program LOGIS2. Figures 12.16(a and b) are the time series of period 3 and period 4 cycles, respectively. Figure 12.16(c) is a chaotic time series. The chaotic nature of Figure 12.16(c) is shown by its first-iterate map in Figure 12.17. This map shows the chaotic attractor covering the entire phase space available to it.

The power spectra of the time series in Figure 12.16 are shown in Figure 12.18. These power spectra are much simpler than those of nonlinear ODEs, such as in Figures 12.8–12.10. For difference equations, the bandwidth limited Nyquist frequency $\nu_{Ny} = 0.5$, and the power spectra show peaks at the fundamental frequency $\nu_{ts} = \nu_{ps} = 1/n$, where n is the period. Accordingly, the power spectra of the logistic equation show peaks at $1/n$, $2/n$, and so on, up to $\nu_{Ny} = 0.5$. For a period 3 cycle, we expect one peak at $\frac{1}{3}$, as observed. For a period 4 cycle, we expect peaks at 0.25 and 0.5, as observed. However, for chaos we find a broad-band power spectrum with many closely spaced peaks. It should be noted that the kind of chaos discussed here is also called *deterministic chaos*, because in principle, at least, the final state of the system can be predicted from the deterministic logistic equation. Deterministic chaos should be distinguished from random or stochastic chaos that arises when the deterministic system is subjected to external noise.

Having discussed period doubling and the approach to chaos, we are now in a position to understand the experimental studies of the BZ reaction and the Oregonator model.

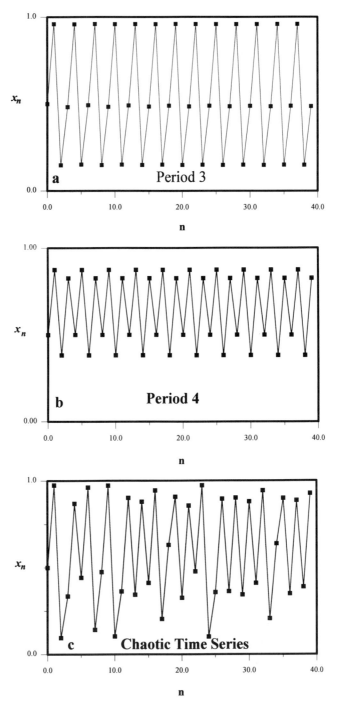

Figure 12.16 Time series for the logistic equation. (a) Period 3 time series for $a = 3.84$. (b) Period 4 time series for $a = 3.5$. (c) Chaotic time series for $a = 3.9$.

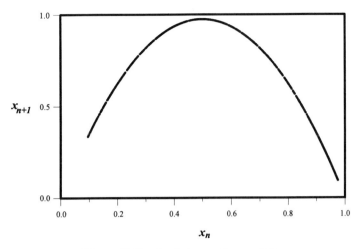

Figure 12.17 Logistic map for $a = 3.9$.

12.5.5   Universality in Chaos

In a deterministic system, because of sensitivity to initial conditions, unpredictable behavior at long times may arise. That is the phenomenon of chaos. In principle, the deterministic system is totally predictable. But in order to be able to do that, we may need to specify the initial conditions to an unreasonable number of digits. Even a small error in the initial conditions may change the behavior. It can be shown that, in dynamic systems involving differential equations, the phase space has to be at least three-dimensional to have chaotic behavior. For difference equations, there is no such dimensional constraint.

Sensitivity to initial conditions can be made more precise. Suppose we have two identical systems that are started at two points in phase space, which are very close to each other. Now let these two systems evolve. Define some measure of separation of two orbits in phase space. Sensitivity to initial conditions means that the two orbits will separate from each other in an exponential way on the average. Note that what we have given here is the definition usually adopted by physicists. Mathematicians do not require exponential growth of separation. They just require that the orbits separate.

To discuss universality in chaos, we go back to the logistic equation and look at the values of parameters a at which period doublings occur, say $a = 3, 3.45, 3.54\ldots$. Now look at the ratio of the numerical differences between consecutive doubling parameters, say corresponding to n and $n + 1$. It was shown by Feigenbaum (Feigenbaum, 1978) that this ratio tends to a constant, now called *Feigenbaum's constant* ($= 4.669201\cdots$), when $n \to \infty$. The constant is universal, because it appears as the limit, not just in the logistic equation but in all maps of any function that follows the period-

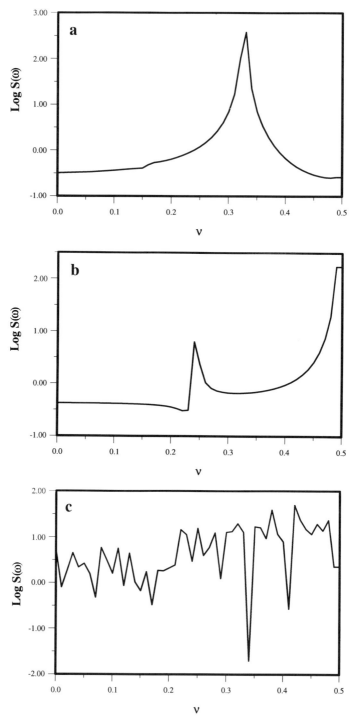

Figure 12.18   The power spectra of the logistic map time series of Figure 12.16 for curve a where $a = 3.84$, curve b where $a = 3.5$, and curve c where $a = 3.9$.

doubling route to chaos and has just one quadratic maximum. For other kinds of maxima, such as cubic and quartic, there are other Feigenbaum constants. These constants remind one of the universality in critical phenomena, where the values of the critical exponents depend on very few qualitative properties of the system. For a study of the universality in chaos by renormalization group methods, see Creswick et al. (Creswick et al., 1992).

12.6   THE BELOUSOV–ZHABOTINSKII REACTION

12.6.1   The Oregonator

The Oregonator is a scheme to model the BZ reaction. It is a simplification of the FKN mechanism. According to Field and Noyes (Field and Noyes, 1974), the Oregonator mechanism is

$$A + Y \rightleftharpoons X \tag{O 1}$$

$$X + Y \rightleftharpoons P \tag{O 2}$$

$$B + X \rightleftharpoons 2X + Z \tag{O 3}$$

$$2X \rightleftharpoons Q \tag{O 4}$$

$$Z \rightleftharpoons fY \tag{O 5}$$

where the overall reaction is $A + 2B \rightleftharpoons P + Q$. Here X, Y, Z are intermediates and f is a stoichiometric factor. The autocatalytic step is (O 3). This mechanism does not show chaos. Oregonators that show chaos have been known for over 10 years (Ringland and Turner, 1984; Richetti and Arenodo, 1985; Richetti et al., 1987). Richetti et al. used a seven-variable Oregonator to account for the approach to chaos in the BZ reaction. It has not been possible to find a three-variable Oregonator that shows chaos. For a discussion of the three-variable Oregonator, see Scott (Scott, 1993, pp. 61–66).

With this brief discussion of the Oregonator, we are now in a position to discuss the large amount of experimental work that has been done on the BZ reaction.

12.6.2   Experimental Observations

By the early 1980s, several experimental studies of the BZ reaction showed periodic, quasiperiodic, and aperiodic time series (e.g, Hudson and Mankin, 1981; Roux et al., 1983). The aperiodic behavior for low flow rates was shown to be deterministic chaos (Turner et al., 1981; Simoyi et al., 1982). The last 10 years have seen detailed studies of the BZ reaction, including the observation of mixed modes, coupled chaotic states, and stochastic chaos. We now turn to a detailed discussion of the BZ reaction in order to elaborate upon these findings.

First, we must determine how many variables are necessary to characterize the BZ reaction. Then we must determine which variables to measure experimentally. If we are to follow the route to chaos, we know that we need at least three variables. The Oregonator mentioned above needs at least seven variables. Fortunately, low-dimensional chaos, requiring only three variables, has been observed in the BZ reaction. However, it is difficult to measure more than one variable such as Br^- or Ce^{4+} in the BZ reaction. It turns out from topological embedding theory (Nayfeh and Balachandran, 1995, Section 7.4) that it is possible to measure only one variable $X(t)$ and the time derivatives dX/dt and d^2X/dt^2 of that variable. Alternatively, the method of time delays can be used. This method requires that we measure a variable X at times $t, t + k, t + 2k$, and so on, where the time delays k must be chosen carefully. The method of time delays has become the method of choice for the BZ studies. The number of time delays represents the effective Euclidean dimension of the attractor. There is no unambiguous method for determining the attractor dimension in a given situation. One method is to increase the dimension till the attractor trajectory does not intersect itself. The book by Nayfeh and Balachandran provides a good discussion of this problem.

The periodic behavior and the approach to chaos of the BZ reaction depend on the temperature, stirring rate, flow rate, and concentration of reactants. Usually, for a given experiment, the temperature, stirring rate, and initial concentrations are fixed, and the flow rate is taken to be the bifurcation parameter. All the bifurcations discussed above have been observed in the BZ reaction. These include primary and secondary Hopf bifurcations, flip bifurcations, and tangent bifurcations. Routes to chaos that have been observed include period doubling via flip bifurcations and intermittency. When intermittency is present, a seemingly periodic state will suddenly show short bursts of chaotic behavior. In the BZ reaction, this is often due to a tangent bifurcation.

A common route to chaos in the low-flow-rate regime (0.1–1.0 mL/min^{-1}) is a primary Hopf bifurcation to give a limit cycle with a frequency f_1 (~ 40 mHz) and then, following that, a secondary Hopf bifurcation to give a torus. The torus gives a quasiperiodic time series as shown in Figure 12.19, obtained by measuring the Ce^{4+} concentration. Figure 12.19(a) shows the time series and power spectrum for a limit cycle. Figure 12.19(b–c) shows the quasiperiodic regime; there is a periodic modulation of the amplitude of the time series and a new frequency f_2 (~ 10 mHz) is apparent in the power spectra, but it goes to zero at the transition to a chaotic attractor. Figure 12.19(d) shows nearly stationary behavior of the time series interrupted by bursts of large

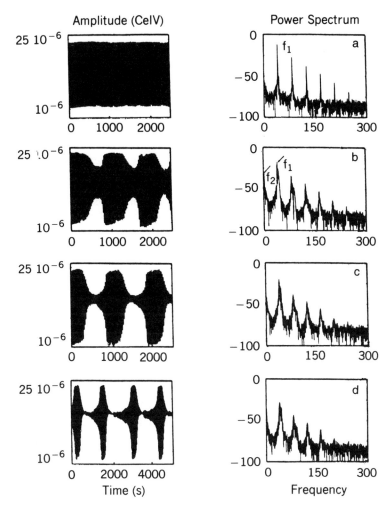

Figure 12.19 Time series and power spectra for the BZ reaction. (a) Flow rate is 0.34 mL/min^{-1}. (b) Flow rate is 0.355 mL/min^{-1}. (c) Flow rate is 0.36 mL/min^{-1}. (d) Flow rate is 0.43 mL/min^{-1}. [Reprinted with permission from F. Argoul, A. Armeodo, P. Richetti, and J.C. Roux, *J. Chem. Phys.*, **86**, 3325 (1987).]

amplitude oscillations. Argoul et al. (Argoul et al., 1987) refer to the behavior in Figure 12.19(d) as the death of the torus.

The torus, as observed in a Poincaré section, is shown evolving in Figure 12.20 as the flow rate increases. Increasing the flow rate beyond 0.45 mL/min^{-1} gives chaos and a fractal torus (not shown here), whose dimensions are stretched and folded till they become fractal. Fractals will be discussed in detail in Chapter 13.

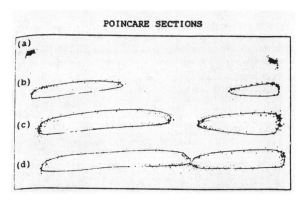

POINCARE SECTIONS

Figure 12.20   A Poincaré section of the torus whose time series are shown in Figure 12.19(a–d). [Reprinted with permission from F. Argoul, A. Armeodo, P. Richetti, and J.C. Roux, *J. Chem. Phys.*, **86**, 3325 (1987).]

Chaotic behavior in the low-flow-rate regime is shown in Figure 12.21, which shows the time series, power spectrum, and one-dimensional map for the BZ reaction. Figure 12.21(a) shows the time series obtained by use of a bromide-sensitive electrode. Figure 12.21(b) shows the power spectrum for this time series, and Figure 12.21(c) is a one-dimensional map obtained from a Poincaré section, which is not shown. The one-dimensional map is a first-return map of the Ce^{4+} concentration. The time series and power spectra are what is expected for chaos. The one-dimensional map, however, is especially informative. It shows a distinct curve, covering its entire phase space, which is typical of deterministic chaos, and is reminiscent of the chaotic logistic map shown in Figure 12.17.

The situation is different in the high-flow-rate regime (3–6 mL/min^{-1}). Here, the chaotic attractor does not show deterministic chaos, but rather shows stochastic chaos arising from statistical fluctuations among close-lying periodic states (Blittersdorf et al., 1992). Evidence for this is shown in Fig.12.22 for the bromide concentration. The time series and power spectra are not much different from those of Figure 12.21. However, the one-dimensional return map of Figure 12.22(c) is considerably different. It shows a random scatter of points, which is similar to the one-dimensional map of uniform white noise shown in Figure 12.23.

It is not easy to distinguish between deterministic and stochastic chaos. The time series and power spectra are not enough; it is necessary to obtain the one-dimensional map. The fractal dimension of the attractor can also be a clue. For the BZ reaction, the attractor of deterministic chaos has a dimension between 2 and 3, whereas the attractor of stochastic chaos has a dimension between 1 and 2 (Blittersdorf et al., 1992).

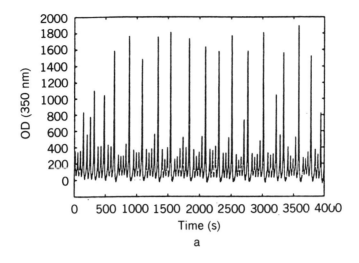

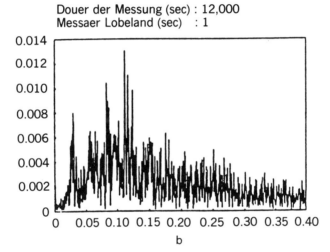

Douer der Messung (sec) : 12,000
Messaer Lobeland (sec) : 1

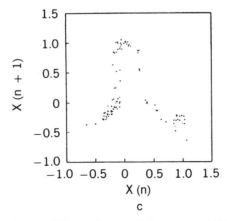

Figure 12.21 A low-flow rate BZ experiment. (a) Time series. (b) Power spectrum. (c) First-return one-dimensional map. [Reprinted with permission from R. Blittersdorf, A.F. Münster, and F.W. Schneider, *J. Phys. Chem.* **96**, 5893 (1992). Copyright ©1992 American Chemical Society.]

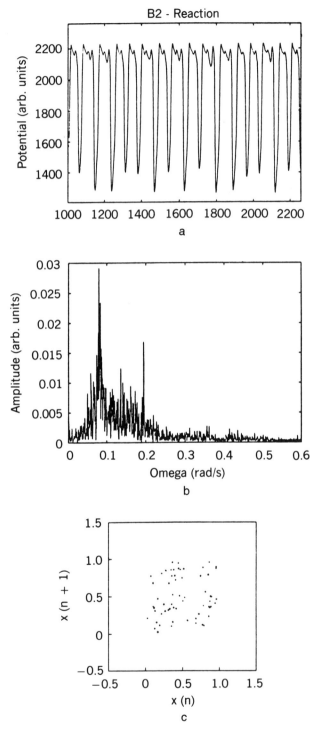

Figure 12.22 A high-flow rate BZ experiment. (a) Time series. (b) Power spectrum. (c) First-return one-dimensional map. [Reprinted with permission from R. Blittersdorf, A.F. Münster, and F.W. Schneider, *J. Phys. Chem.* **96**, 5893 (1992). Copyright ©1992 American Chemical Society.]

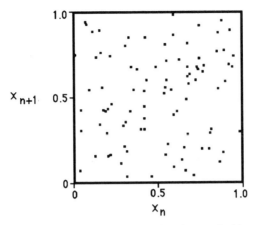

Figure 12.23   First-return one-dimensional map of white noise.

There have been several attempts to model the BZ reaction. The Oregonator was one of the first models and is widely used to study periodicity and chaos. Other models (Petrov et al., 1992; Györgi and Field, 1991) have also given insight into the BZ reaction.

12.6.3   Statistical Mechanics of the BZ Reaction

The requirement that a system be chaotic is that it have a sensitive dependence on initial conditions. Sometimes this gives the attractors of chaotic systems certain unusual properties, and as a result such attractors are known as *strange attractors*. Usually, strange attractors have fractal dimensions. Following convention, we associate "chaos and strangeness" with the dynamics of the system and "fractal" with the geometry of the attractor. Be warned that some authors do not follow this convention.

The BZ reaction is deterministic in that it can be modeled by nonlinear differential equations. However, in the chaotic regime there has been some argument as to whether the chaos is deterministic or stochastic. As seen above, this argument has been settled by an analysis of the experimental time series. Central to the analysis are the Poincaré map, one-dimensional map, and fractal dimension of the attractor. Both deterministic and stochastic chaos have been observed in the BZ reaction.

There have been some attempts to study the bifurcation mechanisms. It is here at the bifurcation points that the system is stochastic, while in between these points the system is deterministic. Similar to the critical point of a continuous phase transition (see Section 3.2.4), fluctuations become very large near bifurcation points. It is possible to apply renormalization group theory (see Section 3.5) to bifurcation points (Drazin, 1993, Section 4.3; Nicolis, 1995, Section 7.5).

Finally, we point out how such studies as the BZ reaction are affecting the foundations of statistical mechanics. For example, Boltzmann invoked his assumption of molecular chaos to account for irreversibility in a gaseous system (see Chapter 8). However, if the gas is chaotic in that two slightly different initial collision trajectories lead to exponential growth between the trajectories, then deterministic chaos might be able to account for irreversibility in place of stochastic assumptions (Baker and Gollub, 1990, p. 142). Irreversibility may also arise from coarse graining as discussed in Section 1.7.3. The coarse graining is automatically built into chaotic systems in the form of the truncation of numbers in a computer or fluctuations in a physical system (Jensen, 1987).

The subject of chaos in nonlinear systems is developing so rapidly that the only way to keep up is the Internet. A guide to resources on the Internet dealing with chaos and nonlinear systems is provided in Appendix A.

12.7   EXERCISES AND PROBLEMS

1. Show that the mechanism (LV 1)–(LV 3) gives the rate equations (12.1).

2. Verify Eq. (12.5).

3. Differentiate Eq. (12.6a) and substitute from Eq. (12.6b) to obtain a second-order differential equation for X. Solve this equation and, using Eq. (12.6a), obtain Y.

4. Set up Eqs. (12.2) in the transformed axis system X, Y and find an exact equation by eliminating the time dependence. Show that the solution of this equation reduces to Eq. (12.9) under the linearity condition $X \ll 1$ and $Y \ll 1$.

5. Use the FORTRAN program LOTKA (or its equivalent) to evaluate the LV equations (12.2) near the critical point $(250/3, 30)$ using the parameters of Figure 12.2 and $x_0 = 30$, $y_0 = 83$. Compare the result with Figure 12.1.

6. Verify the rate equations (12.22).

7. Use the FORTRAN program BRSLTR (or its equivalent) to explore the solutions for $a = 1, b = 5$ and $a = 2, b = 5$. Choose any starting point.

   (a) Construct a limit cycle attractor by use of the FORTRAN program BRSLTR (or its equivalent) such that $a = 2$.

(b) Use the FORTRAN program PWRSPR (or its equivalent) to determine the power spectrum for the limit cycle of part a.

8. Solve the differential equation

$$\frac{dx}{dt} = 2x \qquad (12.38)$$

for $x_0 = 1, t_0 = 0$ and compare it with the values of x obtained from the difference equation

$$x_{n+1} = 2\Delta t x_n + x_n \qquad (12.39)$$

9. Solve the differential equation (12.31).

10. Punch a real number on your calculator. Take its cosine assuming the number to be an angle in radians. Take the cosine of the resulting number. Iterate the procedure. Show that the results converge to $0.73908 \cdots$. Repeat the exercise with other trigonometric functions and their inverses. Explain why you get the results that you do get.

11. Use the FORTRAN program LOGIS1 to obtain the third-iterate maps of the logistic equation for $a = 3.80$ and 3.90. On these maps, draw the diagonal line $x_{n+3} = x_n$ to determine the roots. How many roots are there?

12. Use the FORTRAN program LOGIS2 to obtain the x_{n+1} versus x_n logistic maps for $a = 3.8284$ and $a = 3.8495$. Compare these maps with Figure 12.17.

13

INTRODUCTION TO CELLULAR AUTOMATON MODELS

13.1 INTRODUCTION

This chapter continues the study, which began in Chapter 12, of systems that are far from equilibrium. Such systems are often characterized by self-organization. In the last chapter, such self-organization took the form of increasingly complex periodic behavior via bifurcations. Besides laboratory studies of chemical reactions and turbulence, other examples of self-organization include galaxies, living organisms, and societies. To duplicate such self-organization, John von Neumann used cellular automata. As models of systems far from equilibrium, *cellular automata* appear to be nearly ideal. They illustrate both self-organization and chaos. After presenting a brief history of cellular automata, elementary one-dimensional cellular automata are defined, and the rules for producing three classes of cellular automata are discussed in detail. Examples are given of the three classes of cellular automata.

Two-dimensional cellular automata are then introduced as an extension of one-dimensional ones. It is shown how two-dimensional cellular automata can be used to model hydrodynamics and chemical reactions. These applications are based on the well-known Hardy, de Pazzis, and Pomeau (HPP) model (Hardy et al., 1976). The basic idea behind the HPP model is discussed, and the use of this model to understand diffusion and turbulence is outlined.

Finally, the HPP model is extended to chemical reactions, and it is shown how cellular automata provide a mesoscopic (between the microscopic and the macroscopic) approach to the understanding of limit cycles, bifurcations, period doubling, and the approach to chaos. Examples are drawn from the Selkov and Willamowski–Rössler models to specifically show the influence of fluctuations on these models.

Because cellular automata are self-similar, they often have a fractal nature. To show this fractal nature, fractals and fractal dimension are first defined. The fractal nature of many one-dimensional cellular automata is then demonstrated. Finally, the fractal dimension of Brownian motion and of polymers is discussed.

13.2   ONE-DIMENSIONAL CELLULAR AUTOMATA

13.2.1   A Brief History

The general public became aware of cellular automata in 1970, when Martin Gardner published an account of John Conway's game of "Life" in *Scientific American* (Gardner, 1970). As its name indicates, "Life" shows how populations expand and contract. It is a two-dimensional cellular automaton that can be played with checkers and a checkerboard. However, the field of cellular automata had begun much earlier, with the efforts of John von Neumann in 1948, to prove the possibility of self-replication of machines. Later, von Neumann used cellular automata to model biological reproduction. Since the publication of the game of "Life," the study of cellular automata has expanded rapidly. The seminal work on cellular automata, especially as it relates to statistical mechanics, is that of Stephen Wolfram (Wolfram, 1983). Much of the following discussion is based on Wolfram's article.

13.2.2   Elementary Cellular Automata

Cellular automata are very simple systems that can exhibit a very complex behavior, much as the logistic difference equation does. Cellular automata involve discrete space and time coordinates and are formulated using very simple rules about their evolution. Many physical evolution systems that are described by differential equations can be approximated by cellular automata. For many references to this approach, see Wolfram (1983).

A one-dimensional *cellular automaton* is a line of points, also called cells, each of which is either occupied (labeled 1) or empty (labeled 0), which evolve according to a set of rules. The cellular automaton is completely characterized by the local rules that determine the evolution of a cell and its neighborhood. The simplest cellular automaton is the neighborhood-three automaton, whereby the evolution of a given cell depends on the values of the cell and its two nearest-neighbor cells. Such an automaton is called an *elementary cellular automaton*. There are a total of $2^3 = 8$ possible neighborhoods composed of three cells, each of whose values can be 1 or 0. The possibilities are 111, 110, 101, 100, 011, 010, 001, and 000. Below are shown the cell number and the value of each cell for a system of 24 cells (each neighborhood being equally probable).

Cell No. 1 2 3 4 5 6 7 8 9 10 11 12 13 14 15 16 17 18 19 20 21 22 23 24
Value   1 1 1 1 1 0 1 0 1 1 0 0 0 1 1 0 1 0 0 0 1 0 0 0

If we assume cyclic or periodic boundary conditions, we have an infinite line of cells composed of these repeating units. We note, for example, that the neighborhood of cell No. 1 is 011, that of cell No. 2 is 111, that of cell No. 7 is 010, and that of cell No. 24 is 001.

It remains now to establish rules for the evolution of the values of the cells. We consider the above configuration to be an initial configuration (at time $t = 0$). The local rules (one for each possible neighborhood) can be used to determine the configuration at the next discrete time $t = 1$. The simplest rule is to set each cell equal to zero no matter what its neighborhood, which is called Rule 0 and is shown below.

Neighborhood	111	110	101	100	011	010	001	000
Rule 0	0	0	0	0	0	0	0	0
Rule 7	0	0	0	0	0	1	1	1

The eight local rules comprise the Rule 0 automaton. With this rule, no matter what the initial configuration, all the cells become empty after the first time step. This cellular automaton, which consists of the initial configuration only, is trivial and is an example of a *simple* automaton. In general, all the neighborhood-three rules involve replacing the value of the center cell according to its value and the value of its two neighbors. Thus, Rule 7 produces the following configuration of the 24 cells after the first time step:

Cell No. 1 2 3 4 5 6 7 8 9 10 11 12 13 14 15 16 17 18 19 20 21 22 23 24
$t = 0$ 1 1 1 1 1 0 1 0 1 1 0 0 0 1 1 0 1 0 0 0 1 0 0 0
$t = 1$ 0 0 0 0 0 0 1 0 0 0 0 1 1 0 0 0 1 0 1 1 1 0 1 1

The designation of the rules is that of Wolfram. The number corresponding to the rule is the decimal equivalent of the binary number given by the rule. Hence, Rule 0 is 00000000, Rule 7, shown above, is 00000111, and Rule 15 is 00001111. The relation between the first 15 decimal and binary numbers is shown in Table 13.1. There are a total of $2^8 = 256$ eight-digit binary numbers. In general, there are 2^n n-digit binary numbers. This number can be verified immediately for $n = 4$ from the above binary numbers, which run from 0000, 0001, through 1111.

TABLE 13.1 Numbers in Decimal and Binary Notation

Decimal	0	1	2	3	4	5	6	7	8	9	10	11	12	13	14	15
Binary	0	1	10	11	100	101	110	111	1000	1001	1010	1011	1100	1101	1110	1111

To reduce the number of rules, Wolfram proposed the following two restrictions. First, the binary specification of a rule must end with zero. This specification immediately reduced the number of rules to $2^7 = 128$. The second restriction is that the rules be reflection symmetric, so that 100 and 001, on the one hand, and 110 and 011 on the other, should give rise to the same state. This leaves us with $128/4 = 32$ "legal" rules. Figure 3 of Wolfram's article illustrates the ultimate fate of these 32 elementary cellular automata.

13.2.3 Three Classes of Cellular Automata

There are three classes of elementary cellular automata. For a given disordered initial configuration, Class 1 rapidly evolves to a homogeneous fixed point (all 1's or 0's, which is akin to the equilibrium state), Class 2 rapidly evolves to a period 1 cycle (nonhomogeneous fixed point) or a period 2 cycle, and Class 3 evolves (after transients have died out) to fixed points, multiple-period limit cycles or chaos. These behaviors will be illustrated by the cellular automata Rules 250, 50, and 126, respectively.

Figure 13.1 shows the evolution for cellular automaton Rule 250 (11111010) for 60 cells, 10 time steps, and cyclic boundary conditions. The solid dots represent 1's and the blank spaces represent 0's. A FORTRAN program CELLAUT for generating any of the 256 elementary cellular automata can be found in Appendix D.3. The time evolution of one-dimensional cellular automata illustrates self-organization, which is manifested as inverted triangles. Such behavior is seen in Figure 13.1, where self-organization takes place for

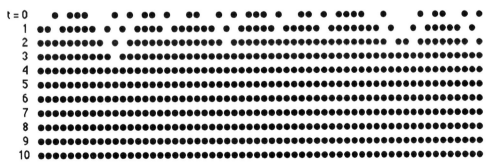

Figure 13.1 Evolution of cellular automaton 250 for 60 cells, 10 time steps, and cyclic boundary conditions.

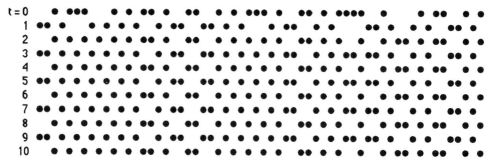

Figure 13.2   Evolution of cellular automaton 50 for 60 cells, 10 time steps, and cyclic boundary conditions.

the first three time steps before the system reaches a homogeneous fixed point of all 1's.

A Class 2 cellular automaton is shown in Figure 13.2 for Rule 50 (00110010). Beginning with time step 3, the automaton goes into a period 2 limit cycle. The period 2 limit cycle is similar to that of the logistic equation discussed in the last chapter. In fact, except for an initial configuration of all 1's or all 0's, cellular automaton 50 eventually goes to a period 2 cycle. The period 2 cycle is the attractor for cellular automaton 50.

Cellular automaton 126 (01111110), a Class 3 cellular automaton, is shown in Figure 13.3. Figure 13.3(a), for an initial configuration of 24 cells, self-organizes for the first 11 transient time steps and then goes into a period 8 limit cycle. In contrast to this, Figure 13.3(b) shows the same cellular automaton except for an initial configuration of 25 cells, an extra empty cell having been added. In Figure 13.3(b), however, there are no repeating configurations for the 36 time steps shown. Figure 13.3(b) effectively represents chaos, since the configurations do not repeat, even when viewed over 500 time steps. There is an analogy here to the logistic map discussed in Chapter 12. If we let the number of cells represent the bifurcation parameter, it is seen that the increase from 24 to 25 cells changes the dynamics of the automaton from a period 8 limit-cycle attractor to a strange attractor. It can be verified by use of the FORTRAN program CELLAUT that changing the initial configuration to 26 cells, say by adding another 0, will produce a period 28 limit cycle, while initial configurations of 22 and 23 cells produce a period 8 limit cycle. Unlike the logistic map, the elementary cellular automata do not show period doublings as an approach to chaos.

There are a number of points to be noted in connection with these one-dimensional cellular automata. First, Figure 13.3(b) is not truly chaotic. In fact, only infinite disordered initial configurations can be chaotic in the true sense of the word, because there are at most 2^N configurations that can be

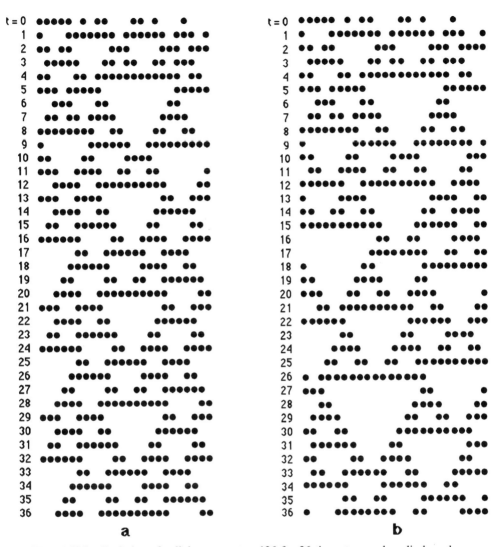

Figure 13.3 Evolution of cellular automaton 126 for 36 time steps and cyclic boundary conditions. (a) 24 cells and (b) 25 cells.

generated for an initial configuration of N cells, each of which are either filled or empty. Hence, there are at most $2^{25} \approx 3.4 \times 10^7$ configurations possible for Figure 13.3(b). By assuming that no configurations repeat for 2^{25} time steps, the time 2^{25} represents the Poincaré recurrence time for the 25-cell automaton. We cannot be sure that a limit cycle will not occur before all 2^{25} configurations have been visited. However, if no recurrences are observed for a sufficiently long time, the dynamics is chaotic for all practical purposes.

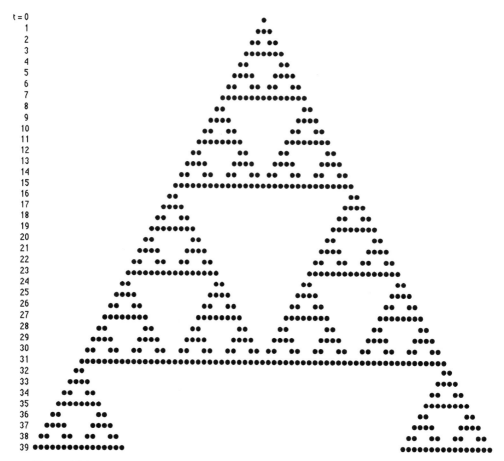

Figure 13.4   Evoltuion of cellular automaton 126 for 39 time steps starting from an infinite configuration.

A second point to be noted is that it is possible to generate true chaotic behavior by considering an infinite initial configuration with only one nonzero (seed) cell, which is shown in Figure 3 of Wolfram's article (Wolfram, 1983). It is also shown in Figure 13.4 for Rule 126. The self-organization is evident in this figure. This figure shows *self-similarity* and has a *fractal dimension*, which will be discussed in Section 13.4.3.

A final point to be noted is that the elementary cellular automata considered here are irreversible in that any given configuration will uniquely determine the future configurations, but the past is unknown, since there may be several ways to get to a given configuration. The irreversibility allows the cellular automata to mimic systems far from equilibrium, where dissipative behavior limits the accessible phase space. It is possible to consider

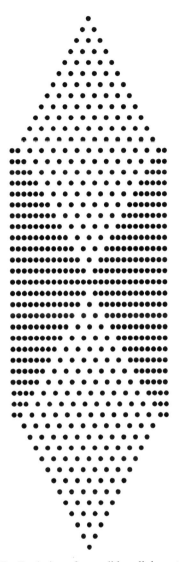

Figure 13.5   Evolution of reversible cellular automaton 90.

reversible one-dimensional cellular automata as shown in Figure 13.5 for Rule 90 (01011010). Figure 13.5 was generated from a seed cell using the FORTRAN program REVCA from Appendix D.13. The triangles generated by the cellular automaton are reversed after 11 time steps. The automaton develops exactly the same way no matter which way time flows.

By considering larger neighborhoods and more values than just 0 and 1, more complex one-dimensional cellular automata can be produced (Wolfram, 1983; Wolfram, 1984). Of more interest for physical applications are two-dimensional cellular automata. Section 13.3 is an introduction to two-dimensional cellular automata.

13.3 TWO-DIMENSIONAL CELLULAR AUTOMATA

13.3.1 Neighborhood Structures

There are two basic neighborhoods for two-dimensional cellular automata. These are the von Neumann and Moore neighborhoods shown in Figure 13.6. For the von Neumann neighborhood, each cell evolves either according to the values of its four neighbors or according to its value and the values of its four neighbors. For the Moore neighborhood, each cell evolves either according to the values of its eight neighbors or according to its value and the values of its eight neighbors. Briefly, the Moore neighborhood includes the diagonal or the second neighbors in addition to the nearest neighbors. To keep the discussion simple, we consider only the von Neumann neighborhood.

To illustrate the use of the von Neumann neighborhood, we consider two states per site, which gives $2^5 = 32$ neighborhoods for which there are a total of $2^{32} \approx 4.3 \times 10^9$ possible rules. Of these rules, we shall consider only one, the so-called sum modulo 2 rule, identified by Packard and Wolfram (Packard and Wolfram, 1985) as code 614. The evolution of this automaton from a single seed site for various time steps is shown in Figure 13.7. Figure 13.7 was generated by use of the FORTRAN program SQRCA listed in Appendix D.14. This automaton is *self-similar* and, hence, has a fractal dimension. The self-similarity can be seen by comparing the configurations at $t = 1, 2,$ and 4, at $t = 3$ and 6, and at $t = 5$ and 10. These figures are identical except for a scaling factor of 2. Self-similar configurations occur for $t = 1, 2, 4, 8, 16, \ldots,$ for $t = 3, 6, 12, 24, 48, \ldots,$ for $t = 5, 10, 20, 40, 80, \ldots,$ and so on. New configurations begin for all odd numbered times, $t = 1, 3, 5, 7, 9,$ and so on.

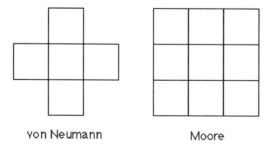

von Neumann   Moore

Figure 13.6   von Neumann and Moore neighborhoods for two-dimensional cellular automata.

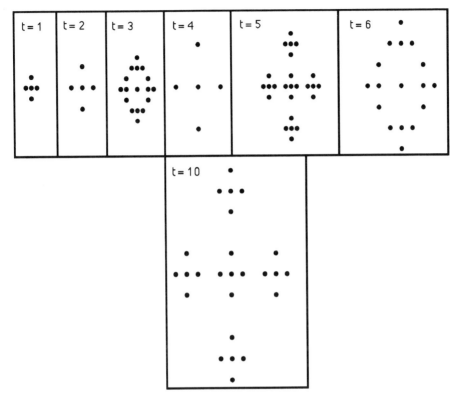

Figure 13.7 Two-dimensional, five neighborhood, sum modulo 2 cellular automaton for time steps 1, 2, 3, 4, 5, 6, and 10.

With this introduction to two-dimensional cellular automata, we are ready to consider their application to the study of turbulence and chemical reactions. It will be seen that the cellular automaton rules must be selected carefully to accurately represent the physical phenomena being portrayed.

13.3.2 The HPP Model

A simple two-dimensional cellular automaton has been described by the HPP model (Hardy et al., 1976). Unlike the von Neumann and Moore neighborhoods, the basic neighborhood here is a square of four sites that can accommodate up to four particles. The four sites define a cell. The particles in a cell have unit velocities in the $+x$, $+y$, $-x$, and $-y$ directions depending on which site they occupy, with the restriction that no two particles in a cell can have identical velocities. The sites are labeled 1–4, respectively, as shown in Figure 13.8.

Figure 13.8   Numbering of sites in a cell.

A cellular automaton consists of an array of basic cells on a square lattice with rules that determine how the particles move among the cells on the lattice. Figure 13.9 shows the occupancy of the 16 basic cells. The sites in the cell are labeled (1234) as indicated in Figure 13.8. If a position is occupied, it is designated by a 1; if it is empty, it is designated by a 0. Thus, the first cell, which is empty, is designated (0000). A designation (0001) indicates a cell containing one particle whose velocity is in the $-y$ direction; (1011) indicates a cell containing three particles whose velocities are in the $+x$, $-x$, and $-y$ directions. The particles are indicated by dots, and the arrows on the dots indicate particle-velocity directions. Because the position in the cell also indicates the velocity direction, the arrows are really superfluous and will hereafter be omitted. An array composed of one or more of the cells in Figure 13.9 on a square lattice at a given time step is called a *configuration*. We shall now define the rules by which a configuration at time t evolves to a configuration at time $t + 1$.

Only two types of events are allowed—a collision within a cell followed by a translation between cells. Two particles within a cell are said to collide if they are in the cells (1010) or (0101), in which cases (1010) becomes (0101) and (0101) becomes (1010) as shown in Figure 13.10. Collisions do not take place in the other 14 basic cells. The particles then translate to the next cell

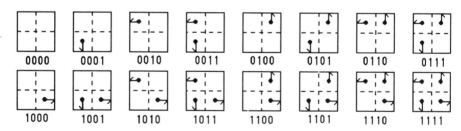

Figure 13.9   The 16 cells of the HPP cellular automaton.

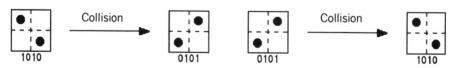

Figure 13.10   The two types of collisions in the HPP cellular automaton.

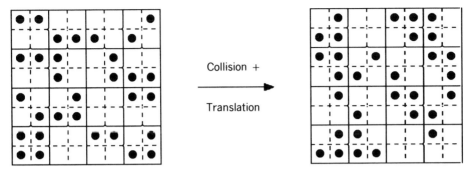

Figure 13.11   An example of a time step in the HPP cellular automaton.

according to their velocities; for example, a $+x$ velocity will proceed in the $+x$ direction to the 1 position in the new cell. The first time step for a 16-cell cellular automaton is shown in Figure 13.11 for cyclic boundary conditions.

13.3.3   Hydrodynamics of the HPP Model

For any configuration, the probability of finding a particle at any one of the four positions of a cell in the configuration, say the ith position, is equal to the number of cells containing a particle at the ith position divided by the total number of cells. This probability can be written $P_i = \sum_c P_c(i)$ for $i = 1$–4, where the sum is over all 16 basic cells, and $P_c(i)$ is the probability of observing one of the 16 basic cells containing a particle in the ith position. For example, for the initial configuration in Figure 13.11, the probability of observing a particle in the 1 position is

$$P_1 = P(1000) + P(1001) + P(1010) + P(1011) + P(1100) + P(1101)$$

$$+ P(1110) + P(1111)$$

$$= 0 + \tfrac{1}{8} + \tfrac{1}{16} + 0 + \tfrac{1}{16} + \tfrac{1}{8} + 0 + \tfrac{1}{16} = \tfrac{7}{16} \tag{13.1}$$

where only those basic cells that allow a particle in the 1 position are considered. In a similar manner, we find that $P_2 = \tfrac{9}{16}$, $P_3 = \tfrac{7}{16}$, and $P_4 = \tfrac{8}{16}$. The total probability must be the average number of particles per basic cell, which is a probability density ρ.

$$\rho = P_1 + P_2 + P_3 + P_4 = \tfrac{31}{16} \tag{13.2}$$

Also, $\sum_{c=1}^{16} P_c = 1$, where the summation extends over all basic cells, and P_c is the probability of finding a given basic cell in the configuration (i.e., the number of times a given basic cell occurs divided by the total number of cells in the configuration). Notice that $0 \leq \rho \leq 4$.

In addition to the probability density for particles, we can define current densities in the x and y directions, J_x and J_y, respectively, as $J_x = P_1 - P_3$ and $J_y = P_2 - P_4$. For example, for the configuration at $t = 0$, $J_x = \frac{7}{16} - \frac{7}{16} = 0$ and $J_y = \frac{9}{16} - \frac{8}{16} = \frac{1}{16}$. Consequently, there is no net velocity along the x axis, but there is a current density in the $+y$ direction.

The fluid conservation equations for mass and velocity are the continuity equation and the Navier–Stokes equation, given by Eqs. (8.46) and (10.56), respectively. If the cellular automaton is to simulate hydrodynamics, then the evolution rules must account for the mass and momentum conservation equations. For the HPP cellular automaton, the mass conservation can be written as

$$\frac{\Delta \rho_m}{\Delta t} = - \left(\frac{\Delta J_x}{\Delta x} + \frac{\Delta J_y}{\Delta y} \right) \tag{13.3}$$

assuming that the particles have unit mass. It is not possible to directly simulate the Navier–Stokes equation, Eq. (10.56), with the HPP cellular automaton owing to the tensor nature of the equation (Wolfram, 1986a). However, Frisch et al. (Frisch et al., 1986) modified the HPP model to accomodate the Navier–Stokes equation. The modified model, known as the FHP model, occupies a hexagonal lattice.

Wolfram (Wolfram, 1986b) discussed the basic theory for modeling hydrodynamic equations with cellular automata. His discussion is beyond the scope of this text, and we refer the interested reader to Wolfram's article. We now turn our attention to the modeling of chemical reactions, especially oscillating chemical reactions, with cellular automata.

13.3.4   Reaction Dynamics on a Cellular Automaton

The cellular automaton model that we shall use is the HPP (or FHP) model. Each chemical species moves on its own HPP (FHP) lattice, there being as many lattices as there are chemical species. The movement on each lattice is independent of all the other lattices. The evolution of the automaton consists of (1) translation of the chemical species as in Section 13.3.2, (2) chemical reaction if two or more species occupy the same cell of their respective lattices, and (3) a rotation of the molecules within a cell (Kapral et al., 1992). The translation step is purely deterministic according to the rules of the HPP or FHP automata. However, the reaction and rotation steps are stochastic, making the automaton a stochastic cellular automaton.

The chemical reaction couples the lattices containing the chemical species. For a continuous-flow stirred tank reactor (CSTR), the stirring gives a binomial distribution of lattice cell occupancy and a binomial probability density (see Section 1.2.1) on each lattice. As in Chapter 12, we shall restrict our discussion to a CSTR. The reaction dynamics is constructed to correspond as closely as possible to the deterministic rate equations. Fluctuations are introduced by use of a master equation as shown below.

Finally, the rotation step randomizes the velocities on each lattice. This randomization is done very simply by randomly choosing a velocity configuration from all possible configurations for each cell. The resulting cellular automaton composed of the translation, reaction, and rotation steps is called a lattice-gas cellular automaton (LGCA).

In general, there are n rate equations of the form (see Section 12.3)

$$\frac{dx_i}{dt} = F_i(x_1, x_2, \ldots, x_n) \qquad (i = 1, 2, \ldots, n) \tag{13.4}$$

for the n concentrations x_i of the chemical species. As we saw in Chapter 12, these are deterministic, macroscopic rate equations. It is assumed that the joint distribution function $P(x_1, x_2, \ldots, x_n, t) \equiv P(x_i, t)$ for observing the concentrations $x_1, x_2, \ldots, x_n$ at time t in a given cell obeys the equation

$$\frac{dP(x_i, t)}{dt} = \sum_{\{x_j\}} [W(x_i|x_j)P(x_j, t) - W(x_j|x_i)P(x_i, t)] \tag{13.5}$$

where $W(x_i|x_j)$ is the transition probability for the concentrations to change from x_j to x_i.

We are usually interested in the mean evolution of the concentrations. So, we define a mean concentration given by

$$\langle x_i \rangle = \sum_{\{x_j\}} x_i P(x_j, t) \tag{13.6}$$

The mean-field rate expression is

$$\frac{d\langle x_i \rangle}{dt} = \sum_{\{x_j, x_k\}} x_i [W(x_j|x_k)P(x_k, t) - W(x_k|x_j)P(x_j, t)] \tag{13.7}$$

Equations such as Eq. (13.7) are called master equations. The left-hand side of this equation is the change in the probability distribution at a given concentration and time. The change comes about as the given concentration increases or decreases owing to chemical reactions. These two processes are described by the two terms in the master equation, respectively. Of course, the sum represents all the possible concentrations. Although we have not used the master equation approach till now, it is used frequently in nonequilibrium statistical mechanics (van Kampen, 1985, Chapter V). The master equation has been derived by projection operator techniques (Zwanzig, 1960).

As noted above, $P(x_i, t)$ is a binomial probability distribution. It remains to determine the transition probabilities $W(x_i|x_j)$. The transition probabilities can be determined to within the constraints imposed by the right-hand side of Eq. (13.4).

$$F(x_i) = \sum_{\{x_j, x_k\}} x_i [W(x_j|x_k)P(x_k, t) - W(x_k|x_j)P(x_j, t)] \qquad (13.8)$$

The transition probability $W(x_i|x_j)$, after having been determined by Eq. (13.8), governs the change in $\{x_i\}$ at each time step of the LGCA.

Recent applications of the LGCA to chemical reactions include the Selkov (Kapral et al., 1992) and Willamowski–Rössler (Wu and Kapral, 1994) reactions. The Selkov reaction mechanism has three reversible steps:

$$A \rightleftharpoons X \qquad (S\ 1)$$

$$X + 2Y \rightleftharpoons 3Y \qquad (S\ 2)$$

$$Y \rightleftharpoons B \qquad (S\ 3)$$

Step (S 2) is the autocatalytic step for the intermediate Y. This mechanism was studied by Richter et al. (Richter et al., 1981) and shows fixed points and limit cycles. The cellular automaton results of Kapral et al. (Kapral et al., 1992) for the Selkov reaction are shown in Figures 13.12 and 13.13. The axes in these figures are the concentrations of intermediates X and Y. The fluctuations in the cellular automaton dynamics are seen quite clearly in the trajectories of Figure 13.12. Compare Figure 13.12, for example, with Figure 12.4(c) for the deterministic Brusselator model.

The Willamowski–Rössler mechanism consists of five reversible steps.

$$A + X \rightleftharpoons 2X \qquad (WR\ 1)$$

$$X + Y \rightleftharpoons 2Y \qquad (WR\ 2)$$

$$B + Y \rightleftharpoons C \qquad (WR\ 3)$$

$$X + Z \rightleftharpoons D \qquad (WR\ 4)$$

$$E + Z \rightleftharpoons 2Z \qquad (WR\ 5)$$

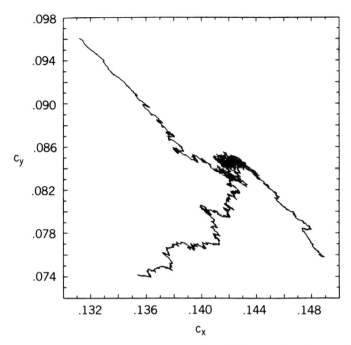

Figure 13.12 Phase portrait in the x, y plane for three different trajectories to a fixed point as determined by the Selkov cellular automaton. [Reprinted with permission from R. Kapral, A. Lawniczak, and P. Masiar, *J. Chem. Phys.*, **96**, 2762 (1992).]

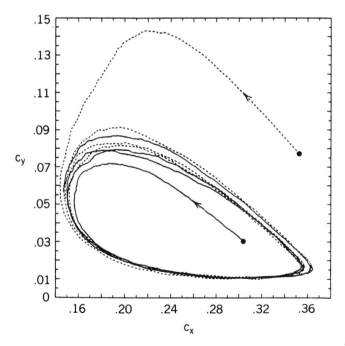

Figure 13.13 Phase portrait in the x, y plane for two different initial trajectories to a limit cycle as determined by the Selkov cellular automaton. [Reprinted with permission from R. Kapral, A. Lawniczak, and P. Masiar, *J. Chem. Phys.*, **96**, 2762 (1992).]

Step (WR 2) is autocatalytic. There are three intermediates, X, Y, and Z, so that this mechanism can exhibit chaos. The first three steps are essentially the LV model of Section 12.3. The WR model shows a series of period doublings leading to chaos (Willamowski and Rössler, 1980; Aguda and Clarke, 1988). The modeling of the WR reaction by a cellular automaton (Wu and Kapral, 1994) is similar to that of the Selkov cellular automaton except that, in addition to limit cycles, period-doubling and chaos can occur. Some results are shown in Figures 13.14 and 13.15. Figure 13.14 shows projections of the three-dimensional phase-space trajectories of a period 2 limit cycle onto the xy plane for different size lattices. As the lattice size increases from 64×64 in Figure 13.14(a) to 1024×1024 in 13.14(e), the effect of the fluctuations is decreased. The fluctuation-free deterministic trajectory is shown in Figure 13.14(f).The same effect is shown in Figure 13.15 for the

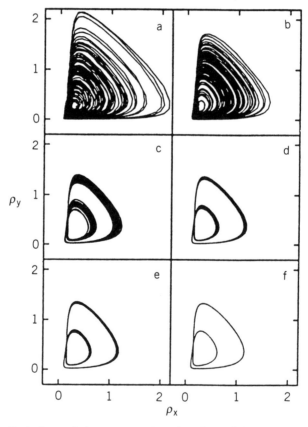

Figure 13.14 Projections of phase-space trajectories of a period 2 limit cycle onto the xy plane for a well-stirred WR cellular automaton. Lattice sizes are (a) 64×64, (b) 128×128, (c) 300×300, (d) 500×500, (e) 1024×1024. The deterministic result is (f). [Reprinted with permission from X.-G. Wu and R. Kapral, *J. Chem. Phys.*, **100**, 5936 (1994).]

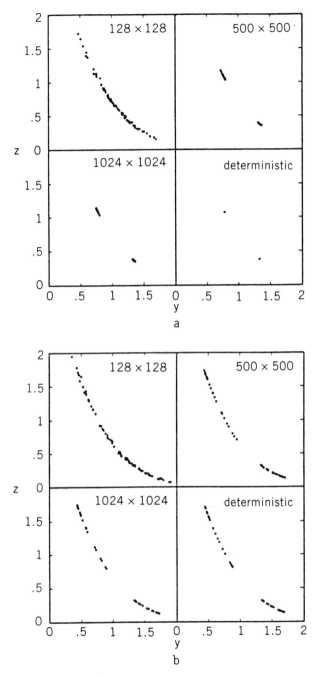

Figure 13.15 Poincaré sections for the WR cellular automaton. (a) Period 2 limit cycle. (b) Chaos. The lattice sizes and the deterministic result are indicated on the figures. [Reprinted with permission from X.-G. Wu and R. Kapral, *J. Chem. Phys.*, **100**, 5936 (1994).]

Poincaré sections in the zy plane. Notice that the fluctuations in the Poincaré sections are restricted to a narrow line, indicating that this line represents the direction of unstable phase-space flow for the attractor. The Poincaré section for a period 2 limit cycle exists in two small regions of the plane as shown for the 1024×1024 lattice size in Figure 13.15(a), whereas the Poincaré section for chaos covers the entire line as shown in Figure 13.15(b). Figure 13.15(b) is reminiscent of the chaotic logistic map of Figure 12.17, which covers the entire phase space available to it.

The cellular automaton approach is well suited to study the effects of fluctuations on limit cycles, bifurcations, and chaos. The LGCA combines the microscopic picture of chemical reactions with the macroscopic deterministic equations of these reactions. The use of the master equation to establish the transition probabilities automatically introduces fluctuations into the microscopic motions. A full MD simulation would be the best approach, but the existence of vastly different time scales, with the collision processes occurring on a much smaller time scale than the overall kinetics, makes a full MD calculation impractical, except for the simplest systems. Also, MD and master equation approaches are not practical, without modifications, when fluctuations become very large, such as near bifuraction points.

13.4   FRACTALS

13.4.1   Introduction

In Section 13.2.3, it was stated that chaotic cellular automata are self-similar and have a fractal dimension. Fractal dimensions were also mentioned in Section 12.6. In this section, we introduce the concept of fractals and fractal dimension and show how fractal dimensions can be determined.

A *fractal* has been defined by mathematicians as a set whose dimension exceeds its expected topological dimension. By a set, we generally mean a point set representing objects or the values of a mathematical function. If we were to pursue the mathematical definition, we would be led to a study of set theory, metric topology, and measure theory (Edgar, 1990). Instead, we shall endeavor to get a feeling for fractals by considering some examples. Fractals are everywhere. Landscapes, rivers, clouds, and the Dow-Jones industrial average are fractals. In science, fractal examples are convection, flow in porous media, molecular topology, weather patterns, and Brownian motion. A fractal is characterized by being scale invariant. That is, a fractal looks the same on any scale; it is said to be *self-similar*. For example, if one were to look at the Dow-Jones industrial average, he could not tell whether it was for 1 day, 1 week, 1 month, or 1 year unless the abscissa was labeled.

Ordinary geometric shapes have topological dimensions of 0 for a point, 1 for a curve, 2 for a surface, and 3 for a three-dimensional object. A curve in two-dimensional Euclidean space can be represented by an equation of the form $y = f(x)$. The curve has a topological dimension of 1 and a length given by

$$s = \int \left[1 + \left(\frac{df}{dy} \right)^2 \right]^{1/2} dx \qquad (13.9)$$

which reflects its Euclidean dimension of 2. Fractal curves, on the other hand, are continuous but nondifferentiable, because the curves are composed of vanishingly small line segments that meet at sharp angles. Since the fractal curves are not differentiable, they have an infinite length. Similar remarks can be made for fractal surfaces and three-dimensional objects. If gaskets or sieves are formed by having holes systematically punched in them, a fractal object such as the Sierpinski gasket (see Mandelbrot, 1982, p. 142; Figure 13.17) is created whose area approaches zero. Likewise, if a solid object has cavities systematically formed in it to give the appearance of Swiss cheese, its volume approaches zero (e.g., the Menger sponge whose surface area is infinite). Like the Cheshire cat, eventually nothing is left but the grin.

13.4.2   Fractal Dimension

Should the dimension of the Sierpinski gasket be classified as one or two? This raises the question of the dimension of a fractal object. The dimension of a fractal object is closely tied to the appearance of the object under scaling. First, consider a line of length L_0. We divide this line into N identical parts, the length of each part being $L = L_0/N$. Let a_0 be the original basis vector for the measurement of the line (i.e., $L_0 = n|a_0|, n > 0$). The new basis vector a_1 is related to the original basis vector by $a_1 = a_0/N$. The new basis is smaller by a factor of $1/N$, which is the scaling factor r. Hence, $L = rL_0 = rn|a_0| = n|a_1|$, and the length L in the a_1 basis is the same multiple of $|a_1|$ as is L_0 in the a_0 basis, which is the gist of the scaling concept. This concept was discussed in Chapter 3 in detail.

For a plane, an original square of length $|a_0|$ on a side has an area a_0^2, which can be divided into N squares. A new area can be defined to be $a_1^2 = a_0^2/N$, or $a_1 = a_0/N^{1/2}$. By extending this argument to three dimensions, $|a_1|^3 = |a_0|^3/N$ or $a_1 = a_0/N^{1/3}$. If we define the scaling factor r as $r = |a_1|/|a_0|$, we get $r = 1/N, 1/N^{1/2}, 1/N^{1/3}$ for one, two, and three dimensions, respectively. In order to define a fractal dimension, the scale factor is written in general as

$$r = \frac{1}{N^{1/D}} \qquad (13.10)$$

where D is called the *fractal dimension*. Actually, what we shall refer to as the fractal dimension is more properly called the *similarity dimension*, since it is defined for a self-similar object. Solving for D,

$$D = -\ln N/\ln r \qquad (13.11)$$

An example of a fractal curve is provided by the Koch curve shown in Figure 13.16. The Koch curve is formed from a straight-line segment (called the initiator) by taking $r = \frac{1}{3}$. The center one-third is removed and replaced by two continuous line segments, each one-third the original length. This process is called the Koch construction, and the resulting Figure 13.16(a) is said to be the generator of the Koch curve. The Koch construction is then repeated iteratively on each straight-line segment till the Koch curve is produced. The second iteration is shown in Figure 13.16(b). In the case of the Koch curve, $r = \frac{1}{3}$ and $N = 4$, which gives $D = \ln 4/\ln 3 \approx 1.26$.

An infinite number of iterations of the Koch construction gives the Koch curve having a fractal dimension of approximately 1.26. Hence, the Koch curve has a higher dimension than that of a topological curve, which is 1, but a smaller dimension than a surface, which is 2. The length of the Koch curve can be written as $L(r) = rN(r)$. Thus, $L(\frac{1}{3}) = \frac{1}{3} \times 4 = \frac{4}{3}, L(\frac{1}{9}) = \frac{1}{9} \times 4^2 = \frac{16}{9}$, and so on. Consequently, $L \to r^n N^n = (r/r^D)^n = (r^{1-D})^n$. To generate the fractal, let $n \to \infty$. Then $L \to \infty$, so that the length of the fractal curve becomes infinite. But since its dimension is less than 2, it does not cover the whole two-dimensional plane.

Another example of fractal dimension is provided by the equilateral triangle Sierpinski gasket shown in the Figure 13.17. The generator, which is the first iteration of the Sierpinski gasket, is shown in Figure 13.17(b). The second iteration is shown in Figure 13.17(c). Since the gasket is two dimensional, we shall be interested in the scaling of the area. The darkened area of the initiator [Figure 13.17(a)] is $A_0 = \sqrt{3}/4$ if the sides are assumed to be of

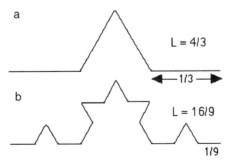

Figure 13.16   The construction of the Koch curve.

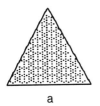
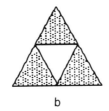
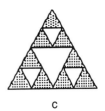

a   b   c

Figure 13.17   The construction of the Sierpinski gasket.

unit length. The total darkened area of the generator is $A_1 = 3\sqrt{3}/16$, where both the base and the height of the triangles have been reduced (scaled) by $\frac{1}{2}$. The total darkened area after the second iteration is $A_2 = 9\sqrt{3}/64$. Thus, $A_1 = (\frac{3}{4})A_0$ and $A_2 = (\frac{9}{16})A_0$. In general, $A_k = (\frac{3}{4})^k A_0$, where k is the number of iterations of the generator. As $k \to \infty$, $A_k \to 0$ and, as stated above, we see that the Sierpinski gasket has zero area. For each application of the generator, $r = \frac{1}{2}$ and $N = 3$, which gives a fractal dimension of $D = \ln 3/\ln 2 \approx 1.58$.

13.4.3   Fractal Nature of Cellular Automaton 126

Starting from an ordered state at a single site with a value 1, cellular automaton 126 generates Figure 13.4. This figure illustrates both self-organization and self-similarity. The figure is characterized by the basic unit consisting of a large triangle surrounded by three smaller congruent triangles. For purposes of determining the fractal dimension, we note that the cellular automaton is one dimensional. Hence, we look at the configurations in space and time of the bases of the triangles. We immediately see that self-similarity requires $N = 3$. The ratio r can be obtained by comparing base lengths: $r =$ base of small triangles : base of large triangle. Beginning at the top of the figure, the successive base ratios are $7 : 15 \approx 0.467$, $15 : 31 \approx 0.484$, $31 : 63 \approx 0.492$. In the long-time limit, the ratio becomes $r = 0.5$. Thus, for an initial single-site state, cellular automaton 126 produces a fractal of dimension $D = \ln 3/\ln 2 \approx 1.58$, the same as the Sierpinski gasket. This dimension is the same for nearly all the one-dimensional neighborhood-three automata (Wolfram, 1983).

Disordered initial configurations, such as the ones considered in Section 13.2.3, do not lead to self-similarity. There is no correlation between values at different sites. The configurations produced correspond to "white noise." Such white noise was discussed in Chapter 9 in connection with Brownian motion. As mentioned above, Brownian motion is fractal and has a fractal dimension. To complete our understanding of Brownian motion, we now discuss the fractal nature of Brownian motion.

13.4.4   Fractal Nature of Brownian Motion

In Chapter 9, we used the Langevin equation as a model for Brownian motion. According to this model, the bath fluctuations y_i relax on a time scale much shorter than that of the particle fluctuations. Consequently, the differences $\Delta y_{t+s} = y_{t+s} - y_t$ occur in a time s, where $s \ll \tau$ and where τ is defined by Eq. (9.29). Einstein's theory requires $s \ll \tau$. Moreover, Eq. (9.29) implies that

$$\langle \Delta z(t + \tau) \rangle \sim \tau^{1/2} \tag{13.12}$$

so that the instantaneous velocity is

$$\dot{z}(t + \tau) = \lim_{\tau \to 0} [z(t + \tau) - z(t)]/\tau = \lim_{\tau \to 0} \tau^{-1/2} \tag{13.13}$$

which becomes infinite. This means that velocity is undefined in the Brownian motion process. Although Brownian motion is continuous, it is not differentiable. In Figure 13.18 the first 10 steps of the $z(t)$ versus t plot of Figure 9.2 are shown. This curve, which is clearly not differentiable, illustrates the fractal nature of Brownian motion.

From Eq. (13.12), we can say that if time is scaled by a factor r, length must be scaled by a factor $r^{1/2}$ (see Feder, 1988, pp. 167–168). Therefore, the Brownian process can be self-similar only if time and length are scaled by different factors. A transformation involving nonuniform scaling is called *affine*, and the curves that are produced under an affine transformation are said to be *self-affine*. Hence, Brownian motion is self-affine. The similarity dimension $D = -\ln N/\ln r$ is not defined for a self-affine fractal. We therefore need a way to define the fractal dimension of a self-affine fractal.

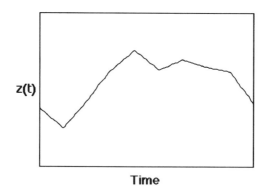

Figure 13.18   The first 10 steps of Figure 9.2.

13.4.5 The Box Dimension

The similarity dimension can be looked upon as the slope of a plot of $-\ln N(r)$ versus $\ln r$. Use is made of this observation in the box-counting method for determining fractal dimensions. The dimension thus obtained is called the box dimension:

$$D_{\mathrm{B}} = - \left[\frac{\ln N(r_2) - \ln N(r_1)}{\ln r_2 - \ln r_1} \right] \qquad (13.14)$$

where $N(r)$ is the minimum number of boxes (usually cubes) necessary to cover the fractal object or trajectory and r is the scaling factor. In two dimensions, squares are used in place of cubes. As the boxes are reduced in size, they approximate the fractal object.

In Figure 13.19, one-dimensional Brownian motion is shown for 2000 time steps and 500 Brownian steps. A grid is placed over the $z(t)$ trajectory, and the squares that contain the trajectory are counted. The number of squares needed to cover the Brownian motion curve is 56. By using Eq. (13.14), we find for Figure 13.19 that $D_{\mathrm{B}} = (\ln 56 - \ln 1)/(\ln 16 - \ln 1) \approx 1.45$. The accepted fractal dimension of Brownian motion is $\frac{3}{2}$ (Mandelbrot, 1982, p. 237).

13.4.6 Fractal Nature of Polymers

The FORTRAN program POLYMER in Appendix D.11 allows one to generate self-avoiding random walks in two dimensions. The program is a myopic

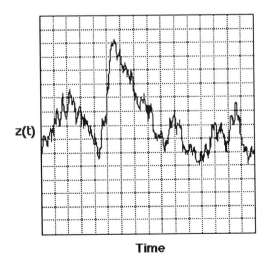

Figure 13.19 One-dimensional Brownian motion.

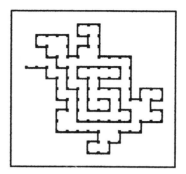

Figure 13.20   Self-similarity of polymers.

SAW (self-avoiding walk) and is not statistically correct; it is intended only to illustrate polymer folding. Figure 13.20 was generated by the program POLY-MER for 100 steps. This figure exhibits self-similarity and a fractal nature as discussed above. Indeed, the fractal nature of the SAW is well established (Mandelbrot, 1982, pp. 329, 330). The random walk, upon which Brownian motion is based, has a fractal dimension $D = 2$ in all dimensions. Hence, for the random walk, the correlation length exponent ν has a value $\frac{1}{2}$, and the fractal dimension can be written $D = 1/\nu$. From Eq. (7.1), $R = aN^\nu = aN^{1/D}$, so that

$$D = \frac{-\ln N}{\ln(R/a)} \tag{13.15}$$

Equation (13.15) is of the same form as Eq. (13.11). We see from Eq. (7.8) that, for the two-dimensional SAW, $D = \frac{4}{3}$.

13.5   EXERCISES AND PROBLEMS

1. For the initial 24-cell configuration given in Section 13.2.2, find the configuration at $t = 1$ for cellular automaton Rule 10.

2. Use the FORTRAN program CELLAUT to show that the time evolution of cellular automaton Rule 150 goes to a period 2 cycle for any finite initial configuration.

3. Use CELLAUT to verify the statement in Section 13.2.3 that initial configurations of 22, 23, and 26 cells produce limit cycles of periods 8, 8, and 28, respectively, for cellular automaton Rule 126.

4. One of the simplest ways to characterize a cellular automaton is by evaluating the average density of occupied sites (i.e., those sites containing a 1). These probability densities are designated ρ_t for the configuration appearing at time step t. For the cellular automaton 126 time evolution in Figure 13.3(a), verify that the probability densities for the first two configurations are $\rho_0 = 0.5$ and $\rho_1 = 0.750$. Find the average probability density for the period 8 limit cycle of cellular automaton 126. For cellular automaton 126, in the limit of an infinite cellular automaton, the average probability density approaches 0.5.

5. Change the initial configuration of Figure 13.3(a) to one with a density of 0.25, and use CELLAUT to find the average probability density for time steps $t = 1 - 20$. Compare your result with Exercise 4. Does the initial configuration have a significant effect on the average probability densities of the configurations generated by the cellular automaton?

6. For the following 24-cell configuration, use the program CELLAUT to follow at least 60 time steps of cellular automaton Rule 18 (00010010).

Cell No. 1 2 3 4 5 6 7 8 9 10 11 12 13 14 15 16 17 18 19 20 21 22 23 24
$t = 0$ 1 1 1 0 1 0 1 0 1 1 0 0 0 1 1 0 1 0 0 1 1 0 0 0

Determine the average probability density for this cellular automaton. How many steps are required before a limit cycle is reached?

7. Use the FORTRAN program REVCA to construct the reversible time evolution for Rule 126 from a seed cell and from a disordered initial configuration. Discuss the reversibility.

8. Use the FORTRAN program SQRCA to construct the time evolution of Figure 13.7 for $t = 7$ and $t = 9$. Are the configurations for $t = 7$ and 9 self-similar? (*Hint:* Set up an 11×11 square array of zeros with a 1 in the middle in the file INPUT. Precede this in the file with the dimension 11 and with the time, either 7 or 9.)

9. Divide the 16 cells used to illustrate the HPP model in the text (Figure 13.11) into quadrants of 4 cells each. Take as your system the 4 cells in the lower right-hand quadrant. Use Eq. (13.3) to find $\Delta \rho_m / \Delta t$. By evaluating the particle probability density directly from Figure 13.11, verify that for the first time step $\Delta \rho_m / \Delta t = -\frac{1}{4}$ for your system.

10. Repeat the previous Exercise for all 16 cells.

11. For the one-dimensional Brownian motion of Figure 13.19, verify that the box dimension is 1.45.

APPENDIX A

RESOURCES ON THE INTERNET AND THE WORLD WIDE WEB

To access the internet and the web, one needs an internet service provider. A typical user of this book will have access to the internet through a university account. Other readers can access the internet with the help of a local or a national service provider. The internet account will typically come with a *web browser*, a *usenet news reader*, and an *ftp (= file-transfer-protocol)* program. The browser is used to connect to web sites, from which one may go to other sites with the help of *links*. Most often, a *search engine* also comes with the browser, and helps in locating the *internet addresses* of various sites of interest. Usually, the search is performed by typing in the *keywords*. But there are so many resources on the internet, that a blind search, such as typing the words *statistical* + *mechanics* is likely to lead to literally tens of thousands of sites. The following is an attempt to help the readers make use of web resources effectively.

Perhaps the best place to start is to visit a USENET newsgroup of interest. For subjects of interest to the readers of this book, the following may be useful: *sci.chem, sci.fractals, sci.mech.fluids, sci.nonlinear, sci.physics.cond-matter, sci.polymers, sci.stat.math*. In each case, it is a good idea to look at the FAQ (frequently-asked-questions), which are available by *ftp* from the *rtfm.mit.edu/pub/usenet/news.answers*. These FAQs do answer the most common questions that are asked again and again by newcomers to the newsgroup and are really very worthwhile. They point to multitudes of printed sources such as books, articles, and, of course other web resources of relevance to the newsgroups. After reading the FAQ, the next step is to visit the newsgroup, and later one may start posting questions and/or answers.

For viewing electronic journals on complexity theory, one may visit the web site *http://journals.wiley.com*. An extensive list of sites dealing with nonlinear science is at *http://cnls-www.lanl.gov/nbt/site.html*. Two FAQs that we have found useful are the *Nonlinear Science* FAQ the *Fractal* FAQ. For general information, the FAQs for *Physics* and *Mathematics* are also quite informative. Latest advances in physics are posted on the *Physics News Update* available at *ftp://ftp.aip.org/archives/physnews*.

APPENDIX B

GENERATING FUNCTIONS

B.1 THE GENERATING FUNCTION CONCEPT

Consider a polynomial such as a Legendre polynomial (Margenau and Murphy, 1956, p. 98),

$$P_l(x) = [1/(2^l l!)]d^l(x^2 - 1)^l/dx^l \tag{B.1}$$

Each value of l has a polynomial of order l associated with it: $P_0(x) = 1, P_1(x) = x, P_2(x) = (3x^2 - 1)/2, P_3(x) = (5x^3 - 3x)/2$, and so on. A generating function for this polynomial is a function $F(x, t)$ such that the coefficients in a Taylor's series expansion of $F(x, t)$ about $t = 0$ are the Legendre polynomials of order l:

$$
\begin{aligned}
F(x,t) &= F(x,0) + [\partial F(x,t)/\partial t]_0 t + [\partial^2 F(x,t)/\partial t^2]_0 t^2/2 + \cdots \\
&= P_0(x) + P_1(x)t + P_2(x)t^2 + \cdots
\end{aligned}
\tag{B.2}
$$

We say that the function $F(x, t)$ has generated the polynomials $P_l(x)$. There is no systematic method for determining the generating function of any given polynomial. For the Legendre polynomial,

$$F(x,t) = (1 - 2xt + t^2)^{-1/2} \tag{B.3}$$

B.2   THE MOMENT GENERATING FUNCTION

The moments of a discrete random variable X are

$$\langle X^n \rangle = \sum_x x^n p(x) \tag{B.4}$$

Equation (B.4) is the nth moment of the random variable X. The zeroth moment is found by setting $n = 0$, the first moment by setting $n = 1$, and so on. A moment generating function of a random variable X is defined to be

$$\langle \exp(Xt) \rangle = \sum_k \exp(x_k t) p(x_k) \tag{B.5}$$

The Taylor's series expansion of $\langle \exp(Xt) \rangle$ about $t = 0$ is readily shown to be

$$\langle \exp(Xt) \rangle = \mu_0 + \mu_1 t + \mu_2 t^2 / 2 + \cdots + \mu_n t^n / n! + \cdots \tag{B.6}$$

where μ_n is the nth moment of X. It is seen that the moments are the coefficients of the Taylor's series expansion of the moment generating function. The moment generating functions for the Gaussian and Poisson distributions are $\exp(\mu t + \sigma^2 t^2 / 2)$ and $\exp[\lambda(e^t - 1)]$, respectively.

Equation (B.5) can be considered to be like the Laplace transform of the probability density. The Fourier transform of the probability density is called the *characteristic function* of X:

$$\langle \exp(iXt) \rangle = \sum_k \exp(ix_k t) p(x_k) \tag{B.7}$$

The Taylor's series expansion of the characteristic function about $t = 0$ is

$$\langle \exp(iXt) \rangle = \mu_0 + \mu_1 it - \mu_2 t^2 / 2 + \cdots + \mu_n (it)^n / n! + \cdots \tag{B.8}$$

so that the characteristic function can also generate moments. For a function symmetric around the origin, such as a Gaussian distribution with zero mean, it can be proved from Eq. (B.7) that the Fourier transform is real, so that all odd moments must vanish.

Another useful function is the log of the characteristic function:

$$\ln\langle \exp(iXt) \rangle = \kappa_1 it - \kappa_2 t^2 / 2 + \cdots + \kappa_n (it)^n / n! + \cdots \tag{B.9}$$

where κ_n is called the nth cumulant. The cumulants are related to the moments by

$$\kappa_1 = \mu_1,$$
$$\kappa_2 = \mu_2 - \mu_1^2, \qquad\qquad (B.10)$$
$$\kappa_3 = \mu_3 - 3\mu_2\mu_1 + 2\mu_1^3.$$

For a Gaussian distribution, κ_3 and all higher cumulants vanish.

APPENDIX C

FERMI'S GOLDEN RULE

C.1 TRANSITION PROBABILITY IN TIME-DEPENDENT PERTURBATION THEORY

For a constant first-order perturbation given by a Hamiltonian $\widehat{\mathcal{H}}_1(t)$ acting between times 0 and t, the probability of a transition from an initial state $|m\rangle$ to a final state $|n\rangle$ is (Dirac, 1958, Section 44)

$$P_{nm} = \left| \left(\frac{-i}{\hbar} \right) \int_0^t \langle n|\widehat{\mathcal{H}}_1(t')|m\rangle \, dt' \right|^2 \tag{C.1}$$

The integral in Eq. (C.1) is

$$\int_0^t \langle n|\widehat{\mathcal{H}}_1(t')|m\rangle \, dt' = \int_0^t \langle n| \exp\left(\frac{iE_n t'}{\hbar} \right) \widehat{\mathcal{H}}_1 \exp\left(-\frac{iE_m t'}{\hbar} \right) |m\rangle \, dt'$$

$$= \langle n|\widehat{\mathcal{H}}_1|m\rangle \int_0^t \exp(i\omega_{nm} t') \, dt'$$

$$= i\langle n|\widehat{\mathcal{H}}_1|m\rangle [1 - \exp(i\omega_{nm} t)]/\omega_{nm} \tag{C.2}$$

where $\langle n|\widehat{\mathcal{H}}_1|m\rangle$ is the Hamiltonian matrix element and $\omega_{nm} = (E_n - E_m)/\hbar$ is the frequency for the $m \rightarrow n$ transition. The transition probability becomes

$$P_{nm} = 4|\langle n|\widehat{\mathcal{H}}_1|m\rangle|^2 \sin^2(\omega_{nm} t/2)/(\hbar^2 \omega_{nm}^2) \tag{C.3}$$

C.2   USE OF THE DELTA FUNCTION

The function $\sin^2(\omega_{nm}t/2)/\omega_{nm}^2$ when plotted versus ω_{nm} is sharply peaked about $\omega_{nm} = 0$ (Schiff, 1955, Section 29), so that nearly all the area under this function is under the central peak. By integrating the function, this area is found to be $\pi t/2$. [For many properties and representations of the delta function, see Pathria (Pathria, 1996, Appendix B).] Thus, in the limit $t \rightarrow \infty$, the function $\sin^2(\omega_{nm}t/2)/\omega_{nm}^2$ may be regarded as a delta function.

$$\delta(\omega_{nm}) = \lim_{t\to\infty}(2/\pi t)\sin^2(\omega_{nm}t/2)/\omega_{nm}^2 \qquad (C.4)$$

Substitution of Eq. (C.4) into Eq. (C.3) gives

$$\lim_{t\to\infty} P_{nm}/t = [2\pi|\langle n|\widehat{\mathcal{H}}_1|m\rangle|^2/\hbar^2]\delta(\omega_{nm}) \qquad (C.5)$$

The transition probability per unit time for $t \rightarrow \infty$ is

$$w_{nm} = dP_{nm}/dt = [2\pi|\langle n|\widehat{\mathcal{H}}_1|m\rangle|^2/\hbar^2]\delta(\omega_n - \omega_m) \qquad (C.6)$$

Equation (C.6) is called *Fermi's golden rule* of time-dependent perturbation theory.

In the case of electromagnetic radiation, the perturbation is harmonic with angular frequency ω, and a similar treatment shows that Eq. (C.6) is now replaced by

$$w_{nm} = [2\pi|\langle n|\widehat{\mathcal{H}}_1|m\rangle|^2/\hbar^2]\delta(\omega_n - \omega_m - \omega) \qquad (C.7)$$

We note that another representation of the delta function is

$$\delta(\omega - \omega_0) = \frac{1}{2\pi}\int_{-\infty}^{+\infty} \exp[it(\omega - \omega_0)]\,dt \qquad (C.8)$$

which enables Eq. (C.7) to be written as

$$w_{nm} = [|\langle n|\widehat{\mathcal{H}}_1|m\rangle|^2/\hbar^2]\int_{-\infty}^{+\infty} \exp[it(\omega_{nm} - \omega)]\,dt \qquad (C.9)$$

APPENDIX D

FORTRAN PROGRAMS

The following FORTRAN 77 programs illustrate concepts discussed in the text. Each program is preceded by an explanation of the program, of the necessary input data and, if necessary, of the output data. The programs have been kept simple; numerous comments have been included as an aid to those who wish to modify the programs. No graphics have been included in the programs, so that the programs can be used directly on mainframes, PCs, and Macs. The output data are in a format such that they can be exported readily to the reader's favorite graphics package, which should be no problem, since there are many graphics packages available for all computers.

The FORTRAN language was selected, rather than Basic, Pascal, or C, because it has been used for so long. It should be possible to copy any of the following programs and have them run on a FORTRAN 77 compiler on a mainframe, PC, Mac, or a workstation.

A source of FORTRAN programs that we have found particularly helpful is a book by Press et al. (Press et al., 1988). Editions of this book have been published for Pascal and C. Many of the programs in this appendix require a random number generator. No random number generators have been included in the programs, since many compilers have random number generators available. For example, Microsoft Fortran Version 5.1 for PC has builtin subroutines SEED and RANDOM that can be used to generate random numbers. The calls in the FORTRAN programs in this book are to the random number generators described by Press et al., references to which will be as PRESS.

D.1 BRNMTN

```
      PROGRAM BRNMTN

C     BROWNIAN MOTION BY INTEGRATION OF GAUSSIAN WHITE NOISE.
C     THIS PROGRAM REQUIRES A GAUSSIAN RANDOM NUMBER GENERATOR SUCH AS
C     GASDEV DESCRIBED IN PRESS.
C     YOU WILL BE PROMPTED FOR THE INPUT DATA, WHICH CONSISTS OF THE
C     NUMBER OF GAUSSIAN WHITE NOISE POINTS (NOT TO EXCEED 2500) AND
C     THE NUMBER OF BROWNIAN MOTION POINTS TO BE CALCULATED (NOT TO
C     EXCEED 500).  THE NUMBER OF GAUSSIAN WHITE NOISE POINTS SHOULD
C     EXCEED THE NUMBER OF BROWNIAN MOTION POINTS BY AT LEAST A FACTOR
C     OF FIVE.
C     THE PROGRAM THEN CALCULATES THE BROWNIAN MOTION TRAJECTORY AND
C     PUTS THESE INTO A FILE LABELED 'OUTPUT.'  THESE POINTS MAY BE
C     EXPORTED TO YOUR FAVORITE GRAPHICS PROGRAM.
C
      IMPLICIT REAL*8(A-H,O-Z)
      DIMENSION Y(2500),Z(500)
      WRITE(*,20)
   20 FORMAT(1X,'SELECT THE NUMBER OF GAUSSIAN WHITE NOISE POINTS AND'/,
     X' BROWNIAN MOTION POINTS.  ',\)
      READ(*,*) N,NBM
      OPEN(3,FILE='OUTPUT', STATUS='NEW')
C     GENERATE GAUSSIAN WHITE NOISE POINTS
      IDUM = -1
      DO 30 I=1,N
   30 Y(I) = 0.0
      DO 35 I=1,NBM
   35 Z(I) = 0.0
      DO 40 I=1,N
   40 Y(I) = GASDEV(IDUM)
C     GENERATE BROWNIAN MOTION POINTS
      M = N/NBM
      MIN = 2
      MAX = M
      DO 70 J=1,NBM
      DO 50 I=MIN,MAX
      Y(I) = Y(I-1)+Y(I)
   50 Z(J) = Y(I)
C     START NEW SUMMATION WITH ORIGINAL Y(I-1)
      MIN = MIN+M
      MAX = MAX+M
   70 CONTINUE
C     BROWNIAN MOTION TRAJECTORY BY SUMMATION OF THE Z POINTS
      ZSUM = 0.0
      DO 90 I=1,NBM
      ZSUM = ZSUM+Z(I)
      WRITE(3,80) ZSUM
   80 FORMAT(E14.5)
   90 CONTINUE
      CLOSE(3)
      WRITE (*,100)
```

```
100 FORMAT(1X,'PRESS RETURN TO TERMINATE.')
    PAUSE
    END
```

D.2 BRSLTR

```
    PROGRAM BRSLTR

C    SOLUTION OF THE BRUSSELATOR BY THE RUNGE-KUTTA METHOD
C    THE EQUATIONS ARE
C                DX/DT = A-(B+1)*X+X^2*Y
C                DY/DT = B*X-X^2*Y
C    WHERE A,B,C ARE PARAMETERS.  THE SOLUTIONS ARE ADVANCED BY A
C    TIME INCREMENT DT.  THE PRINTOUTS OF X AND Y ARE CONTROLLED BY
C    THE NUMBER OF TIME INCREMENTS (NUM) BETWEEN WRITES TO THE OUTPUT
C    FILE.  THIS IS SET BY THE VALUE OF M, WHERE M = 1/NUM*DT.
C    TOO LARGE A TIME INCREMENT WILL CAUSE ERRATIC BEHAVIOR OF THE
C    PROGRAM.  AN INITIAL VALUE OF 0.01 IS RECOMMENDED.
C    WHEN ENTERING VALUES OF PARAMETERS, UPON SEEING THE SCREEN PROMPT,
C    YOU MUST LEAVE A SPACE BETWEEN NUMBERS.
C    THE OUTPUT FILES ARE LABELED 'XRESULT' AND 'YRESULT.'  THESE
C    RESULTS MAY BE EXPORTED TO YOUR FAVORITE GRAPHICS PROGRAM.
C
    F1(X,Y,A,B) = A-(B+1.0)*X+X*X*Y
    F2(X,Y,B) = B*X-X*X*Y
    OPEN(2,FILE='XRESULT', STATUS='NEW')
    OPEN(3,FILE='YRESULT', STATUS='NEW')
    WRITE(*,10)
 10 FORMAT(1X,'SELECT THE NUMBER OF POINTS TO BE CALCULATED, THE VALUE
   XS OF A,B, AND THE',/1X,'INITIAL VALUES OF X AND Y.  ',\)
    READ(*,*) N,A,B,X,Y
    WRITE(*,25)
 25 FORMAT(1X,'SELECT THE TIME INCREMENT AND THE VALUE OF M.   ',\)
    READ(*,*) DT, M
    WRITE(2,40) X
    WRITE(3,40) Y
C    NUM IS A COUNTER FOR TIME INTERVALS BETWEEN OUTPUT WRITES
    NUM = 0
    DO 50 I=1,N
    AK1 = F1(X,Y,A,B)*DT
    AM1 = F2(X,Y,B)*DT
    X1 = X+AK1/2.0
    Y1 = Y+AM1/2.0
    AK2 = F1(X1,Y1,A,B)*DT
    AM2 = F2(X1,Y1,B)*DT
    X2 = X+AK2/2.0
    Y2 = Y+AM2/2.0
    AK3 = F1(X2,Y2,A,B)*DT
    AM3 = F2(X2,Y2,B)*DT
    X3 = X+AK3
    Y3 = Y+AM3
    AK4 = F1(X3,Y3,A,B)*DT
    AM4 = F2(X3,Y3,B)*DT
```

```
      X1 = X+(AK1+2.0*AK2+2.0*AK3+AK4)/6.0
      Y1 = Y+(AM1+2.0*AM2+2.0*AM3+AM4)/6.0
      X = X1
      Y = Y1
      IF(NUM.EQ.0) GO TO 35
      IF(NUM*M*DT.EQ.1) GO TO 30
      GO TO 45
   30 NUM = 0
   35 WRITE(2,40) X
      WRITE(3,40) Y
   40 FORMAT(E14.5)
   45 NUM = NUM + 1
   50 CONTINUE
      CLOSE(2)
      CLOSE(3)
      WRITE(*,60)
   60 FORMAT(1X,'PRESS RETURN TO TERMINATE')
      PAUSE
      END
```

D.3   CELLAUT

```
      PROGRAM CELLAUT

C     CELLULAR AUTOMATA
C     THIS PROGRAM CALCULATES ONE-DIMENSIONAL CELLULAR AUTO-
C     MATA USING THE NEIGHBOR-THREE RULES OF [WOLFRAM, 1983].
C     THE INPUT CONSISTS OF THE NUMBER OF SITES
C     (CELLS) N IN THE CA PLUS 2 TO TAKE INTO ACCOUNT THE ENDS FOR
C     CYCLIC BOUNDARY CONDITIONS, THE NUMBER OF TIME VALUES (INCLUDING
C     TIME ZERO) NT, AND THE ORIGINAL CONFIGURATION IX(NT,N) COMPRISED
C     OF 1'S AND 0'S.  THE VALUES OF N, NT, AND THE INITIAL
C     CONFIGURATION MUST BE ENTERED IN A FILE LABELED 'INPUT.'
C     SEE THE PROGRAM FOR THE FORMAT OF THE INPUT DATA.
C     THE CA RULES IN THE PROGRAM MUST BE CHANGED FOR EACH NEW CA AND
C     THE PROGRAM MUST BE RECOMPILED FOR EACH NEW CA.
C     EACH TIME N AND OR NT ARE CHANGED, THE PROGRAM MUST BE MODIFIED
C     AND RECOMPILED.  THESE CHANGES ARE INDICATED IN THE PROGRAM
C     BY *****.
C     THE VALUE OF N MUST NOT EXCEED 100 AND THE VALUE OF NT MUST NOT
C     EXCEED 500.
C     THE CA APPEAR DIRECTLY ON THE SCREEN.
C
      IMPLICIT REAL*8(A-H,O-V), INTEGER(I-N)
      DIMENSION IX(500,100)
      OPEN(2, FILE='INPUT',STATUS='OLD')
C     READ IN THE NUMBER OF CELLS (+2) AND THE NUMBER OF TIME STEPS
      READ(2,*) N,NT
C     READ IN THE INITIAL CONFIGURATION.
      READ(2,*) (IX(1,J), J=1,N)
      DO 120 I=1,NT-1
      DO 118 J=2,N-1
C*****THE CA RULES MUST BE SPECIFIED BELOW.  THE NEIGHBORHOODS ARE
```

```
C*****LABELED.  IF DESIRED, THE CA NUMBER CAN BE RECORDED IN THE
C*****NEXT STATEMENT.
C     CELLULAR AUTOMATA RULES FOR CA 126
      IF(IX(I,J).EQ.1) GO TO 108
C     NEIGHBORHOOD 101
      IF(IX(I,J-1).EQ.1.AND.IX(I,J+1).EQ.1) IX(I+1,J) = 1
C     NEIGHBORHOOD 100
      IF(IX(I,J-1).EQ.1.AND.IX(I,J+1).EQ.0) IX(I+1,J) = 1
C     NEIGHBORHOOD 001
      IF(IX(I,J-1).EQ.0.AND.IX(I,J+1).EQ.1) IX(I+1,J) = 1
C     NEIGHBORHOOD 000
      IF(IX(I,J-1).EQ.0.AND.IX(I,J+1).EQ.0) IX(I+1,J) = 0
      GO TO 118
C     NEIGHBORHOOD 111
  108 IF(IX(I,J-1).EQ.1.AND.IX(I,J+1).EQ.1) IX(I+1,J) = 0
C     NEIGHBORHOOD 110
      IF(IX(I,J-1).EQ.1.AND.IX(I,J+1).EQ.0) IX(I+1,J) = 1
C     NEIGHBORHOOD 011
      IF(IX(I,J-1).EQ.0.AND.IX(I,J+1).EQ.1) IX(I+1,J) = 1
C     NEIGHBORHOOD 010
      IF(IX(I,J-1).EQ.0.AND.IX(I,J+1).EQ.0) IX(I+1,J) = 1
  118 CONTINUE
      IX(I+1,1) = IX(I+1,N-1)
      IX(I+1,N) = IX(I+1,2)
  120 CONTINUE
      DO 125 I=1,NT
      DO 127 J=1,N
      IF(IX(I,J).EQ.0) GO TO 128
      WRITE(*,130)
  130 FORMAT(''\)
      GOTO 127
  128 WRITE(*,129)
  129 FORMAT(' '\)
  127 CONTINUE
      WRITE(*,126)
  126 FORMAT(' ')
  125 CONTINUE
      CLOSE(2)
C     SCREEN REMARK
      WRITE(*,160)
  160 FORMAT(1X,'PRESS RETURN TO TERMINATE')
      PAUSE
      END
```

D.4 GSSWTN

```
      PROGRAM GSSWTN

C     GAUSSIAN WHITE NOISE
C     THIS PROGRAM CALCULATES GAUSSIAN WHITE NOISE.
C     THIS PROGRAM REQUIRES A GAUSSIAN RANDOM NUMBER GENERATOR SUCH AS
C     GASDEV DESCRIBED IN PRESS.
C     INPUT CONSISTS OF THE NUMBER OF WHITE NOISE POINTS (N).
```

```
C     THERE IS NO RESTRICTION ON THE NUMBER OF POINTS.
C     THE OUTPUT IS IN A FILE NAMED 'RESULT.'
C
C
      IMPLICIT REAL*8(A-H,O-Z)
      WRITE(*,20)
   20 FORMAT(1X,'SELECT THE NUMBER OF GAUSSIAN WHITE NOISE POINTS.  ',\)
      READ(*,*) N
      OPEN(3,FILE='RESULT',STATUS='NEW')
      IDUM = -1
      DO 60 I=1,N
      X = GASDEV(IDUM)
      WRITE(3,50) X
   50 FORMAT(E14.5)
   60 CONTINUE
      CLOSE(3)
      WRITE(*,70)
   70 FORMAT(1X,'PRESS RETURN TO TERMINATE.')
      PAUSE
      END
```

D.5 ISING

```
      PROGRAM ISING

C     TWO-DIMENSIONAL ISING LATTICE BY MONTE CARLO METHOD
C     THIS FORTRAN PROGRAM CALCULATES THE SHORT-RANGE ORDER
C     PARAMETER F1 (FRACTION OF NEAREST-NEIGHBOR SITES OCCUPIED
C     BY AN ANTIPARALLEL PAIR OF SPINS) AND E/NJ.  INPUT CONSISTS OF THE
C     LATTICE DIMENSION N (WHERE TOTAL NUMBER OF SITES IS N * N),
C     THE REDUCED COUPLING CONSTANT R, AND THE NUMBER OF ITERATIONS
C     NIT.  THE COUNTER NCOUNT RECORDS THE NUMBER OF ITERATIONS.
C     EACH ITERATION CONSISTS OF THE METROPOLIS MC METHOD APPLIED
C     SUCCESSIVELY TO EACH OF THE NXN SITES.  THE TOTAL NUMBER OF
C     CONFIGURATIONS USED IN THE CALCULATION OF F1 IS N * N * NIT.
C     A RANDOM NUMBER GENERATOR, SUCH AS RAN2 FROM PRESS
C     MUST BE SUPPLIED.
C     THE OUTPUT ON THE SCREEN IS THE ORDER PARAMETER.  IN A FILE
C     LABELED 'RESULT' CAN BE FOUND VALUES FOR E/NJ AFTER EACH
C     ITERATION.
C     AT THE END OF THE CALCULATION, YOU WILL BE PROMPTED AS TO WHETHER
C     YOU WISH TO DO ANOTHER CALCULATION.  IF YOU DO, THE RESULTS FOR
C     EACH CALCULATION WILL BE SAVED IN THE 'RESULT' FILE.
C
      IMPLICIT REAL*8(A-H,O-Z)
      CHARACTER*3 ANS
      DIMENSION IA(50,50),SUMF1(1000)
      OPEN(3,FILE='RESULT',STATUS='NEW')
C     DIMENSION OF LATTICES IS N; REDUCED COUPLING CONSTANT R=J/KT
    1 WRITE(*,5)
    5 FORMAT(1X,'PLEASE TYPE IN THE LATTICE DIMENSION, J/KT, AND NUMBER
     X OF ITERATIONS; LEAVE A'/' SPACE BETWEEN NUMBERS.    ',\)
      READ(*,*) N,R,NIT
```

```
      WRITE(*,6)
    6 FORMAT(1X,'CALCULATION IS IN PROGRESS.')
      NCOUNT = 0
      DO 200 IJ=1,NIT
      NCOUNT = NCOUNT+1
      IF(NCOUNT.EQ.1) GO TO 7
      F1 = F1*N*N*(NCOUNT-1)
      GO TO 12
    7 F1 = 0.0
      IDUM = -1
C     SET ALL SPINS EQUAL TO 1 TO BEGIN
    9 DO 10 I=1,N
      DO 10 J=1,N
   10 IA(I,J) = 1
   12 K = 0
C     COUNT NUMBER OF +1,+1 NEIGHBOR PAIRS
   15 NPP = 0
      DO 20 J=1,N
      IF(IA(1,J).NE.IA(N,J).OR.IA(1,J).EQ.0) GO TO 20
      NPP = NPP+1
   20 CONTINUE
      DO 30 I=1,N
      IF(IA(I,1).NE.IA(I,N).OR.IA(I,1).EQ.0) GO TO 30
      NPP = NPP+1
   30 CONTINUE
      DO 40 I=1,N
      DO 40 J=1,N-1
      IF(IA(I,J).NE.IA(I,J+1).OR.IA(I,J).EQ.0) GO TO 40
      NPP = NPP+1
   40 CONTINUE
      DO 50 I=1,N-1
      DO 50 J=1,N
      IF(IA(I,J).NE.IA(I+1,J).OR.IA(I,J).EQ.0) GO TO 50
      NPP = NPP+1
   50 CONTINUE
C     COUNT NUMBER OF +1 SPINS
      NP = 0
      DO 60 I=1,N
      DO 60 J=1,N
      IF(IA(I,J).EQ.0) GO TO 60
      NP = NP+1
   60 CONTINUE
      SS = 4.0*NPP-8.0*NP+2.0*N*N
      IF(K.GT.0) GO TO 65
      E1 = -R*SS
      GO TO 66
   65 E2 = -R*SS
      DELE = E2-E1
      GO TO 85
C     REVERSE SPIN AT SITE I,J
   66 L = 0
   67 L = L+1
      M = 0
   68 M = M+1
```

```
      IF(IA(L,M)) 70,70,80
   70 IA(L,M) = 1
      GO TO 82
   80 IA(L,M) = 0
   82 K = K+1
      NP1 = NP
      NPP1 = NPP
      GO TO 15
   85 IF(DELE) 90,90,100
   90 E1 = E2
      GO TO 140
C     DELTA E IS GT ZERO; CHECK BOLTZMANN DISTRIBUTION
  100 IDUM = IDUM-1
      RN = RAN2(IDUM)
      BOL = DEXP(-DELE)
      IF(RN.LT.BOL) GO TO 130
C     NON-BOLTZMANN DISTRIBUTION---RESET TO OLD VALUES
      IF(IA(L,M)) 110,110,120
  110 IA(L,M) = 1
      NP = NP1
      NPP = NPP1
      GO TO 140
  120 IA(L,M) = 0
      NP = NP1
      NPP = NPP1
      GO TO 140
  130 E1 = E2
  140 F1 = (4.0*NP-2.0*NPP)/(2.0*N*N)+F1
      NP1 = NP
      NPP1 = NPP
      IF(M.NE.N) GO TO 68
      IF(L.NE.N) GO TO 67
      F1 = F1/(N*N*NCOUNT)
      SUMF1(IJ) = -2.0*(1.0 - 2.0*F1)
  200 CONTINUE
      WRITE(3,210) (SUMF1(IJ), IJ=1,NIT)
  210 FORMAT(E12.5)
      WRITE(*,150) F1
  150 FORMAT(1X,'THE ORDER PARAMETER IS',E12.5)
      WRITE(*,155)
  155 FORMAT(1X,'DO YOU WISH TO CONTINUE WITH A NEW CONFIGURATION? (yes
     XOR no) ',\)
      READ(*,157) ANS
  157 FORMAT(A3)
      IF(ANS.EQ.'yes') GO TO 1
      CLOSE(3)
      WRITE(*,160)
  160 FORMAT(1X,'PRESS RETURN TO TERMINATE')
      PAUSE
      END
```

D.6 LJMDFOR1D

```
      PROGRAM LJMDFOR1D

C     THIS PROGRAM PERFORMS AN MD SIMULATION ON A ONE-DIMENSIONAL
C     LENNARD-JONES FLUID MODEL. THE PROGRAM USES THE RANDOM NUMBER
C     GENERATOR RAN2 FROM PRESS. CARE SHOULD BE TAKEN TO CHANGE THE
C     FUNCTION AND ALL THE REAL VARIABLES IN IT INTO DOUBLE PRECISION.
C     THE VARIABLES HAVE EASY TO UNDER3TAND
C     NAMES. REDUCED UNITS HAVE BEEN USED. THE NTH POSITION DERIVATIVE
C     HAS BEEN MULTIPLIED BY (DELTA)**N/N!. NO CUTOFF CORRECTIONS HAVE
C     BEEN APPLIED.

      IMPLICIT REAL*8(A-H, O-Z)
      COMMON /XOF/ X0(300), F(300)
      COMMON /X123/ X1(300), X2(300), X3(300)
      COMMON /NSIDEL/ N, SIDEL, SIDEL2
      COMMON /SUMS/ V2SUM, ENSUM, VISUM
      COMMON /ENERVIR/ ENERGY, VIRIAL
      OPEN( UNIT=6, FILE='MDOUT.OUT')
C     GIVE THE VALUES OF THE INPUT PARAMETERS.
      NEQUI=5000
      NCALC=20
      NPROD=15000
      N=300
      AN=N
      DELTA=0.005D0
      DELTA2=DELTA*DELTA
      DELTA3=0.5D0*DELTA2
      TEMP=2.10D0
      DENSITY=0.4D0
      SIDEL=AN/DENSITY
      SIDEL2=0.5D0*SIDEL
      DELTAX=SIDEL/(AN + 1.0D0)
      DO 10 I=1, N
      X2(I)=0.0D0
      X3(I)=0.0D0
   10 F(I)=0.0D0
      V2SUM=0.0D0
      ENSUM=0.0D0
      VISUM=0.0D0
      X0(1)=DELTAX
C     ASSIGN INITIAL POSITIONS
      DO 20 I=2, N
   20 X0(I)=X0(I-1) + DELTAX
      SUMV=0.0D0
      WRITE(*,25)
   25 FORMAT(1X,'PLEASE TYPE IN RANDOM NUMBER SEED INTEGER. '\)
C     ASSIGN RANDOM VELOCITIES
      READ(*,*)IDUM
      DO 30 I=1,N
      XX = RAN2(IDUM)
      X1(I) = 2.0D0*XX-1.0D0
   30 SUMV=SUMV + X1(I)
```

```
         SUMV2=0.0D0
         DO 40 I=1, N
         X1(I) = X1(I) - SUMV/AN
   40    SUMV2 = SUMV2 + X1(I)*X1(I)
         SCALE=DSQRT(AN*DELTA2*TEMP/SUMV2)
C        SCALE THE VELOCITIES TO GET THE GIVEN REDUCED TEMPERATURE
         DO 50 I=1, N
   50    X1(I) = X1(I)*SCALE
C        START THE EQUILIBRATION PHASE
         CALL FORCEETC
         DO 60 I=1, N
   60    X2(I) =F(I)*DELTA3
         DO 70 NST=1, NEQUI
         CALL PREDCOR(DELTA3)
         IF(MOD(NST,NCALC).EQ.0)THEN
         CALL CALC(DENSITY, SQV, DELTA2, NCALC, NST)
         SCALE=DSQRT(AN*DELTA2*TEMP/SQV)
         DO 80 I=1, N
   80    X1(I) = X1(I)*SCALE
         END IF
   70    CONTINUE
C        START THE PRODUCTION RUN
         V2SUM=0.0D0
         ENSUM=0.0D0
         VISUM=0.0D0
         DO 90 NST=1, NPROD
         CALL PREDCOR(DELTA3)
         IF(MOD(NST,NCALC).EQ.0)CALL CALC(DENSITY, SQV,
       1 DELTA2, NCALC, NST)
   90    CONTINUE
         STOP
         END


         SUBROUTINE FORCEETC
C        THIS SUNROUTINE CALCULATES THE FORCES, POTENTIAL ENERGIES AND THE
C        VIRIAL FOR A GIVEN CONFIGURATION.
         IMPLICIT REAL*8(A-H, O-Z)
         COMMON /X0F/ X0(300), F(300)
         COMMON /NSIDEL/ N, SIDEL,SIDEL2
         COMMON /ENERVIR/ ENERGY, VIRIAL
         ENERGY=0.0D0
         VIRIAL=0.0D0
         DO 10 I=1, N-1
         R=X0(I)-X0(I+1)
         IF(R.GT.SIDEL2)R=R-SIDEL
         IF(R.LT.-SIDEL2)R=R+SIDEL
         R2=R*R
         R6=R2*R2*R2
         R61=1.0D0/R6
         EN=4.0D0*R61*(R61-1.0D0)
         FORCE=24.0D0*R61*(2.0D0*R61-1.0D0)/R
         F(I)=F(I)+FORCE
         F(I+1)=F(I+1)-FORCE
         ENERGY=ENERGY+EN
```

```
   10  VIRIAL=VIRIAL+FORCE*R
       R=X0(N)-X0(1)
       IF(R.GT.SIDEL2)R=R-SIDEL
       IF(R.LT.-SIDEL2)R=R+SIDEL
       R2=R*R
       R6=R2*R2*R2
       R61=1.0D0/R6
       EN=4.0D0*R61*(R61-1.0D0)
       FORCE=24.0D0*R61*(2.0D0*R61-1.0D0)/R
       F(N)=F(N)+FORCE
       F(1)=F(1)-FORCE
       ENERGY=ENERGY+EN
       VIRIAL=VIRIAL+FORCE*R
       RETURN
       END


       SUBROUTINE PREDCOR(DELTA3)
C      THIS SUBROUTINE APPLIES THE GEAR'S THIRD ORDER PREDICTOR CORRECTOR
C      ALGORITHM.
       IMPLICIT REAL*8(A-H, O-Z)
       COMMON /XOF/ X0(300), F(300)
       COMMON /X123/ X1(300), X2(300), X3(300)
       COMMON /NSIDEL/ N, SIDEL, SIDEL2
       DO 10 I=1, N
       X0(I)=X0(I)+X1(I)+X2(I)+X3(I)
       X1(I)=X1(I)+2.0D0*X2(I)+3.0D0*X3(I)
       X2(I)=X2(I)+3.0D0*X3(I)
   10  F(I)=0.0D0
       CALL FORCEETC
       DO 20 I=1, N
       DIFF=DELTA3*F(I)-X2(I)
       X0(I)=X0(I)+DIFF/6.0D0
       X1(I)=X1(I)+5.0D0*DIFF/6.0D0
       X2(I)=X2(I)+DIFF
       X3(I)=X3(I)+DIFF/3.0D0
       IF(X0(I).GT.SIDEL)X0(I)=X0(I)-SIDEL
   20  IF(X0(I).LT.0.0D0)X0(I)=X0(I)+SIDEL
       RETURN
       END


       SUBROUTINE CALC(DENSITY, SQV, DELTA2, NCALC, NST)
C      THIS SUBROUTINE CALCULATES THE PHYSICAL PROPERTIES USING THEIR
C      DEFINITIONS, AS NEEDED.
       IMPLICIT REAL*8(A-H, O-Z)
       COMMON /X123/ X1(300), X2(300), X3(300)
       COMMON /NSIDEL/ N, SIDEL, SIDEL2
       COMMON /SUMS/ V2SUM, ENSUM, VISUM
       COMMON /ENERVIR/ ENERGY, VIRIAL
       SQV=0.0D0
       AN=N
       DO 10 I=1, N
   10  SQV=SQV+X1(I)*X1(I)
       V2SUM=V2SUM+SQV
       ENSUM=ENSUM+ENERGY
```

```
          VISUM=VISUM+VIRIAL
          TEMPCALC=SQV/(AN*DELTA2)
          PRES=DENSITY*(TEMPCALC+VIRIAL/AN)
          EPOT=ENERGY/AN
          ETOT=0.5D0*TEMPCALC +EPOT
          POINTS=DFLOAT(NST/NCALC)
          TAVE=V2SUM/(AN*POINTS*DELTA2)
          EAVE=ENSUM/(AN*POINTS)
          PAVE=DENSITY*(TAVE+VISUM/(AN*POINTS))
          WRITE(6,*)'NST,TEMPCALC,TAVE,EPOT,EAVE,PRES,PAVE,ETOT'
          WRITE(6,*)NST,TEMPCALC,TAVE,EPOT,EAVE,PRES,PAVE,ETOT
          RETURN
          END
```

D.7 LOGIS1

```
          PROGRAM LOGIS1
C         THE LOGISTIC MAP
C         THIS PROGRAM GENERATES POINTS FOR THE LOGISTIC MAP
C                     X(N+1) = AX(1-X)
C         WHERE A IS THE BIFURCATION PARAMETER.  AT THE PROMPT, SELECT 1 FOR
C         THE FIRST ITERATE, 2 FOR THE SECOND ITERATE, 3 FOR THE THIRD
C         ITERATE, AND 4 FOR THE FOURTH ITERATE.
C         THE VALUES OF X ARE TO BE FOUND IN A FILE TITLED 'RESULT.'
C
          DIMENSION X(101),Y(101)
          OPEN(2,FILE='RESULT',STATUS='NEW')
          WRITE(*,10)
   10     FORMAT(1X,'SELECT THE VALUE OF THE CONTROL PARAMETER.   ',\)
          READ(*,*) A
          WRITE(*,20)
   20     FORMAT(1X,'SELECT THE ITERATIVE MAP.   ',\)
          READ(*,*) N
          X(1) = 0.0
          GO TO (25,35,45,55) N
C         FIRST ITERATE MAP
   25     DO 30 I=1,100
          X(I+1) = X(I)+0.01
   30     Y(I) = A*X(I)*(1.0-X(I))
          GO TO 65
C         SECOND ITERATE MAP
   35     DO 40 I=1,100
          X(I+1) = X(I)+0.01
          Y(I) = A*X(I)*(1.0-X(I))
   40     Y(I) = A*Y(I)*(1.0-Y(I))
          GO TO 65
C         THIRD ITERATE MAP
   45     DO 50 I=1,100
          X(I+1) = X(I)+0.01
          Y(I) = A*X(I)*(1.0-X(I))
          Y(I) = A*Y(I)*(1.0-Y(I))
   50     Y(I) = A*Y(I)*(1.0-Y(I))
```

```
      GO TO 45
30 NUM = 0
35 WRITE(2,40) X
   WRITE(3,40) Y
40 FORMAT(E14.5)
45 NUM = NUM + 1
50 CONTINUE
   CLOSE(2)
   CLOSE(3)
   WRITE(*,60)
60 FORMAT(1X,'PRESS RETURN TO TERMINATE')
   PAUSE
   END
```

D.10 MCINT

```
      PROGRAM MCINT

C     MONTE CARLO INTEGRATION
C     THIS PROGRAM INTEGRATES 1/(1+X*X) USING THE RANDOM NUMBER
C     GENERATOR RAN2(IDUM) FROM PRESS. THE PROGRAM ASKS
C     FOR THE INPUT, WHICH IS THE NUMBER OF SAMPLE POINTS
C     (RAMDOM NUMBERS) IN THE EVALUATION
C     OF THE INTEGRAL.
C     THE OUTPUT IS TO THE SCREEN.
C     NOTE HOW MANY POINTS MUST BE SAMPLED TO APPROACH THE TRUE VALUE OF
C     2.65163.
C
      IMPLICIT REAL*8(A-H,O-Z)
      CHARACTER*3 ANS
    3 WRITE (*,5)
    5 FORMAT(1X,'ENTER THE NUMBER OF SAMPLE POINTS DESIRED.    ',\)
      READ (*,*) N
      WRITE (*,7)
    7 FORMAT(1X,'THE PROGRAM IS RUNNING.')
      IDUM = -1
      AINT = 0.0
      DO 10 I=1,N
      X = -4.0+8.0*RAN2(IDUM)
C     THE FOLLOWING LINE IS THE FUNCTION, WHICH MAY BE CHANGED.
      F = 1.0/(1.0+X*X)
      AINT = AINT+F
   10 CONTINUE
      AINT = AINT*8.0/N
   20 FORMAT(6E12.5)
      WRITE (*,12) AINT
   12 FORMAT(1X,'THE VALUE OF THE INTEGRAL IS',2X,E10.5)
      WRITE (*,25)
   25 FORMAT(1X,'DO YOU WISH TO DO ANOTHER CALCULATION? (yes OR no) ',\)
      READ (*,30) ANS
   30 FORMAT(A3)
      IF (ANS.EQ.'yes') GO TO 3
      WRITE (*,35)
```

```
   35 FORMAT(1X,'PRESS RETURN TO TERMINATE')
      PAUSE
      END
```

D.11 POLYMER

```
      PROGRAM POLYMER

C     THIS FORTRAN PROGRAM ILLUSTRATES POLYMER FOLDING BY GENERATING A
C     SELF-AVOIDING RANDOM WALK.  THE POLYMER MODEL IS THE BEAD-SPRING
C     MODEL.  THE PROGRAM PROMPTS FOR THE NUMBER OF BEADS AND FOR THE
C     SEED VALUE FOR THE RANDOM NUMBER GENERATOR.
C     THE NUMBER OF BEADS CANNOT EXCEED 500.
C     IN SOME CASES THE POLYMER CHAIN MUST BE TERMINATED BECAUSE IT
C     CANNOT GROW FARTHER.  SUCH CASES ARE INDICATED ON THE SCREEN AND
C     THE POLYMER IS DRAWN AS FAR AS IT CAN GO.
C     THIS IS NOT A STATISTICAL SELF-AVOIDING WALK, SINCE THE MOVES ARE
C     NOT ALL EQUALLY PROBABLE.  IT IS INTENDED ONLY TO SHOW POLYMER
C     FOLDING.
C     THIS PROGRAM REQUIRES A RANDOM NUMBER GENERATOR SUCH AS RAN2 FOUND
C     IN PRESS.
C     THIS PROGRAM WRITES THE OUTPUT TO FILES 'RESULTX' AND 'RESULTY.'
C     THE OUTPUT FILES CAN BE EXPORTED TO A PLOTTING ROUTINE, OR THE
C     RESULTS CAN BE PLOTTED ON GRAPH PAPER.
C
C
      DIMENSION IX(500),IY(500)
      OPEN(2,FILE='RESULTX',STATUS='NEW')
      OPEN(3,FILE='RESULTY',STATUS='NEW')
C     NUMBER OF BEADS IS N
      WRITE(*,5)
    5 FORMAT(1X,'PLEASE TYPE IN THE NUMBER OF BEADS.  '\)
      READ(*,*) N
      WRITE(*,10)
   10 FORMAT(1X,'PLEASE TYPE IN THE VALUE OF THE RANDOM NUMBER SEED.  '\
     X)
      READ(*,*) IDUM
C     SET UP INITIAL SITES
      L = 1
      IX(1) = 100
      IY(1) = 100
      IX(2) = 100+L
      IY(2) = 100
C     BEGIN THE RANDOM WALK
      DO 150 I=3,N
  100 A = RAN2(IDUM)
  102 NCOUNT = 0
      IF(A.LT.0.25) GO TO 115
      IF(A.LT.0.5) GO TO 125
      IF(A.LT.0.75) GO TO 135
  105 NCOUNT = NCOUNT+1
      IF(NCOUNT.GT.4) GO TO 500
      DO 110 K=1,I-1
```

```
110 IF(IX(K).EQ.IX(I-1).AND.IY(K).EQ.IY(I-1)-L) GO TO 115
    IX(I) = IX(I-1)
    IY(I) = IY(I-1)-L
    GO TO 150
115 NCOUNT = NCOUNT+1
    IF(NCOUNT.GT.4) GO TO 500
    DO 120 K=1,I-1
120 IF(IX(K).EQ.IX(I-1)-L.AND.IY(K).EQ.IY(I-1)) GO TO 125
    TX(I) = IX(I-1)-L
    IY(I) = IY(I-1)
    GO TO 150
125 NCOUNT = NCOUNT+1
    IF(NCOUNT.GT.4) GO TO 500
    DO 130 K=1,I-1
130 IF(IX(K).EQ.IX(I-1).AND.IY(K).EQ.IY(I-1)+L) GO TO 135
    IX(I) = IX(I-1)
    IY(I) = IY(I-1)+L
    GO TO 150
135 NCOUNT = NCOUNT+1
    IF(NCOUNT.GT.4) GO TO 500
    DO 140 K=1,I-1
140 IF(IX(K).EQ.IX(I-1)+L.AND.IY(K).EQ.IY(I-1)) GO TO 105
    IX(I) = IX(I-1)+L
    IY(I) = IY(I-1)
    GO TO 150
C     NCOUNT > 4; CHAIN IS STUCK.   REEVALUATE PREVIOUS POSITION.
500 MCOUNT = 0
    IF(A.LT.0.25) GO TO 510
    IF(A.LT.0.5) GO TO 520
    IF(A.LT.0.75) GO TO 530
540 MCOUNT = MCOUNT+1
    IF(MCOUNT.GT.4) GO TO 590
    DO 505 K=1,I-1
505 IF(IX(K).EQ.IX(I-2).AND.IY(K).EQ.IY(I-2)-L) GO TO 510
    GO TO 550
510 MCOUNT = MCOUNT+1
    IF(MCOUNT.GT.4) GO TO 590
    DO 515 K=1,I-1
515 IF(IX(K).EQ.IX(I-2)-L.AND.IY(K).EQ.IY(I-2)) GO TO 520
    GO TO 560
520 MCOUNT = MCOUNT+1
    IF(MCOUNT.GT.4) GO TO 590
    DO 525 K=1,I-1
525 IF(IX(K).EQ.IX(I-2).AND.IY(K).EQ.IY(I-2)+L) GO TO 530
    GO TO 570
530 MCOUNT = MCOUNT+1
    IF(MCOUNT.GT.4) GO TO 590
    DO 535 K=1,I-1
535 IF(IX(K).EQ.IX(I-2)+L.AND.IY(K).EQ.IY(I-2)) GO TO 540
    GO TO 580
550 IX(I-1) = IX(I-2)
    IY(I-1) = IY(I-2)-L
    GO TO 102
560 IX(I-1) = IX(I-2)-L
```

```
      IY(I-1) = IY(I-2)
      GO TO 102
  570 IX(I-1) = IX(I-2)
      IY(I-1) = IY(I-2)+L
      GO TO 102
  580 IX(I-1) = IX(I-2)+L
      IY(I-1) = IY(I-2)
      GO TO 102
  590 WRITE(*,595)
  595 FORMAT('CHAIN GROWTH TERMINATED')
      GO TO 205
  150 CONTINUE
C     WRITE FINAL CONFIGURATION TO FILES "RESULTX" AND "RESULTY"
  205 WRITE(2,645) (IX(I),I=1,N)
      WRITE(3,645) (IY(I),I=1,N)
  645 FORMAT(I5)
      CLOSE(2)
      CLOSE(3)
      WRITE(*,650)
  650 FORMAT(1X,'PRESS RETURN TO TERMINATE')
      PAUSE
      END
```

D.12 PWRSPR

```
      PROGRAM PWRSPR
C     POWER SPECTRUM
C     THE INPUT FOR THE POWER SPECTRUM CONSISTS OF A TIME SERIES,
C     EITHER CALCULATED OR MEASURED IN AN EXPERIMENT.  THESE DATA
C     ARE IN A FILE NAMED 'DATA.'  THE NUMBER OF DATA POINTS
C     OF THE TIME SERIES MUST BE PROVIDED ACCORDING TO THE PROMPT.
C     THE OUTPUT IS THE POWER SPECTRUM IN A FILE NAMED 'RESULT.'
C     THERE IS NO WINDOW PROVIDED FOR THE FOURIER TRANSFORM, SO THERE
C     MAY BE LEAKAGE IN THE TRANSFORM AND THE POWER SPECTRUM.
C     ALSO, CONSTANT COEFFICIENTS SUCH AS THE TIME INCREMENT AND THE
C     TOTAL TIME HAVE BEEN OMITTED.
C     THE OUTPUT DATA MAY BE EXPORTED TO YOUR FAVORITE GRAPHICS PROGRAM.
C
      IMPLICIT REAL*8(A-H,O-Z)
      DIMENSION X(2700),Y(2700),YIMAG(2700)
      DATA PI/3.141593/
      OPEN(2,FILE='DATA',STATUS='OLD',FORM='FORMATTED')
      OPEN(3,FILE='RESULT',STATUS='NEW',FORM='FORMATTED')
      WRITE(*,10)
   10 FORMAT(1X,'SELECT THE NUMBER OF POINTS FOR THE POWER SPECTRUM.  TH
     XIS SHOULD AGREE WITH',/1X,'THE OUTPUT (LESS ONE) FROM THE TIME SER
     XIES.  ',\)
      READ(*,*) N
      READ (2,20) (X(I), I=1,N)
   20 FORMAT (E14.5)
C     FOURIER TRANSFORM
      DO 50 J=1,N
      Y(J) = 0.0
```

```
      YIMAG(J) = 0.0
      DO 40 I=1,N
      Y(J) = X(I)*COS(2.0*PI*J*(I-1)/(N+1))+Y(J)
   40 YIMAG(J) = X(I)*SIN(2.0*PI*J*(I-1)/(N+1))+YIMAG(J)
   50 CONTINUE
      DO 60 J=1,N
      Y(J) = Y(J)*Y(J)
      YIMAG(J) = YIMAG(J)*YIMAG(J)
   60 Y(J) = Y(J)+YIMAG(J)
      WRITE(3,20) (Y(J), J=1,N)
      CLOSE(2)
      CLOSE(3)
      WRITE(*,70)
   70 FORMAT(1X,'PRESS RETURN TO TERMINATE')
      PAUSE
      END
```

D.13 REVCA

```
      PROGRAM REVCA

C     REVERSIBLE CELLULAR AUTOMATA
C     THIS PROGRAM CALCULATES ONE-DIMENSIONAL REVERSIBLE
C     CELLULAR AUTOMATA USING THE NEIGHBOR-THREE RULES OF
C     [WOLFRAM, 1983]. THE INPUT CONSISTS OF THE NUMBER OF
C     SITES (CELLS) N IN THE CA PLUS TWO TO TAKE INTO ACCOUNT THE ENDS
C     FOR CYCLIC BOUNDARY CONDITIONS, THE NUMBER OF TIME VALUES
C     (INCLUDING TIME ZERO) NT, AND THE ORIGINAL CONFIGURATION IX(NT,N)
C     COMPRISED OF 1 OR 0.  THE VALUES OF N, NT, AND THE INITIAL
C     CONFIGURATION MUST BE ENTERED IN A FILE LABELED 'INPUT.'
C     SEE THE PROGRAM FOR THE FORMAT OF THE INPUT DATA.
C     THE CA RULES IN THE PROGRAM MUST BE CHANGED FOR EACH NEW CA AND
C     THE PROGRAM MUST BE RECOMPILED FOR EACH NEW CA.
C     THE VALUE OF N MUST NOT EXCEED 100 AND THE VALUE OF NT MUST NOT
C     EXCEED 500.
C     THE CA APPEAR DIRECTLY ON THE SCREEN.
C
      IMPLICIT REAL*8(A-H,O-V), INTEGER(I-N)
      DIMENSION IX(500,100),IY(500,100)
      OPEN(2, FILE='INPUT',STATUS='OLD')
C     READ IN THE NUMBER OF CELLS (+2) AND THE NUMBER OF TIME STEPS
      READ(2,*) N,NT
C     READ IN THE INITIAL CONFIGURATION.
      READ(2,*) (IX(1,J), J=1,N)
      DO 20 J=2,N-1
C*****THE CA RULES MUST BE SPECIFIED BELOW.  THE NEIGHBORHOODS ARE
C*****LABELED.  IF DESIRED, THE CA NUMBER CAN BE RECORDED IN THE
C*****NEXT STATEMENT.
C     CELLULAR AUTOMATA RULES FOR CA 126
      IF(IX(1,J).EQ.1) GO TO 15
C     NEIGHBORHOOD 101
      IF(IX(1,J-1).EQ.1.AND.IX(1,J+1).EQ.1) IX(1+1,J) = 1
C     NEIGHBORHOOD 100
```

```
          IF(IX(1,J-1).EQ.1.AND.IX(1,J+1).EQ.0) IX(1+1,J) = 1
C         NEIGHBORHOOD 001
          IF(IX(1,J-1).EQ.0.AND.IX(1,J+1).EQ.1) IX(1+1,J) = 1
C         NEIGHBORHOOD 000
          IF(IX(1,J-1).EQ.0.AND.IX(1,J+1).EQ.0) IX(1+1,J) = 0
          GO TO 20
C         NEIGHBORHOOD 111
       15 IF(IX(1,J-1).EQ.1.AND.IX(1,J+1).EQ.1) IX(1+1,J) = 0
C         NEIGHBORHOOD 110
          IF(IX(1,J-1).EQ.1.AND.IX(1,J+1).EQ.0) IX(1+1,J) = 1
C         NEIGHBORHOOD 011
          IF(IX(1,J-1).EQ.0.AND.IX(1,J+1).EQ.1) IX(1+1,J) = 1
C         NEIGHBORHOOD 010
          IF(IX(1,J-1).EQ.0.AND.IX(1,J+1).EQ.0) IX(1+1,J) = 1
       20 CONTINUE
          IX(1+1,1) = IX(1+1,N-1)
          IX(1+1,N) = IX(1+1,2)
          DO 120 I=2,NT-1
          DO 118 J=2,N-1
C*****THE CA RULES MUST BE SPECIFIED BELOW.  THE NEIGHBORHOODS ARE
C*****LABELED.  IF DESIRED, THE CA NUMBER CAN BE RECORDED IN THE
C*****NEXT STATEMENT.
C     CELLULAR AUTOMATA RULES FOR CA 126
          IF(IX(I,J).EQ.1) GO TO 108
C         NEIGHBORHOOD 101
          IF(IX(I,J-1).EQ.1.AND.IX(I,J+1).EQ.1) IY(I+1,J) = 1
C         NEIGHBORHOOD 100
          IF(IX(I,J-1).EQ.1.AND.IX(I,J+1).EQ.0) IY(I+1,J) = 1
C         NEIGHBORHOOD 001
          IF(IX(I,J-1).EQ.0.AND.IX(I,J+1).EQ.1) IY(I+1,J) = 1
C         NEIGHBORHOOD 000
          IF(IX(I,J-1).EQ.0.AND.IX(I,J+1).EQ.0) IY(I+1,J) = 0
C         ADD LINE NT-2 MOD 2 (DO NOT CHANGE THE NEXT FOUR STATEMENTS.)
          IF(IX(I-1,J).EQ.0.AND.IY(I+1,J).EQ.0) IX(I+1,J) = 0
          IF(IX(I-1,J).EQ.0.AND.IY(I+1,J).EQ.1) IX(I+1,J) = 1
          IF(IX(I-1,J).EQ.1.AND.IY(I+1,J).EQ.0) IX(I+1,J) = 1
          IF(IX(I-1,J).EQ.1.AND.IY(I+1,J).EQ.1) IX(I+1,J) = 0
          GO TO 118
C         NEIGHBORHOOD 111
      108 IF(IX(I,J-1).EQ.1.AND.IX(I,J+1).EQ.1) IY(I+1,J) = 0
C         NEIGHBORHOOD 110
          IF(IX(I,J-1).EQ.1.AND.IX(I,J+1).EQ.0) IY(I+1,J) = 1
C         NEIGHBORHOOD 011
          IF(IX(I,J-1).EQ.0.AND.IX(I,J+1).EQ.1) IY(I+1,J) = 1
C         NEIGHBORHOOD 010
          IF(IX(I,J-1).EQ.0.AND.IX(I,J+1).EQ.0) IY(I+1,J) = 1
C         ADD LINE NT-2 MOD 2 (DO NOT CHANGE THE NEXT FOUR STATEMENTS.)
          IF(IX(I-1,J).EQ.0.AND.IY(I+1,J).EQ.0) IX(I+1,J) = 0
          IF(IX(I-1,J).EQ.0.AND.IY(I+1,J).EQ.1) IX(I+1,J) = 1
          IF(IX(I-1,J).EQ.1.AND.IY(I+1,J).EQ.0) IX(I+1,J) = 1
          IF(IX(I-1,J).EQ.1.AND.IY(I+1,J).EQ.1) IX(I+1,J) = 0
      118 CONTINUE
C     ADD CYCLIC BOUNDARY CONDITIONS
          IX(I+1,1) = IX(I+1,N-1)
```

```
      IX(I+1,N) = IX(I+1,2)
  120 CONTINUE
      DO 125 I=1,NT
      DO 127 J=1,N
      IF(IX(I,J).EQ.0) GO TO 128
      WRITE(*,130)
  130 FORMAT(''\)
      GOTO 127
  128 WRITE(*,129)
  129 FORMAT(' '\)
  127 CONTINUE
      WRITE(*,126)
  126 FORMAT(' ')
  125 CONTINUE
      CLOSE(2)
C     SCREEN REMARK
      WRITE(*,160)
  160 FORMAT(1X,'PRESS RETURN TO TERMINATE')
      PAUSE
      END
```

D.14 SQRCA

```
      PROGRAM SQRCA

C     SQUARE CELLULAR AUTOMATON
C     THIS PROGRAM FOLLOWS THE EVOLUTION OF FIVE-NEIGHBOR, SUM MODULO 2
C     RULE FOR CA 614.  CYCLIC BOUNDARY CONDITIONS ARE USED.  THE (1,1),
C     (1,N), (N,1), AND (N,N) ELEMENTS ARE NEVER USED AND CAN BE READ IN
C     AS 0.
C     THE INPUT IS IN A FILE LABELED 'INPUT' AND CONTAINS THE SIZE OF
C     THE CA, THE NUMBER OF TIME STEPS, AND THE INITIAL CONFIGURATION.
C     SEE THE PROGRAM FOR THE FORMAT OF THE INPUT DATA.
C     THE OUTPUT IS DIRECTLY TO THE SCREEN.
C
      IMPLICIT REAL*8(A-H,O-V), INTEGER(I-N)
      DIMENSION IX(100,100),IY(100,100)
      OPEN(2, FILE='INPUT',STATUS='OLD')
C     READ IN THE SIZE OF THE CA (+2) AND THE NUMBER OF TIME STEPS (NT)
      READ(2,*) N,NT
C     READ IN THE INITIAL CONFIGURATION
C     THE ARRAY IS READ IN WITH THE COLUMN INDEX INCREASING MOST RAPIDLY
      READ(2,*) ((IX(I,J), J=1,N), I=1,N)
      DO 100 IT=1,NT
      DO 30 I=2,N-1
      DO 20 J=2,N-1
      IY(I,J) = IX(I,J-1)+IX(I,J)+IX(I,J+1)+IX(I-1,J)+IX(I+1,J)
   20 IY(I,J) = MOD(IY(I,J),2)
   30 CONTINUE
C     SET UP NEW CONFIGURATION
      DO 60 I=2,N-1
      DO 50 J=2,N-1
   50 IX(I,J) = IY(I,J)
```

```
   60 CONTINUE
C     IMPOSE CYCLIC BOUNDARY CONDITIONS
      DO 80 I=2,N-1
      IX(I,1) = IX(I,N-1)
   80 IX(I,N) = IX(I,2)
      DO 90 J=2,N-1
      IX(1,J) = IX(N-1,J)
   90 IX(N,J) = IX(2,J)
  100 CONTINUE
C     PRINT OUT IN A SQUARE ARRAY
      DO 140 I=2,N-1
      DO 110 J=2,N-1
      IF(IX(I,J).EQ.0) GO TO 120
      WRITE(*,115)
  115 FORMAT(' '\)
      GOTO 110
  120 WRITE(*,125)
  125 FORMAT(' '\)
  110 CONTINUE
      WRITE(*,130)
  130 FORMAT(' ')
  140 CONTINUE
      CLOSE(2)
C     SCREEN REMARK
      WRITE(*,160)
  160 FORMAT(1X,'PRESS RETURN TO TERMINATE')
      PAUSE
      END
```

BIBLIOGRAPHY

[Aguda and Clarke, 1988] B. D. Aguda and B. L. Clarke, *J. Chem. Phys.*, **89**, 7428 (1988).

[Alder and Wainwright, 1957–1964] B. J. Alder and T. E. Wainwright, *J. Chem. Phys.*, **27**, 1208 (1957); **31**, 459 (1959); **33**, 1439 (1960); **40**, 2724 (1964).

[Alder and Wainwright, 1962] B. J. Alder and T. E. Wainwright, *Phys. Rev.*, **127**, 359 (1962).

[Alder and Wainwright, 1970] B. J. Alder and T. E. Wainwright, *Phys. Rev.*, **A 1**, 18 (1970).

[Allen and Tildesley, 1989] M. P. Allen and D. J. Tildesley, *Computer Simulation of Liquids*, Oxford University Press, Oxford, 1989.

[Anderson, 1972] P. W. Anderson, *Science*, **177**, 393 (1972).

[Anderson, 1984] P. W. Anderson, *Basic Notions of Condensed Matter Physics*, Benjamin, Menlo Park, CA, 1984.

[Anderson, 1994] P. W. Anderson, *A Career in Theoretical Physics*, World Scientific, Singapore, 1994.

[Anderson et al., 1995] M. H. Anderson, J. R. Ensher, M. R. Matthews, C. E. Weiman, and E. A. Cornell, *Science*, **269**, 198 (1995).

[Argoul et al., 1987] F. Argoul, A. Armeodo, P. Richetti, and J. C. Roux, *J. Chem. Phys.*, **86**, 3325 (1987).

[Baker and Agard, 1994] D. Baker and D. A. Agard, *Biochemistry*, **33**, 7505 (1994).

[Baker and Gollub, 1990] G. L. Baker and J. P. Gollub, *Chaotic Dynamics*, Cambridge University Press, Cambridge, UK, 1990.

[Barker and Henderson, 1976] J. A. Barker and J. Henderson, *Rev. Mod. Phys.*, **48**, 588 (1976).

[Barker and Watts, 1969] J. A. Barker and R. O. Watts, *Chem. Phys. Lett.*, **3**, 144 (1969).

[Baxter, 1989] R. J. Baxter, *Exactly Solved Models in Statistical Mechanics*, Academic, New York, 1989.

[Berne, 1985] B. J. Berne, in *Multiple Time Scales*, J. U. Brackbill and B. I. Cohen, Eds., Academic, New York, 1985, pp. 419–436.

[Berne and Harp, 1970] B. J. Berne and G. D. Harp, *Adv. Chem. Phys.*, **17**, 63 (1970).

[Berne and Pecora, 1976] B. J. Berne and R. Pecora, *Dynamic Light Scattering*, Wiley, New York, 1976.

[Binder, 1995] K. Binder, Ed., *The Monte Carlo Method in Condensed Matter Physics. Topics in Applied Physics*, Springer-Verlag, Berlin, 1995, Vol. 71, 2nd corrected and updated edition.

[Binder and Young, 1986] K. Binder and A. P. Young, *Rev. Mod. Phys.*, **58**, 801 (1986).

[Bixon, 1973] M. Bixon, *J. Chem. Phys.*, **58**, 1459 (1973).

[Blittersdorf et al., 1992] R. Blittersdorf, A. F. Münster, and F. W. Schneider, *J. Phys. Chem.*, **96**, 5893 (1992).

[Box and Mueller, 1958] G. E. P. Box and M. E. Mueller, *Ann. Math. Stat.*, **29**, 610 (1958).

[Branden and Tooze, 1991] C. Branden and J. Tooze, *Introduction to Protein Structure*, Garland, New York, 1991.

[Bray, 1921] W. C. Bray, *J. Am. Chem. Soc.*, **43**, 1262 (1921).

[Bray and Liebhafsky, 1931] W. C. Bray and H. A. Liebhafsky, *J. Am. Chem. Soc.*, **53**, 38 (1931).

[Brush, 1967] S. G. Brush, *Rev. Mod. Phys.*, **39**, 883 (1967).

[Bryngelson and Wolynes, 1987] J. D. Bryngelson and P. G. Wolynes, *Proc. Natl. Acad. Sci. USA*, **84**, 7524 (1987).

[Bryngelson and Wolynes, 1990] J. D. Bryngelson and P. G. Wolynes, *Biopolymers*, **30**, 177 (1990).

[Callen, 1985] H. B. Callen, *Thermodynamics and an Introduction to Thermostatics*, Wiley, New York, 1985, 2nd ed.

[Carmeli and Nitzan, 1982] B. Carmeli and A. Nitzan, *Phys. Rev. Lett.*, **51**, 233 (1982).

[Chan and Dill, 1991] H. S. Chan and K. A. Dill, *J. Chem. Phys.*, **95**, 3775 (1991).

[Chan and Dill, 1993] H. S. Chan and K. A. Dill, *Phys. Today*, **46**, No. 2, Feb. 1993, p. 24.

[Chandrasekhar, 1943] S. Chandrasekhar, *Rev. Mod. Phys.*, **15**, 1 (1943).

[Cheung, 1977] P. S. Y. Cheung, *Mol. Phys.*, **33**, 519 (1977).

[Cho et al., 1996] C. H. Cho, S. Singh, and G. W. Robinson, *Phys. Rev. Lett.*, **76**, 1651 (1996).

[Churchill, 1958] R. V. Churchill, *Operational Mathematics*, McGraw-Hill, New York, 1958, 2nd ed.

[Ciccotti et al., 1987] G. Ciccotti, D. Frendel, and I. R. McDonald, Eds., *Simulation of Liquids and Solids*, North-Holland, Amsterdam, 1987.

[Ciccotti and Hoover, 1986] G. Ciccotti and W. G. Hoover, Eds., *Proc. Int. School Phys. Enrico Fermi*, North-Holland, Amsterdam, 1986, Course 97.

[Cole, 1967] G. H. A. Cole, *The Statistical Theory of Classical Simple Dense Fluids*, Pergamon, New York, 1967.

[Creswick et al., 1992] R. J. Creswick, H. A. Farach, and C. P. Poole, Jr., *Introduction to Renormalization Group Methods in Physics*, Wiley, New York, 1992.

[de Gennes, 1979] P. G. de Gennes, *Scaling Concepts in Polymer Physics*, Cornell University Press, Ithaca, NY, 1979.

[Dirac, 1958] P. A. M. Dirac, *The Principles of Quantum Mechanics*, Oxford University Press, London, 1958, 4th ed.

[Domb and Green, 1972–1983] C. Domb and M. S. Green, Eds., *Phase Transitions and Critical Phenomena*, Academic, New York, 1972–1983, Vols. 1–6.

[Domb and Lebowitz, 1984–] C. Domb and J. L. Lebowitz, Eds., *Phase Transitions and Critical Phenomena*, Academic, New York, 1984–, Vols. 7–.

[Doob, 1942] J. L. Doob, *Ann. Math.*, **43**, 351 (1942).

[Drazin, 1993] P. G. Drazin, *Nonlinear Systems*, Cambridge University Press, Cambridge, 1993.

[Edgar, 1990] G. A. Edgar, *Measure, Topology, and Fractal Geometry*, Springer-Verlag, Berlin, 1990.

[Edwards and Anderson, 1975] S. F. Edwards and P. W. Anderson, *J. Phys.*, **F 5**, 965 (1975).

[Einstein, 1956] A. Einstein, in *Investigations on the Theory of the Brownian Movement*, R Fürth, Ed., Dover, New York, 1956.

[Einstein, 1925] A. Einstein, *Berliner Ber.*, **1**, 3 (1925).

[Evans and Morriss, 1983] D. J. Evans and G. P. Morriss, *Phys. Lett.*, **98A**, 433 (1983).

[Eyring, 1980] H. Eyring, *Ber. Bunsenges. Phys. Chem.*, **86**, 348 (1980).

[Feder, 1988] J. Feder, *Fractals*, Plenum, New York, 1988.

[Feigenbaum, 1978] M. J. Feigenbaum, *J. Stat. Phys.*, **19**, 25 (1978).

[Feller, 1968] W. Feller, *An Introduction to Probability Theory and Its Applications*, Wiley, New York, 1968, Vols. I and II.

[Fiebig and Dill, 1993] K. M. Fiebig and K. A. Dill, *J. Chem. Phys.*, **98**, 3475 (1993).

[Field and Burger, 1985] R. J. Field and M. Burger, Eds., *Oscillations and Traveling Waves in Chemical Systems*, Wiley, New York, 1985.

[Field and Noyes, 1974] R. J. Field and R. M. Noyes, *J. Chem. Phys.*, **60**, 1877 (1974).

[Field et al., 1972] R. J. Field, E. Körös, and R. M. Noyes, *J. Am. Chem. Soc.*, **94**, 8649 (1972).

[Fisher, 1967] M. E. Fisher, *Rept. Prog. Phys.*, **30**, 615 (1967).

[Fisher, 1969] M. E. Fisher, *J. Phys. Soc. Jpn. (Suppl.)*, **26**, 44–45 (1969).

[Fisher, 1974] M. E. Fisher, *Rev. Mod. Phys.*, **46**, 597 (1974).

[Fisher, 1983] M. E. Fisher, in *Lecture Notes in Physics*, F. J. W. Hahne, Ed., Springer-Verlag, New York, 1983, Vol. 186, pp. 1–139.

[Fleming, 1986] G. R. Fleming, *Chemical Applications of Ultrafast Spectroscopy*, Oxford University Press, New York, 1986.

[Flory, 1953] P. Flory, *Principles of Polymer Chemistry*, Cornell University Press, Ithaca, 1953.

[Ford et al., 1965] G. W. Ford, M. Kac, and P. Mazur, *J. Math. Phys.*, **6**, 504 (1965).

[Frauenfelder and Wolynes, 1994] H. Frauenfelder and P. G. Wolynes, *Phys. Today*, **47**, No. 2, Feb. 1994, p. 58.

[Frenkel, 1946] J. Frenkel, *Kinetic Theory of Liquids*, Oxford University Press, Oxford, 1946.

[Frisch et al., 1986] U. Frisch, B. Hasslacher, and Y. Pomeau, *Phys. Rev. Lett.*, **56**, 1505 (1986).

[Furrow, 1985] S. D. Furrow, in *Oscillations and Traveling Waves in Chemical Systems*, R. J. Field and M. Burger, Eds., Wiley, New York, 1985.

[Gardner, 1970] M. Gardner, *Sci. Am.*, **223**, 120 (1970).

[Go and Taketomi, 1978] N. Go and H. Taketomi, *Proc. Natl. Acad. Sci. USA*, **75**, 559 (1978).

[Go and Taketomi, 1979] N. Go and H. Taketomi, *Int. J. Peptide Protein Res.*, **13**, 235 (1979).

[Goldenfeld, 1992] N. Goldenfeld, *Lectures on Phase Transitions and the Renormalization Group*, Addison-Wesley, Reading, MA, 1992.

[Goldstein, 1980] H. Goldstein, *Classical Mechanics*, Addison-Wesley, Reading, MA, 1980, 2nd ed.

[Gordon, 1965] R. G. Gordon, *J. Chem. Phys.*, **43**, 1307 (1965).

[Greenspoon and Pathria, 1974] S. Greenspoon and R. K. Pathria, *Phys. Rev. A*, **9**, 2103 (1974).

[Griffiths, 1972] R. B. Griffiths, in *Phase Transitions and Critical Phenomena*, C. Domb and M. S. Green, Eds., Academic, New York, 1972, Vol. 1, pp. 7–109.

[Grote and Hynes, 1980] R. F. Grote and J. T. Hynes, *J. Chem. Phys.*, **73**, 2715 (1980).

[Grote and Hynes, 1982] R. F. Grote and J. T. Hynes, *J. Chem. Phys.*, **77**, 3736 (1982).

[Györgi and Field, 1991] L. Györgyi and R. J. Field, *J. Phys. Chem.*, **95**, 6594 (1991).

[Hänggi et al., 1990] P. Hänggi, P. Talkner, and M. Borkovec, *Rev. Mod. Phys.*, **62**, 251 (1990).

[Haile, 1992] J. M. Haile, *Molecular Dynamics Simulation*, Wiley, New York, 1992.

[Hansen and McDonald, 1986] J. P. Hansen and I. R. McDonald, *Theory of Simple Liquids*, Academic, London, 1986.

[Hardy et al., 1976] J. Hardy, O. de Pazzis, and Y. Pomeau, *Phys. Rev. A*, **13**, 1949 (1976).

[Heermann and Burkitt, 1995] D. W. Heermann and A. N. Burkitt, in *The Monte Carlo Method in Condensed Matter Physics. Topics in Applied Physics*, K. Binder, Ed., Springer-Verlag, Berlin, 1995, Vol. 71, 2nd corrected and updated edition, pp. 53–74.

[Hill, 1987] T. L. Hill, *Statistical Mechanics*, Dover, New York, 1987.

[Hirschfelder et al., 1966] J. O. Hirschfelder, C. F. Curtiss, and R. B. Bird, *Molecular Theory of Gases and Liquids*, Wiley, New York, 1966.

[Hoover, 1984] W. G. Hoover, *Phys. Today*, **37**, No. 1, Jan. 1984, p. 44.

[Huang, 1987] K. Huang, *Statistical Mechanics*, Wiley, New York, 1987, 2nd ed.

[Hudson and Mankin, 1981] J. L. Hudson and J. C. Mankin, *J. Chem. Phys.*, **74**, 6171 (1981).

[Hynes, 1985] J. T. Hynes, in *Theory of Chemical Reaction Dynamics*, M. Baer, Ed., CRC, Boca Raton, FL, 1985, Vol. IV, pp. 171–235.

[Isihara, 1971] A. Isihara, *Statistical Physics*, Academic, New York, 1971.

[Ising, 1925] E. Ising, *Z. Phys.*, **31**, 253 (1925).

[Jauch, 1968] J. M. Jauch, *Foundations of Quantum Mechanics*, Addison-Wesley, Reading, MA, 1968.

[Jensen, 1987] R. V. Jensen, *Am. Sci.*, **75**, 168 (1987).

[Kadanoff, 1966] L. P. Kadanoff, *Physics*, **2**, 263 (1966).

[Kadanoff et al., 1967] L. P. Kadanoff, W. Götze, D. Hamblen, R. Hecht, E. A. S. Lewis, V. V. Palciauskas, M. Rayl, J. Swift, D. Aspnes, and J. Kane, *Rev. Mod. Phys.*, **39**, 395 (1967).

[Kapral et al., 1992] R. Kapral, A. Lawniczak, and P. Masiar, *J. Chem. Phys.*, **96**, 2762 (1992).

[Kausmann, 1959] W. Kausmann, *Adv. Protein Chem.*, **14**, 1 (1959).

[Kirkwood, 1935] J. G. Kirkwood, *J. Chem. Phys.*, **3**, 300 (1935).

[Kirkwood, 1951] J. G. Kirkwood, in *Phase Transformations in Solids*, R. Smoluchowski, J. E. Mayer, and W. A. Weyl, Eds., Wiley, New York, 1951.

[Kittel, 1969] C. Kittel, *Thermal Physics*, Wiley, New York, 1969.

[Kittel, 1983] C. Kittel, *Introduction to Solid State Physics*, Wiley, New York, 1983, 6th ed.

[Klein, 1922] O. Klein, *Ark. Mat. Astr. Fys.*, **16** No. 5, 1 (1922).

[Kleppner, 1996] D. Kleppner, *Phys. Today*, **49**, No. 8, 11 (1996).

[Kramers, 1940] H. A. Kramers, *Physica*, **7**, 284 (1940).

[Kubo, 1957] R. Kubo, *J. Phys. Soc. Jpn.*, **12**, 570 (1957).

[Kuhn, 1934] W. Kuhn, *Kolloid-Z*, **68**, 2 (1934).

[Landau, 1995] D. P. Landau, in *The Monte Carlo Method in Condensed Matter Physics. Topics in Applied Physics*, K. Binder, Ed., Springer-Verlag, Berlin, 1995, Vol. 71, 2nd corrected and updated edition, pp. 23–51.

[Landau and Lifshitz, 1970] L. D. Landau and E. M. Lifshitz, *Statistical Physics*, Addison-Welsey, Reading, MA, 1970.

[Landau and Lifshitz, 1975] L. D. Landau and E. M. Lifshitz, *Fluid Mechanics*, Pergamon, Oxford, 1975.

[Levine, 1983] I. N. Levine, *Quantum Mechanics*, Allyn and Bacon, Boston, 1983, 3rd ed.

[Levine, 1995] I. N. Levine, *Physical Chemistry*, McGraw-Hill, Boston, 1995, 4th ed.

[Levinthal, 1968] C. Levinthal, *J. Chim. Phys.*, **65**, 44 (1968).

[Lieb and Mattis, 1966] E. H. Lieb and D. C. Mattis, *Mathematical Physics in One Dimension*, Academic, New York, 1966.

[Lifshitz, 1969] I. M. Lifshitz, *Sov. Phys.—JETP*, **28**, 1280 (1969).

[Lifshitz et al., 1978] I. M. Lifshitz, A. Y. Grosberg, and A. R. Khokhlov, *Rev. Mod. Phys.*, **50**, 683 (1978).

[Ma, 1976] S.-K. Ma, *Phys. Rev. Lett.*, **37**, 461 (1976).

[Ma, 1985] S.-K. Ma, *Statistical Mechanics*, World Scientific, Singapore, 1985.

[Madras and Slade, 1993] N. Madras and G. Slade, *The Self-Avoiding Walk*, Birkhäuser, Boston, 1993.

[Mandelbrot, 1982] B. Mandelbrot, *The Fractal Geometry of Nature*, Freeman, New York, 1982.

[Margenau and Murphy, 1956] H. Margenau and G. M. Murphy, *The Mathematics of Physics and Chemistry*, Van Nostrand, New York, 1956, 2nd ed.

[McQuarrie, 1976] D. A. McQuarrie, *Statistical Mechanics*, Harper & Row, New York, 1976.

[Metropolis et al., 1953] N. Metropolis, A. W. Rosenbluth, M. N. Rosenbluth, A. H. Teller, and E. Teller, *J. Chem. Phys.*, **21**, 1087 (1953).

[Mori, 1965] H. Mori, *Prog. Theor. Phys.*, **33**, 423 (1965); **34**, 399 (1965).

[Münster, 1969] A. Münster, *Statistical Thermodynamics*, Springer-Verlag, New York, 1969, Vol. I.

[Nayfeh and Balachandran, 1995] A. H. Nayfeh and B. Balachandran, *Applied Nonlinear Dynamics*, Wiley, New York, 1995.

[Nicolis, 1971] G. Nicolis, *Adv. Chem. Phys.*, **19**, 209 (1971).

[Nicolis, 1995] G. Nicolis, *Introduction to Nonlinear Science*, Cambridge University Press, Cambridge, 1995.

[Nicolis and Prigogine, 1977] G. Nicolis and I. Prigogine, *Self-organization in Nonequilibrium Systems*, Wiley, New York, 1977.

[Nordholm and Zwanzig, 1975] S. Nordholm and R. Zwanzig, *J. Stat. Phys.*, **13**, 347 (1975).

[Nosé, 1984] S. Nosé, *Mol. Phys.*, **52**, 255 (1984).

[Onuchic et al., 1995] J. N. Onuchic, P. G. Wolynes, A. Luthey-Schulten, and N. D. Socci, *Proc. Natl. Acad. Sci. USA*, **92**, 3626 (1995).

[Ornstein and Zernike, 1914] L. S. Ornstein and F. Zernike, *Proc. Acad. Sci. Amsterdam*, **17**, 793 (1914).

[Packard and Wolfram, 1985] N. H. Packard and S. Wolfram, *J. Stat. Phys.*, **38**, 901 (1985).

[Paine and Scheraga, 1985] G. H. Paine and H. A. Scheraga, *Biopolymers*, **24**, 1391 (1985).

[Pais, 1982] A. Pais, *'Subtle is the Lord. . . '*, Oxford University Press, New York, 1982.

[Pathria, 1996] R. K. Pathria, *Statistical Mechanics*, Butterworth-Heinemann, Oxford, 1996, 2nd ed.

[Petrov et al., 1992] V. Petrov, S. K. Scott, and K. Showalter, *J. Chem. Phys.*, **97**, 6191 (1992).

[Pollak, 1986] E. Pollak, *J. Chem. Phys.*, **85**, 865 (1986).

[Pollak et al., 1989] E. Pollak, H. Grabert, and P. Hänggi, *J. Chem. Phys.*, **91**, 4073 (1989).

[Press et al., 1988] W. H. Press, B. P. Flannery, S. A. Teukolsky, and W. T. Vetterling, *Numerical Recipes*, Cambridge University Press, New York, 1988.

[Prigogine, 1980] I. Prigogine, *From Being to Becoming*, Freeman, New York, 1980.

[Prigogine and Stengers, 1984] I. Prigogine and I. Stengers, *Order out of Chaos*, Bantam, Toronto, 1984.

[Rahman, 1964] A. Rahman, *Phys. Rev.*, **A 136**, 405 (1964); *Phys. Rev. Lett.*, **12**, 575 (1964).

[Rahman and Stillinger, 1971] A. Rahman and F. H. Stillinger, *J. Chem. Phys.*, **55**, 3336 (1971).

[Reif, 1965] F. Reif, *Fundamentals of Statistical and Thermal Physics*, McGraw-Hill, New York, 1965.

[Rice and Gray, 1965] S. A. Rice and P. Gray, *The Statistical Mechanics of Simple Liquids*, Interscience, New York, 1965.

[Richards, 1980] E. G. Richards, *An Introduction to Physical Properties of Large Molecules in Solution*, Cambridge University Press, Cambridge, 1980.

[Richetti and Arenodo, 1985] P. Richetti and A. Arenodo, *Phys. Lett. A*, **109**, 359 (1985).

[Richetti et al., 1987] P. Richetti, J. C. Roux, F. Argoul, and A. Arneodo, *J. Chem. Phys.*, **86**, 3339 (1987).

[Richter et al., 1981] P. H. Richter, P. Rehmus, and J. Ross, *Prog. Theor. Phys.*, **66**, 385 (1981).

[Ringland and Turner, 1984] J. Ringland and J. S. Turner, *Phys. Lett. A*, **105**, 93 (1984).

[Robertson, 1993] H. P. Robertson, *Statistical Thermophysics*, Prentice Hall, Englewood Cliffs, NJ, 1993.

[Robinson et al., 1996] G. W. Robinson, S.-B. Zhu, S. Singh, and M. W. Evans, *Water in Biology, Chemistry and Physics*, World Scientific, New Jersey, 1996.

[Rouse, 1953] P. E. Rouse, Jr., *J. Chem. Phys.*, **21**, 1272 (1953).

[Roux et al., 1983] J. C. Roux, R. H. Simoyi, and H. L. Swinney, *Physica*, **8D**, 257 (1983).

[Sakurai, 1985] J. J. Sakurai, *Modern Quantum Mechanics*, Benjamin/Cummings, Menlo Park, CA, 1985.

[Salsburg et al., 1953] Z. W. Salsburg, R. W. Zwanzig, and J. G. Kirkwood, *J. Chem. Phys.*, **21**, 1098 (1953).

[Schiff, 1955] L. I. Schiff, *Quantum Mechanics*, McGraw-Hill, New York, 1955, 2nd ed.

[Schrödinger, 1944] E. Schrödinger, *What is Life?*, Cambridge University Press, Cambridge, 1944.

[Schwarz, 1966] L. Schwarz, *Mathematics for the Physical Sciences*, Addison-Wesley, Reading, MA, 1966.

[Scott, 1993] S. K. Scott, *Chemical Chaos*, Oxford University Press, Oxford, 1993.

[Sherrington and Kirkpatrick, 1975] D. Sherrington and S. Kirkpatrick, *Phys. Rev. Lett.*, **35**, 104 (1975).

[Simoyi et al., 1982] R. H. Simoyi, A. Wolf, and H. L. Swinney, *Phys. Rev. Lett.*, **49**, 245 (1982).

[Singh and Pathria, 1985] S. Singh and R. K. Pathria, *Phys. Rev.*, **B31**, 4483 (1985).

[Singh and Robinson, 1994] S. Singh and G. W. Robinson, *J. Chem. Phys.*, **100**, 6640 (1994).

[Singh and Robinson, 1995] S. Singh and G. W. Robinson, *J. Chem. Phys.*, **103**, 4920 (1995).

[Singh et al., 1992, 1994] S. Singh, R. Krishnan, and G. W. Robinson, *Phys. Rev. Lett.*, **68**, 2608 (1992); *Phys. Rev. E*, **49**, 2540 (1994).

[Stanley, 1971] H. E. Stanley, *Introduction to Phase Transitions and Critical Phenomena*, Oxford University Press, New York, 1971.

[Straub et al., 1985, 1986] J. E. Straub, M. Borkovec, and B. Berne, *J. Chem. Phys.*, **83**, 3172 (1985); **84**, 1788 (1986).

[Swendsen, 1979] R. H. Swendsen, *Phys. Rev. Lett.*, **42**, 859 (1979).

[Swendsen, 1982] R. H. Swendsen, in *Real Space Renormalization. Topics in Current Physics*, T. W. Burkhardt and J. M. J. van Leeuwen, Eds., Springer-Verlag, Berlin, 1982, Vol. 30, pp. 57–86.

[Swendsen et al., 1995] R. H. Swendsen, J.-S. Wang, and A. M. Ferrenberg, in *The Monte Carlo Method in Condensed Matter Physics. Topics in Applied Physics*, K. Binder, Ed., Springer-Verlag, Berlin, 1995, Vol. 71, 2nd corrected and updated edition, pp. 75–91.

[Takahashi, 1942] H. Takahashi, *Proc. Phys.-Math. Soc. Jpn.*, **24**, 60 (1942).

[Taketomi et al., 1975] H. Taketomi, Y. Ueda, and N. Go, *Int. J. Peptide Protein Res.*, **7**, 445 (1975).

[Tolman, 1979] R. C. Tolman, *The Principles of Statistical Mechanics*, Dover, New York, 1979.

[Tonks, 1936] L. Tonks, *Phys. Rev.*, **50**, 955 (1936).

[Tufillaro et al., 1992] N. B. Tufillaro, T. Abbot, and J. Reilly, *An Experimental Approach to Nonlinear Dynamics and Chaos*, Addison-Wesley, Redwood City, 1992.

[Turner et al., 1981] J. S. Turner, J. C. Roux, W. D. McCormick, and H. L. Swinney, *Phys. Lett.*, **85A**, 9 (1981).

[van der Zwan and Hynes, 1983, 1984] G. van der Zwan and J. T. Hynes, *J. Chem. Phys.*, **78**, 4174 (1983); *Chem. Phys.*, **90**, 21 (1984).

[van Kampen, 1985] N. G. van Kampen, *Stochastic Processes in Physics and Chem istry*, North-Holland, Amsterdam, 1985.

[Verlet, 1967] L. Verlet, *Phys. Rev.*, **159**, 98 (1967).

[Vicentini-Missoni, 1971] M. Vicentini-Mossini, in *Proc. Int. School Phys. Enrico Fermi*, M. S. Green, Ed., North-Holland, Amsterdam, 1971, Vol. 51, pp. 157–187.

[Weeks et al., 1971] J. D. Weeks, D. Chandler, and H. C. Andersen, *J. Chem. Phys.*, **54**, 5237 (1971).

[Widom, 1965] B. J. Widom, *J. Chem. Phys.*, **43**, 3892, 3898 (1965).

[Willamowski and Rössler, 1980] K.-D. Willamowski and O. E. Rössler, *Z. Natur-forsch.*, **35a**, 317 (1980).

[Wilson, 1979] K. G. Wilson, *Sci. Am.*, **241**, 158 (1979).

[Wilson, 1985] K. G. Wilson, in a report by M. M. Waldrop, in *Science*, **228**, 568 (1985).

[Wolfram, 1983] S. Wolfram, *Rev. Mod. Phys.*, **55**, 601 (1983).

[Wolfram, 1984] S. Wolfram, *Physica*, **10D**, 1 (1984).

[Wolfram, 1986a] S. Wolfram, *Theory and Applications of Cellular Automata*, World Scientific, Singapore, 1986.

[Wolfram, 1986b] S. Wolfram, *J. Stat. Phys.*, **45**, 471 (1986).

[Wood, 1968, 1970] W. W. Wood, *J. Chem. Phys.*, **48**, 415 (1968); **52**, 729 (1970).

[Wood and Jacobson, 1957] W. W. Wood and J. D. Jacobson, *J. Chem. Phys.*, **27**, 1207 (1957).

[Wood and Parker, 1957] W. W. Wood and F. R. Parker, *J. Chem. Phys.*, **27**, 720 (1957).

[Wu and Kapral, 1994] X.-G. Wu and R. Kapral, *J. Chem. Phys.*, **100**, 5936 (1994); *Phys. Rev. E*, **50**, 3560 (1994).

[Zasada and Pathria, 1976] C. S. Zasada and R. K. Pathria, *Phys. Rev. A*, **14**, 1269 (1976).

[Zernike and Prins, 1927] F. Zernike and J. A. Prins, *Z. Phys.*, **41**, 184 (1927).

[Zhabotinskii, 1964] A. M. Zhabotinskii, *Biofizika*, **9**, 306 (1964).

[Zimm, 1956] B. H. Zimm, *J. Chem. Phys.*, **24**, 269 (1956).

[Zwanzig, 1960] R. Zwanzig, *J. Chem. Phys.*, **33**, 1338 (1960).

[Zwanzig, 1961] R. Zwanzig, *Phys. Rev.*, **124**, 983 (1961).

[Zwanzig, 1964] R. Zwanzig, *Physica*, **30**, 1109 (1964).

[Zwanzig, 1973] R. Zwanzig, *J. Stat. Phys.*, **9**, 215 (1973).

[Zwanzig, 1974] R. Zwanzig, *J. Chem. Phys.*, **60**, 2717 (1974).

INDEX